Care and Repair
of Advanced
Composites

Other SAE books on this topic:

Design of Durable, Repairable, and Maintainable Aircraft Composites (Order No. AE-27)

For more information or to order this book, contact SAE at 400 Commonwealth Drive, Warrendale, PA 15096-0001; (412) 776-4970; fax (412) 776-0790; e-mail: publications@sae.org.

Care and Repair
of Advanced
Composites

by
Keith B. Armstrong
and
Richard T. Barrett

Society of Automotive Engineers, Inc.
Warrendale, Pa.

Library of Congress Cataloging-in-Publication Data

Armstrong, K. B.
    Care and repair of advanced composites / by K. B. Armstrong
and R. T. Barrett.
        p. cm.
    Includes bibliographical references and index.
    ISBN 0-7680-0047-5 (alk. paper)
    1. Composite materials—Repairing. 2. Composite materials—
Bonding. 3. Airplanes—Materials. 4. Airframes—Maintenance and
repair. I. Barrett, R. T. II. Title.
TA418.9.C6A76 1997
620.1'18'0288—dc21

97-35791

CIP

Cover photos (from top to bottom): De Havilland Mosquito, Beechcraft Starship, B-2 Stealth Bomber, Bell-Boeing V22 Osprey, and Boeing 777.

This book is dedicated to the members of the original International Air Transport Association (IATA) Task Force (1988–1991) set up to standardize composite repairs, for which the first author acted as chairman, and to the members of the various Task Groups of the Commercial Aircraft Composite Repair Committee (CACRC) and especially the Training Task Group and its chairmen over several years, first Ed Herman of Douglas, followed by Rick Barrett of Continental, and then by Ralph Edwards of Northwest. Work from the Design, NDI, and Repair Techniques Task Groups is included in this book and is gratefully acknowledged. We particularly wish to record the great debt we all owe to Henk Lodewijk of KLM. When the IATA Task Force completed its work and produced IATA Document DOC:GEN:3043, it made several recommendations. Henk had the vision and enthusiasm to persuade three committees (ATA, IATA, and SAE) to combine to form the CACRC and to implement the recommendations of the IATA Task Force. This book is one of the many products of that vision.

Acknowledgments

The first author especially wishes to thank E.I. Du Pont de Nemours and Company for support for attendance at the early Training Task Group Meetings of the Commercial Aircraft Composite Repair Committee (CACRC) through the good offices and personal interest of Dr. Rene Pinzelli of its Geneva office and Ginger Gupton of Wilmington, who is now with her own company, Vantage Marketing. More recently, support from the Defence Evaluation and Research Agency (DERA), Farnborough (United Kingdom), and British Aerospace Systems and Equipment Ltd. allowed attendance at later meetings. Without their interest and support, neither this book nor the *Composite and Metal Bonding Glossary* would have been possible.

Our thanks also are extended to Mike Hoke of Abaris Training Inc. for many of the diagrams and to Matt Waterson of Aerobond (United Kingdom) for the forms in Chapter 12. Dave Bunt of British Airways (BA) kindly reviewed the book and made many helpful comments. Ian Fidler of BA supplied some diagrams, and the first author would like to thank BA for the 24 years of experience on which this book is based. Many companies have contributed photographs and diagrams, and we have individually acknowledged them.

Our thanks are extended to Chris Mekern for scanning and improving some of the drawings and to Bert Groenewoud for arranging the text around them. The support of KLM by giving them the time to do this is greatly appreciated. Because of a burglary at his home, Chris lost his computer and the disks for most of the work he had done; therefore, the majority of the scanning had to be redone by our local printer, Chris Bond, at very short notice.

Finally, our thanks are extended to Philip Griggs, Ralph Edwards, and Patricia Morrison for their time in reviewing the book and in making valuable comments before the final version was prepared for printing.

Table of Contents

Foreword

This book is a detailed work of reference, drawing from many industry sources around the world. It is a comprehensive, in-depth look at the principal materials and processes involved in manufacturing, inspecting, and repairing advanced composite aircraft structures. It should also be useful to those working with composites in non-aerospace industries.

This book becomes even more valuable considering the background of the authors. Dr. Keith Armstrong is extremely well respected for his many years of experience and leadership in advanced composite aircraft structures. He was involved in the development of carbon fiber/Nomex® honeycomb floor panels for the VC-10 airliner with Graham Bevan of Rolls-Royce Composite Materials and now with British Aerospace. These were the first carbon fiber composites to fly on commercial aircraft. Almost all flooring used in commercial aircraft today meets specifications that have evolved from these early composite structures. Dr. Armstrong developed many new methods in the infancy of the industry during his 24 years with British Airways. He led the effort to standardize composite repair as chairman of the IATA Composite Repair Task Force from 1988 to 1991. He continues to be a key player in the effort to improve composites education via his ongoing participation in the Training Task Group of the SAE/IATA/ATA Commercial Aircraft Composite Repair Committee (CACRC). As part of this work, he compiled a glossary of terms, published by SAE as AIR 4844, "Composites and Metal Bonding Glossary."

Dr. Armstrong possesses a unique combination of theoretical and practical knowledge. He earned an M.Sc. (1978) and a Ph.D. (1991) in Adhesion Science and Technology from The City University in London, England. He also has spent many years performing simple but valuable hands-on experiments such as wedge testing of composite and aluminum alloy samples after years of immersion in water. These have now been published and are listed as references in the book.

This book is co-authored by Richard (Rick) Barrett. Rick's many years of hands-on experience have proven invaluable in ensuring that this book covers not only the technical theory but also the practical aspects of composites. Formerly a Composites Shop Manager with Continental Airlines, Rick has spent many years performing, managing, and training hands-on repair of damaged composite parts. Rick's leadership as chairman of the CACRC Training Task Group from 1992 to 1994 helped to establish the syllabus for this book.

This book will become a standard by which future efforts are judged. All of us in this evolving field owe the authors of this book a debt of gratitude, not only for this work but also for Dr. Armstrong's many fundamental contributions to composites technology and for Rick Barrett's ideas and enthusiasm which made this project possible.

Mike Hoke, President
Abaris Training Inc.
Reno, Nevada

Preface

This book is intended to meet the background information needs of composite repair technicians who repair aircraft composite components and structures for return to airline service. It supplements the practical work required to achieve the necessary level of skill in performing these repairs. The syllabus was developed by the Training Task Group and approved by the Commercial Aircraft Composite Repair Committee (CACRC). It has been published by the Society of Automotive Engineers (SAE) as SAE AIR 4938 Composite and Bonded Structure Technician/Specialist Training Document and covers materials, handling and storage, design aspects, damage assessment, repair techniques, manufacturers' documents, fastening systems, tools and equipment, health and safety aspects, and care of composite parts. Chapter 15 covers metal bonding, partly because some composite parts include metal reinforcements and fittings and partly because the term "composite" is sometimes used to mean any combination of two or more materials such as a skin and honeycomb core, for example. Repair design engineers should also find the book useful.

This book is designed primarily for aerospace use, but these notes have been compiled for all composite repair technicians, regardless of the type of vehicle or component they have to repair. Although some of the materials used may be either different from, or less strictly regulated than, those used in the aerospace industries, we hope that this book will be useful to the marine, automobile, railway, wind power generation, and sporting goods industries and others using composite components. The general principles involved are very similar for all composite materials, regardless of their end use.

A book of this size and topic range cannot cover any subject in great detail; therefore, more extensive references should be studied when necessary. For detailed specific repair procedures, the Original Equipment Manufacturers (OEMs) Structural Repair Manuals (SRMs), Overhaul Manuals (OHMs), etc., should be used. SAE (CACRC) produced Aerospace Recommended Practices (ARPs) will be included in these manuals as alternative means of compliance as opportunity arises during normal amendment procedures and as these ARPs become available.

Metal bonding has been covered in Chapter 15 because there are occasions when metal parts are bonded to composites, and the Design Guide in Chapter 16 documents experience and shows where small design improvements could extend the service lifetimes of composite parts. The support of SAE for all of the CACRC work is greatly appreciated. This book has been written to meet the CACRC syllabus, and it has been reviewed by some CACRC members. The book has not been approved by the CACRC; therefore, the views expressed are those of the authors and do not necessarily reflect the opinions of the committee as a whole or of the organizations that sponsor the committee.

The purpose of training is to improve performance and to provide leadership for others. We hope that the following thoughts will inspire further effort.

"Education expands the boundaries of our known ignorance faster than it expands the boundaries of our knowledge."

If we wish to be good leaders, we have a vast amount of study and research to do.

As an old poem states,

1. *He that knows and knows that he knows is wise, follow him,*
2. *He that knows not and knows he knows not is willing to learn, teach him,*
3. *He that knows not and knows not that he knows not is a fool, shun him.*

Let us all try to become at least "Type 2" and strive to reach "Type 1."

Chapter 1

Introduction to Composites and Care of Composite Parts

This book is presented in the same sequence as the ATA/IATA/SAE syllabus of training recommended for composite repair technicians. With additions, it is also suitable for repair design engineers. Chapter 15, Metal Bonding, has been included partly because the definition of "composite" can be interpreted to include metal-skinned honeycomb panels and partly because some composite parts have metal fittings or reinforcements which must be treated before bonding. Chapter 16 covers a number of the problems experienced in service, some of which may also be applicable to metallic sandwich panels, and offers suggestions for design improvements. The Commercial Aircraft Composite Repair Committee (CACRC) considered it necessary to provide training for a wider range of staff and recommended that aircrew, managers, aircraft mechanics, storekeepers, and delivery drivers should be given some knowledge of composites. Extracts from this book may be useful for this purpose.

1.1 Definition of Composites

Composite materials are quite different from metals. Composites are combinations of materials differing in composition or form where the individual constituents retain their separate identities and do not dissolve or merge together. These separate constituents act together to give the necessary mechanical strength or stiffness to the composite part. Today this definition usually refers to fibers as reinforcement in a resin matrix, but it can also include metal-skinned honeycomb panels, for example.

Reinforced concrete is a good example of a composite material. The steel and concrete retain their individual identities in the finished structure. However, because they work together, the steel carries the tension loads and the concrete carries the compression loads. Although not covered by this book, metal and ceramic matrix composites are being studied intensively. Composites in structural applications have the following characteristics:

- They generally consist of two or more physically distinct and mechanically separable materials.

- They are made by mixing the separate materials in such a way as to achieve controlled and uniform dispersion of the constituents.

- Mechanical properties of composites are superior to and in some cases uniquely different from the properties of their constituents. This is clearly seen with glass-reinforced plastics (GRP). In the case of GRP, the epoxy resin is a relatively weak, flexible, and brittle material, and although the glass fibers are strong and stiff, they can be loaded in tension only as a bare fiber. When combined, the resin and fiber give a strong, stiff composite with excellent toughness characteristics.

1.2 History of Composite Materials

A composite is, by definition, something made from two or more components—in our case here, a fiber and a resin. Composites are not a new idea. Moses floated down the Nile in a basket made from papyrus reeds coated with pitch. Papyrus is a form of paper with a visible fibrous reinforcement; therefore, it would not have been difficult to make a waterproof basket from it. From ancient times, it was known that bricks were stronger if filled with chopped straw. African "mud" huts were reinforced with grasses and thin sticks. The Butser Hill farm project shows that woven sticks, bonded with a mixture of cow dung and mud, were used to build house walls in England in 1500 B.C. It would be interesting to know how the correct mix ratio for the cow dung and the mud was determined! The lath and plaster walls in old English houses were a form of composite. Although the concept is old, the materials have changed. Carbon, aramid, and glass fibers are very expensive compared to straw, and epoxy resins are costly compared to a mixture of cow dung and mud! Fortunately, the performance for a given weight is much higher. There also are natural composites such as wood. The structure of a tree consists of long, strong cellulose fibers bonded together by a protein-like substance called lignin. The fibers that run up the trunk and along the branches are thus aligned by nature in the optimum way to resist the stresses experienced from gravity and wind forces. Large radii are provided at the trunk-to-branch and branch-to-branch joints to reduce stress concentrations at high-load points.

1.3 Advantages and Disadvantages of Composites

Composite parts have both advantages and disadvantages when compared to the metal parts they are being used to replace.

1.3.1 Advantages of Composites

1. A higher performance for a given weight leads to fuel savings. Excellent strength-to-weight and stiffness-to-weight ratios can be achieved by composite materials. This is usually expressed as strength divided by density and stiffness (modulus) divided by density. These are so-called "specific" strength and "specific" modulus characteristics.

2. Laminate patterns and ply buildup in a part can be tailored to give the required mechanical properties in various directions.

3. It is easier to achieve smooth aerodynamic profiles for drag reduction. Complex double-curvature parts with a smooth surface finish can be made in one manufacturing operation.

4. Part count is reduced.

5. Production cost is reduced. Composites may be made by a wide range of processes.

6. Composites offer excellent resistance to corrosion, chemical attack, and outdoor weathering; however, some chemicals are damaging to composites (e.g., paint stripper), and new types of paint and stripper are being developed to deal with this. Some thermoplastics are not very resistant to some solvents. Check the data sheets for each type.

1.3.2 Disadvantages of Composites

1. Composites are more brittle than wrought metals and thus are more easily damaged. Cast metals also tend to be brittle.

2. Repair introduces new problems, for the following reasons:

    • Materials require refrigerated transport and storage and have limited shelf lives.

    • Hot curing is necessary in many cases, requiring special equipment.

    • Curing either hot or cold takes time. The job is not finished when the last rivet has been installed.

3. If rivets have been used and must be removed, this presents problems of removal without causing further damage.

4. Repair at the original cure temperature requires tooling and pressure.

5.   Composites must be thoroughly cleaned of all contamination before repair.

6.   Composites must be dried before repair because all resin matrices and some fibers absorb moisture.

1.3.3 Advantages of Thermoset Resin Composites

Thermoset resin composites have advantages and disadvantages when compared to thermoplastic resin composites. The advantages of thermoset resin composites over thermoplastic resin composites include the following:

1.   Thermosets will cure at lower temperatures than most thermoplastics will melt. Therefore, thermosets can be manufactured at lower temperatures than thermoplastics.

2.   Two-part systems can be cured at room temperature, and their cure can be speeded by heating to approximately 80°C (176°F).

3.   A range of curing temperatures, particularly with epoxy systems, allows repair at lower temperatures than the original cure.

4.   Tooling can be used at lower temperatures than with thermoplastics.

5.   Chemical resistance is generally good, but check for resistance to any chemicals that may come into contact with the part. For example, some epoxies are more resistant to chemicals than others.

1.3.4 Disadvantages of Thermoset Resin Composites

1.   Slow to process (cold store/thaw/cure).

2.   Relatively low toughness, environmental performance, and strength.

3.   Can be health hazards.

4.   Slow to repair.

1.3.5 Advantages of Thermoplastic Resin Composites

The advantages of thermoplastic resin composites over thermoset resin composites include the following:

1.   Thermoplastic resin composites are much tougher than thermosets and offer fast processing times and good environmental performance, except against certain solvents in some cases. Again, check each material and its response to each solvent likely to be encountered.

2. No health hazards.

3. More closely match fiber performance.

4. Good fire/smoke performance (interiors and fuel tanks and engine parts).

5. Good fatigue performance.

6. Primary structure usage.

7. High temperature uses polyetheretherketone (PEEK) 250 to 300°C (482 to 572°F).

8. Commercial applications include helicopter rotor blades, some high-strength interior parts, and fairing panels on civil aircraft.

9. Future possibility of resin transfer molding (RTM) around reinforcing fiber or use in conventional application mode (i.e., pre-preg stacking). Single crystal growth versions could be used for engine parts.

1.3.6 Disadvantages of Thermoplastic Resin Composites

1. Cost.

2. New process methods.

3. Long-term fatigue characteristics unknown.

4. Temperature to melt for repairs is very high in some cases. This could cause serious problems for *in-situ* repairs to primary or secondary structures, especially if being done near fuel tanks or hydraulic systems.

5. Polyimides suffer microcracking (Ref. 1.1).

1.4 Applications of Composites to Modern Aircraft, Yachts, Cars, and Trains

1.4.1 Early Aircraft Structures

Early aircraft were composite-based structures because they were built from wood, which is a composite material comprising a cellulose/lignin mixture that gives wood its excellent strength-to-weight performance and properties of resilience and damage resistance. However, wood is subject to deterioration by moisture-induced decay and attack from fungal growths. By the 1930s, wooden aircraft structures began to be replaced by stressed-skin, monocoque aluminum alloy structures.

One notable exception to the trend toward all-metal aircraft structures was the
De Havilland Mosquito, which was an all-composite wooden aircraft in which adhesively
bonded sandwich structures were used extensively, consisting of plywood skins on a balsa
wood core. The Mosquito was a successful aircraft, and more than 8,000 were built
between 1940 and 1946.

The first documented use of fibrous composite aerospace structures was an experimental
Spitfire fuselage made in the early 1940s by Aero Research Ltd. at Duxford near Cambridge, England. This structure was made from untwisted flax fibers impregnated with
phenolic resin and formed into 150-mm (6-in.) wide unidirectional tape ribbons which were
then assembled into 0°/90° ply sheets that were hot-pressed to their final form. These
precured sheets were assembled on the airframe structure with conventional riveting techniques. This fuselage was tested at RAE (now DERA) Farnborough and was found to meet
all the structural requirements of a conventional duralumin aluminum alloy structure.

1.4.2 Modern Aircraft Structures and Other Applications

The introduction of glass fibers in the 1940s and their post-war use with polyester and
epoxy resins laid the foundations for the aerospace composites industry as we know it
today. The new, higher performance fibers introduced in the 1960s, such as boron (1966),
carbon (1968), aramid fiber (1972), and high-performance polyethylene (1987), have accelerated this trend. Projections for the next decade indicate that the following aircraft types
will use composites to a large extent.

Projected Use of Composites

Type of Aircraft	Percentage Usage
Rotorcraft and corporate aircraft	60–80% of structure weight
Fighter and attack aircraft	40–50% of structure weight
Large commercial and military	20–30% of structure weight

Composites started being used in modern aircraft for tertiary structure, such as fairings;
however, composites now are used for secondary and primary structures, in some cases
whole wings. Filament-wound composite fuselages are being seriously studied for medium
to large aircraft, and in some cases prototypes have been built. They are already flying in
some corporate aircraft. Obviously, the quality of composite parts and repairs to these parts
is becoming a matter of great importance, and that is one of the reasons for producing this
book. Composites are here to stay. All of the major airframe manufacturers are heavily
engaged in further efforts to capitalize on the exceptional strength-to-weight ratio and other
benefits to be obtained with composites. However, to obtain and retain these benefits,

airframe manufacturers must make these materials more repairable and airlines must learn to take good care of them, to avoid needless damage, and to become skilled and equipped for repair when necessary.

The first Boeing 707 had approximately 2% composite construction. In contrast, today's Boeing 767 has approximately 35%, and the new Beech Starship is the first certified general aviation aircraft to have a structure built entirely of composites. The Bell-Boeing V-22 tilt rotor aircraft is also the first of its kind to be made entirely of composite materials.

Some aircraft using considerable amounts of composites are:

Aerospatiale	ATR 42 and 72 and others in this series
Airbus	A-300, A-310, A-319, A-320, A-321, A-330, and A-340
Boeing	737, 747, 757, 767, and 777
British Aerospace	146, RJ Series, Jetstream Series
Fokker	F-50, F-70, and F-100
McDonnell Douglas	DC-10, MD-11, MD-12 project, and MD-80, 83, and 90

Composites are also being used extensively in modern racing cars, yachts, railway trains, and sporting equipment such as skis.

For illustrations of some applications of composites, see Figures 1.1 through 1.7.

1.5 Care of Composite Parts

A great opportunity exists to save money, time, and trouble by taking good care of composite parts, components, and structures. Composite construction offers the opportunity for parts integration, i.e., to make a large part in one piece rather than to assemble it from a large number of smaller parts. See Ref. 1.2. A composite part may be made from a very large number of tape or fabric plies all bonded at the same time; however, when finished, it is only one large part, having one assembly number and requiring only one record card. This means that the larger a part is and the more expensive it is, the less likely it is to be held as a spare. Consequently, if damage or deterioration occurs, the choice must be made between an expensive spare that is not in stock or a repair that may be difficult and time consuming. Many large parts also have many bolt holes, and these are not always jig located. This means that if a new part is fitted and the old part repaired, then the holes in that part may not fit the next airplane. Problems of this type are common and mean that these parts must be repairable quickly so that they can be refitted to the airplane from which they came. All repairs cause problems that airlines would prefer not to have; therefore, the value of taking good care of composites is very high, especially if the parts are large, expensive, or not easily obtained when required.

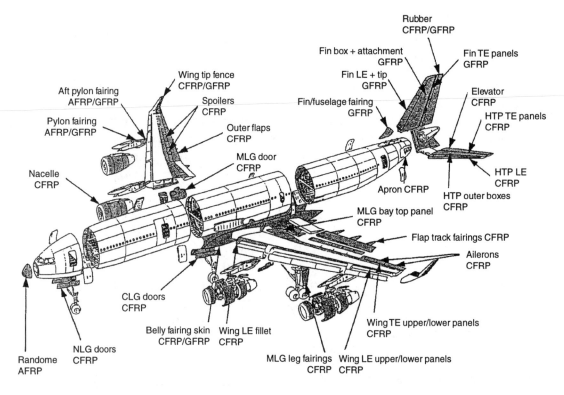

Fig. 1.1 Use of composites on the Airbus A-340. (Courtesy of Airbus Industrie.)

1.5.1 Sources of Damage to Composite Parts

Unfortunately, there are many potential sources of damage to composite parts, and most of these apply regardless of the material of construction. They can be listed under two basic headings: physical or chemical damage.

Physical or Chemical Damage: These types of damage can occur from the sources listed below and probably from a few more.

*   Collision with ground service vehicles, ladders, and other ground equipment.

*   Taxiing into other aircraft or stationary objects.

*   In severe storms, hail impact can cause extensive damage, even if the aircraft is on the ground.

*   Lightning strike.

*   Sand erosion.

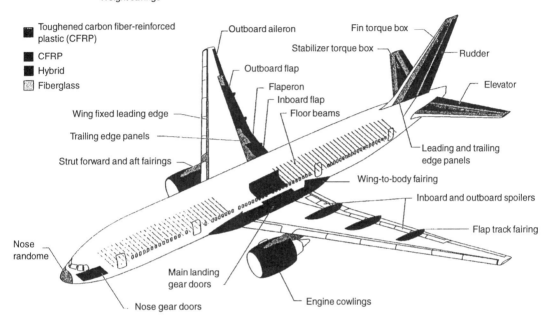

777 composite materials
- Toughened material for improved damage resistance and damage tolerance, and parts are designed for simple, bolted, or bonded repairs
- Corrosion and fatigue resistant
- Weight savings

■ Toughened carbon fiber-reinforced plastic (CFRP)
■ CFRP
■ Hybrid
▨ Fiberglass

Outboard aileron
Fin torque box
Stabilizer torque box
Rudder
Outboard flap
Elevator
Flaperon
Inboard flap
Wing fixed leading edge
Floor beams
Trailing edge panels
Leading and trailing edge panels
Strut forward and aft fairings
Wing-to-body fairing
Inboard and outboard spoilers
Flap track fairing
Nose randome
Main landing gear doors
Nose gear doors
Engine cowlings

Fig. 1.2 Use of composites on the Boeing 777. (Courtesy of Boeing Commercial Airplane Group.)

- Careless use of screwdrivers and/or overtightening of screws can damage countersunk holes.

- Damage can result from bad handling. Some long, thin panels can be fractured by bad handling. They must be supported by several people along their length in the same way as a person with a back injury being carefully moved onto a stretcher. In addition, corners are easily broken off if parts are moved without protection. As mentioned in a later chapter, composite parts should be stored in their boxes until they are removed for use on an aircraft. They should not be removed from their boxes to save storage space as soon as they are received. The purpose of good packaging is to have the part reach its final destination in good condition. This can be achieved only if the packaging remains in place until the last possible moment.

- Damage in transit can result from incorrect, inadequate, or insufficient padding or packaging. Trailers and vehicles must be softly sprung. Staff must appreciate the value and importance of the parts they are moving.

- Store these parts under good conditions, particularly avoiding excessive heat.

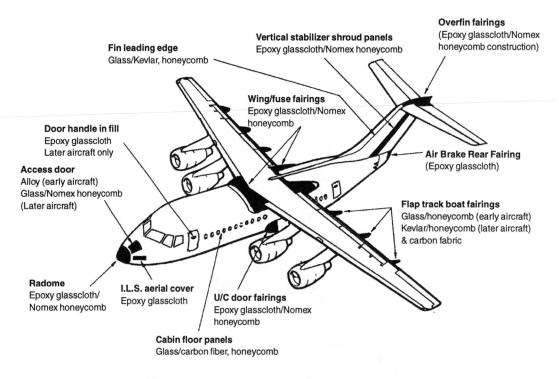

Overfin fairings
(Epoxy glasscloth/Nomex honeycomb construction)

Vertical stabilizer shroud panels
Epoxy glasscloth/Nomex honeycomb

Fin leading edge
Glass/Kevlar, honeycomb

Wing/fuse fairings
Epoxy glasscloth/Nomex honeycomb

Door handle in fill
Epoxy glasscloth
Later aircraft only

Access door
Alloy (early aircraft)
Glass/Nomex honeycomb
(Later aircraft)

Air Brake Rear Fairing
(Epoxy glasscloth)

Flap track boat fairings
Glass/honeycomb (early aircraft)
Kevlar/honeycomb (later aircraft)
& carbon fabric

Radome
Epoxy glasscloth/
Nomex honeycomb

I.L.S. aerial cover
Epoxy glasscloth

U/C door fairings
Epoxy glasscloth/Nomex
honeycomb

Cabin floor panels
Glass/carbon fiber, honeycomb

*Fig. 1.3 Use of composites on British Aerospace (Avro) 146/RJ Series.
(Courtesy of Avro International.)*

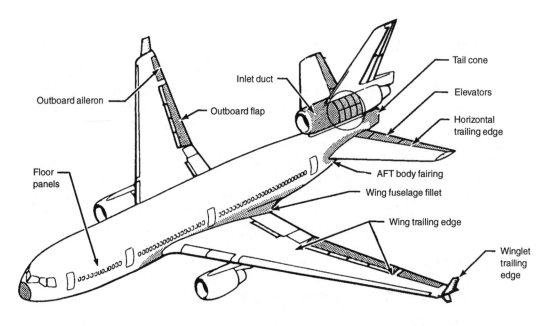

Tail cone

Inlet duct

Outboard aileron

Elevators

Outboard flap

Horizontal
trailing edge

Floor
panels

AFT body fairing

Wing fuselage fillet

Wing trailing edge

Winglet
trailing
edge

Fig. 1.4 Use of composites on McDonnell Douglas MD-11. (Courtesy of McDonnell Douglas.)

Fig. 1.5 Composites in a racing yacht. (Courtesy of Du Pont.)

Fig. 1.6 Composites in a sports car. (Courtesy of Du Pont.)

Fig. 1.7 Composites in a modern train. (Courtesy of SNCF—French Railways.)

• Dropped tools can cause damage. If personnel are working at some height above composite parts, these parts should be covered with thick foam-rubber padding that would be sufficient to offer protection from any tools that may be dropped.

• Chemical damage can occur. Paint stripper, Skydrol, other hydraulic fluids, and any other damaging fluids should be removed from composite parts immediately if accidental spillage occurs. The area should be thoroughly washed with warm soapy water to remove any residual chemicals after first removing most of it with clean rags or absorbent paper. On the inside faces of fairings and panels, which can be contaminated with hydraulic fluid due to small leaks, a protective coating is not always provided. In addition, these panels may have only one layer of skin, and fluid penetration into the honeycomb is likely.

• Overheat damage can occur mainly to engine cowlings if engine components or accessories overheat and occasionally in wheel bay areas if brakes have been overheated. Reflective aluminum layers and heat shields must be maintained in good condition. It is a sad fact that more repairs are made to composites because of damage resulting from lack of foresight or experience at the design stage and lack of care in handling than because of deterioration or unavoidable damage such as severe hail impact and lightning strikes. Even these are, to some extent, avoidable by cooperation between pilots and ground controllers. In one sense, this is a compliment to composite parts, but it is not encouraging to think that most damage is inflicted by airline or associated personnel. Room for improvement exists in the care of composites to avoid the need for repair. Chapter 16 shows that detail design can be improved in many areas.

Many of these problems could be resolved by improved staff training and more care on the flight line.

Deterioration in Normal Service: If less important items were being described, this would probably be dismissed almost flippantly as FW&T (fair wear and tear). Although this is always true to a degree, it is fair to say that anything that significantly affects the safety, serviceability, or maintenance cost of an airplane should be considered for design improvement if the type has enough aircraft with sufficient remaining service life to justify the effort. This type of improvement is usually brought about by the issue of a manufacturer's Service Bulletin, if the customer service department has received enough complaints via its service representatives based at major airports or with the large airline maintenance bases. This type of improvement depends on airlines maintaining good records of problems and forwarding this information to the manufacturer by the normal channels. It often requires reports from more than one airline to achieve action, but it is essential for airlines to do this. Good reporting is particularly necessary if the parts in question can be repaired fairly easily. If the manufacturer receives no orders for spares and no complaints, it follows that nothing will be done.

Deterioration can occur in many ways:

- Rain, hail, or sand erosion

- Cyclic heat, especially in engine cowlings

- Weathering (moisture absorption) and ultraviolet (UV) radiation

- Hydraulic oil damage to paint and resin and adhesive bonds if allowed to remain on composite parts

- Vibration causing wear at bolt holes

- Frequent panel removal for inspection or replenishment, causing wear at bolt holes

- Careless use of screwdrivers when removing panels

- Use of excessive torque when tightening bolts in composite parts, which can split the edge or cause delamination around the hole

- Minor damage that goes unnoticed

- Minor lightning strikes causing pinholes in radomes that allow moisture penetration over a period of time

- Paint or sealant that comes off and is not replaced quickly

1.5.2 Avoidance of Damage and Reduction of Deterioration in Service

Airlines and other users can minimize in-service damage and deterioration in the following ways:

1. Maintain protective paint schemes in good condition.

2. Coat any dry spots found in composite parts with resin as soon as possible. Dry spots must be given a good brush coat of a high-quality, low-viscosity laminating resin, preferably with a high glass transition temperature (T_g) and low water uptake. This should be done under vacuum pressure and with mild heat (approximately 80°C [176°F] maximum).

3. Replace any missing sealant quickly after drying the area.

4. Repair oil leaks promptly (hydraulic or engine oil), and wash all oil contamination from composite parts as soon as possible.

5. Train staff to close all doors and access panels carefully to prevent slamming damage and to report any doors that are difficult to close immediately so that rerigging or other adjustments can be made before damage is done. Any doors and access panels that are not properly flush should also be reported so that adjustments can be made to reduce edge erosion damage and fuel wastage due to avoidable drag.

6. Ensure that any loose bolt inserts are removed, replaced, and correctly bonded.

7. Use the correct torque values when fitting bolts in composite parts.

8. Repair damage of a minor nature quickly. This can prevent repairs from becoming larger if the damage grows or water seepage occurs.

9. Seal any allowable damage immediately and in accordance with the SRM to prevent moisture penetration or damage growth while repairs are organized.

10. Check erosion-resistant coatings on radomes, leading edges of wings, tailplanes, fins, and helicopter rotor blades at regular intervals, and replace or repair them as required.

11. Check radomes for moisture content at regular intervals, and dry and repair as required.

12. Check lightning protection systems regularly and replace any defective parts.

13. Report repetitive defects to the manufacturer and request modification action to correct the problem and thus prevent recurrence.

14. Design and apply local repairs or improvements if they have, or have access to, a design authority to do so.

15. Encourage all personnel to report any deterioration or damage, however small, as soon as it is found. This includes accidental damage caused by the person reporting it. It is important to safety and economics of operation that airlines should not embarrass or penalize staff for doing this, because everyone makes mistakes at times and often when cutting corners to keep an aircraft on schedule.

16. Store parts in the boxes in which they are delivered until they are needed on the aircraft or assembled to a component. Note that the value of air space and floor space in the stores, although not insignificant, is unlikely to compare with the cost of finding that a part transported to the aircraft without a box is damaged and requires repair or replacement when it is urgently needed. There is a tendency among storekeepers to take everything out of its packaging to save space. This is understandable but should be discouraged for all parts that can suffer damage in transit and especially for easily damaged parts such as composite components and aircraft windows.

17. Conduct very minor modifications. In the case of panels or fairings contaminated by hydraulic fluid, local minor modification is possible. The problem usually has two causes:

    • No protection has been provided.

    • The inner skin may have only one fabric layer, and this may allow fluid penetration into the honeycomb, especially if the resin content is low.

    Several solutions are possible:

    • Fit a protective layer after careful cleaning of the inner skin. This layer may be made from a range of materials. A Tedlar plastic coating or a layer of aluminum foil (Speedtape) may be applied, or a brush coating of polysulfide or polythioether sealant may be used. These will prevent direct contact between the fluid and the component inner skin.

    • After cleaning the existing skin, additional layers of fabric may be applied with a suitable epoxy resin to bring the total number of layers to three. The additional layers may be thin ones. Three layers is the minimum number to ensure that fluid penetration of the skin is unlikely.

18. Use care in rigging of parts that require correct and careful adjustment. Another point where care may be profitable is in ensuring the correct rigging of items such as control surfaces and particularly undercarriage doors. The manufacture of jigs for this purpose

has been found to be worth the cost. On one well-remembered occasion with which one of the authors is familiar, an undercarriage door was sent to the hangar for fitting after many manhours had been spent repairing it. A shift change was in progress, and the incoming shift thought that the rigging checks and adjustments on the new door had been done by the previous shift. A retraction test was called for, and it was quickly discovered that the door had not been rigged; it was almost instantly damaged to a far worse degree than it had been when sent to the workshop the first time. The workshop staff members were not amused. The only good that arose from this incident was that the supervisor was so embarrassed that he measured several aircraft and designed a jig to check and ensure that the rigging was correct. To date, no similar incidents have occurred. Another good reason for rigging doors and access panels correctly and flush with the skin is that drag is reduced and a useful amount of fuel is saved on each flight.

19. Ensure that drain holes are clear of dirt and debris and that they are redrilled if covered during repairs.

20. Ensure that any repairs do not encroach on necessary mechanical clearance required by moving parts.

21. High-pressure washing of aircraft by untrained personnel may be acceptable for metal structures but has proved damaging to composite parts. Composite honeycomb panels suffer the most from this treatment, as any holes or delaminations are penetrated by the high pressures employed, which significantly increases the amount of damage. Staff training is required, together with a reduction of the pressure used, if damage is to be avoided when cleaning composite parts by this method. Composite parts are usually expensive. Although in many cases repairs are not difficult, any resin or adhesive requires at least an hour to cure, even if it is a hot-curing system. These systems require heater blankets and vacuum bags, and they take time to set up before curing can commence. Room-temperature curing systems can be warm cured in a few hours, but both methods usually take longer than riveting. A bonded repair cannot fly until the resin is cured. A riveted repair may require time to complete, but it can fly as soon as the last rivet is fitted (unless sealants are needed). It is easier to take care of a part, whatever its form of construction, than it is to do a repair. Following the old principle that "a stitch in time saves nine," always make permanent repairs as soon as possible after damage is found, before the damaged area can increase in size. On the same basis, a permanent repair should always be made in preference to a temporary repair if time and other factors permit. (However, it must be recognized that there will be times when a honeycomb sandwich type fairing panel is damaged and the only way to meet a service deadline is to drill a large number of holes and bolt a plate on the outside with large washers on the inside. If this makes a large wide-bodied jet take off on or near schedule with 400 passengers, thereby saving 400 hotel bills and an upset schedule the following day, then the decision will have been a sound economic one, even if the panel is scrapped.) A temporary repair must be removed before a permanent repair can

be made, and further damage is usually done in the process, which results in the second repair being larger and nearer to the permitted limits. Parts in need of repair are unlikely to having packing cases available for transport to and from hangars or work-shops. The best possible provision should be made for covering, padding, physical support, and careful handling during transport to and from the point of repair and any other necessary journeys such as visits to the paint shop. Damage on the way to or from the paint shop is not unknown. This provision should include training for the staff involved and should give them an appreciation of the value of the parts they are handling and of the work they do.

22. One further action by which airlines could minimize long-term damage to the entire external surface of the aircraft is to adopt a new paint system mentioned in more detail in Chapter 15. This is the new scheme tested on aluminum surfaces where the chro-mate or other corrosion-resistant primer remains on the aircraft for its entire life. This is painted with an intercoat and a topcoat, which are both capable of being stripped by benzyl alcohol. This is a more environmentally friendly solvent than methylene chlo-ride, which was commonly used for many years. Benzyl alcohol does not damage the basic primer; therefore, paint can be removed an almost unlimited number of times without removing the primer and the underlying oxide layer produced by anodizing. This anodic treatment provides a good bonding surface for the primer and also gives corrosion resistance to the aluminum alloy skin. Trials indicate that this primer and paint system may also be used on composite surfaces. If this proves as successful as expected, then this paint topcoat, intercoat, and primer system will meet the protective needs of composite components extremely well. Adoption of this paint system by airlines will make a great contribution to the successful maintenance of composite parts, which are progressively becoming a large part of the external surface areas of modern aircraft.

1.6   References

1.1   Phillips, Leslie N. [ed.], *Design with Advanced Composite Materials,* The Design Council, Springer-Verlag, London, ISBN 0-85072-238-1, 1989.

1.2   Armstrong, K.B., "Parts Integration—Advantages and Problems," Carbon Fibers III, Third International Conference, 8–10 October 1985, Kensington, London, UK, The Plastics and Rubber Institute (now part of The Institute of Materials), 1985, pp. 15/1–15/6.

Chapter 2

Materials

2.1 Fiber Reinforcement

Many types of fiber reinforcement are available today, and a number of them are described in detail in this chapter. Before examining the individual fibers, it is worth considering some of the characteristics common to all fiber-reinforced materials.

2.1.1 Comparison of High-Performance Fibers and Common Metals

Fiber manufacturers often promote their fibers by showing graphs of their strength and stiffness in comparison with metallic materials. This is a rather unfair comparison because the fibers are rarely used as isolated fibers working in tension. More commonly, fibers are combined with a resin in a 0° unidirectional laminate form or in a 0/90° woven fabric based composite form. An infinite number of orientations are possible; however, in practice, only a few are used.

The following shows how the actual performance drops from the pure fiber to a woven laminate:

Single Fiber: Very high strength (say, x)

Unidirectional Tape: At 50% volume fraction (say, x/2)

Cross-Plied Unidirectional Tape: Only half of the fiber is in the loading direction (strength now x/4)

Woven Fabric: Because of crimping of the fibers, strength is now less than x/4, especially in compression

Similar effects are found with stiffness.

A point that must always be remembered is that composites, like plywood, have considerable strength in the fiber directions but little transverse strength. Metals have similar strength in all directions.

It is interesting, and important from a design viewpoint, that a woven carbon fiber fabric laminate has a tensile modulus approximately the same as aluminum alloy, when the above factors are taken into account. This statement applies to a low-modulus (LM) carbon fiber. The values for a high-modulus (HM) carbon fiber and ultrahigh-modulus (UHM) carbon fiber would be higher. Because of its lower density, carbon fiber continues to have an advantage or it would not be used; however, the advantage is much less than the single fiber strength would suggest. If a quasi-isotropic laminate is required to approximate as closely as possible the properties of a metal sheet, which are the same in all directions, then

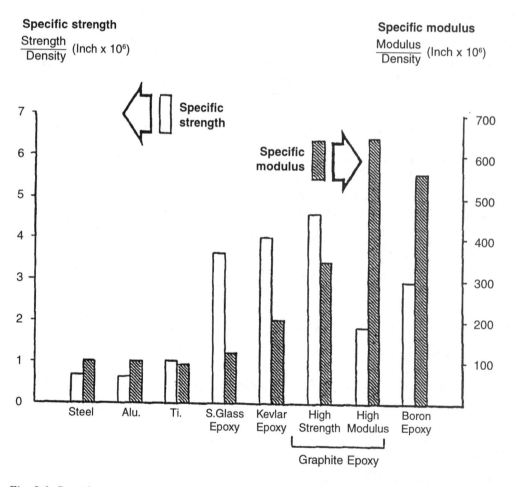

Fig. 2.1 Specific strength and modulus. (Courtesy of Aero Consultants [United Kingdom] Ltd.)

the strength in any one direction will be even lower. If we examine the properties on this basis by plotting tensile strength divided by density (specific tensile strength) against tensile modulus divided by density (specific tensile modulus) for three forms of material (i.e., fiber alone, 0° unidirectional laminate, and 0/90° woven fabric laminate), we obtain a more realistic idea of the relative performance of these materials. See Figure 2.1. For specific mechanical properties, see Figure 2.2 and Tables 2.1 and 2.2.

It is also useful to look at the relative costs of these various high-performance fibers to make a cost comparison. See Figure 2.3.

We see from the chart that high-strength carbon fiber prices and aramid (Kevlar) prices seem to be leveling off with increasing volume of production of carbon fiber and two more producers for aramids. It is hoped that prices will stabilize or even fall in the next few years. However, due to reduced defense budgets around 1993–1994, many suppliers may be struggling because of reduced production with fixed overhead. Thus, increased prices could result.

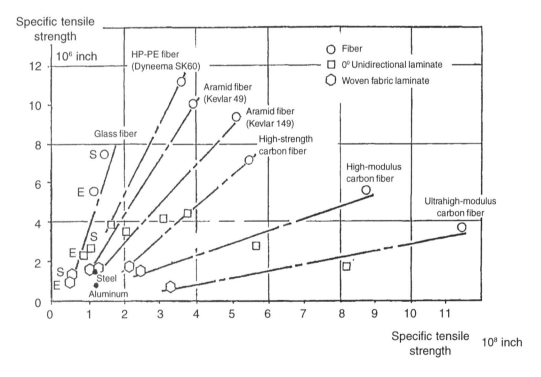

Fig. 2.2 Specific mechanical properties. (Courtesy of Aero Consultants [United Kingdom] Ltd.)

Table 2.1
Mechanical Properties of Advanced Composite Fibers

Fiber Type	Fiber Diameter		Fiber Density		Tensile Strength		Tensile Modulus	
	inch x 10⁻³	µm	lb/in³	g/cm³	ksi	GPa	msi	GPa
E Glass	0.3–0.6	8–14	0.092	2.54	500	3.45	10.5	72.4
S Glass	0.3–0.6	8–14	0.090	2.49	665	4.58	12.5	86.2
HP-Polyethylene[1]	0.4–0.47	10–12	0.035	0.97	392	2.70	12.6	87.0
Aramid (Kevlar 49)[2]	0.47	12	0.052	1.44	525	3.62	19.0	130.0
Aramid (Kevlar 149)[2]	0.47	12	0.052	1.44	503	3.47	27.0	186.2
High-Strength Carbon (T300)[3]	0.28	7.0	0.063	1.76	514	3.53	33.6	230.0
High-Modulus Carbon (HM400)[4]	0.26	6.5	0.067	1.87	450	3.10	58.7	405.0
Ultrahigh-Modulus Carbon (GY 80)[5]	0.33	8.4	0.071	1.96	270	1.86	83.0	572.0
Boron Filament	2–8	50–203	0.094	2.60	500	3.44	59.0	406.7
Silicon Carbide Filament (SiC on Carbon)	5.6	142	0.11	3.04	611	4.21	60.0	413.6

Manufactuters trade names/references are as follow:

1. Dyneema reference (Dyneema SK60)
2. Du Pont registered trademark
3. Toray Industries reference
4. Courtaulds Grafil reference
5. BASF reference

The above values are taken from manufacturers' literature as typical characteristic properties. A wide range of values has been published, and this information should be taken as an approximate guide.

2.1.2 Glass Fibers

Glass fiber, usually called fiberglass, is as its name implies—a fiber made of glass. Approximately 90 miles of glass filament can be run from a single glass sphere no larger than a common marble. This fiber possesses tremendous strength for its weight and is fairly resistant to all but the strongest acids or alkalis. It will not rust or burn. The chemical resistance of cured glass-reinforced plastic (GRP) laminates depends on many factors, including the type of glass, the type of resin, the surface finish on the glass, edge sealing of

Table 2.2
Fiber Properties

Fiber	Length, μm	Diameter, μm	Density, g/cm³	Tensile Strength, psi	Modulus, psi	Continuous Temperature, °F
PBI	Continuous	11–21	1.43	65,000	800,000	840
LCP	Continuous	23	1.4	360,000–450,000	12–15 million	550–600
PE	Continuous	27–38	0.97	375,000–430,000	17–25 million	230
Carbon	Continuous	7	1.66	350,000–450,000	33-55 million	1000
S Glass	Continuous	7	2.50	500,000–660,000	13 million	600–700
Aramid	Continuous	12	1.44	400,000	10–25 million	500
Boron	Continuous	100–140	2.50	510,000	58 million	3600
Quartz	Continuous	9	2.2	500,000	10 million	1920
SiC Fibers, Polymer Precursor	Continuous	10–20	2.3–2.6	400,000	28 million	2370
SiC Monofilaments by CVD	Continuous	140	3.05	500,000–600,000	60 million	2000
SiC Whiskers	20–50	0.3–1.0	3.19	450,000–2 million	58–100 million	2900
VLS SiC Whiskers	50,000	5	3.05	1.2 million	84 million	3270
SiO Whiskers	0.2–2.5	0.002–0.05	2.3	1 million		
Alumina	Continuous	20	3.9	200,000–300,000	55 million	1800
85% Alumina/15% Silica	Continuous	17	3.25	210,000	28 million	2280
50% Alumina/50% Silica	750	2.5	2.7	250,000	15 million	2300
72% Alumina/28% Silica	750	3.0	3.05	120,000	22 million	3000
Alumina/Boria/Silica	Continuous	8–12	2.7–3.0	250,000–300,000	22–33 million	2500
$Si_xN_yC_z$	Continuous	25–30		200,000	29 million	
HPZ SiNC	Continuous	8–15	2.3–2.5	350,000–420,000	25–35 million	2000
Titanium Based	Continuous	8–10	2.3–2.5	400,000	28 million	2370

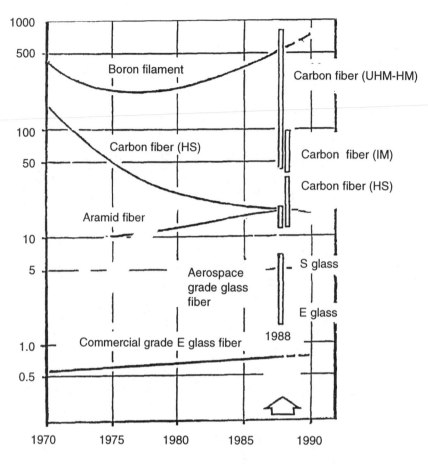

Note: These figures are intended only as an approximate indication of relative prices because exact fiber price will depend on fiber grade/tex/filament count

Fig. 2.3 Advanced composite fibers—price trend. (Courtesy of Aero Consultants [United Kingdom] Ltd.)

the laminate, and the use of gel coats. Stress reduces this resistance considerably. The resin protects the fiber from the acid. (See Refs. 2.1, 2.2, and 2.3). The threadlike fibers may be woven into a cloth as any other fiber. The fiberglass used as reinforcement for thermosetting resins in aircraft applications is available as a cloth in many different weights and weaves, as a loose mat, and as loose strands of fiberglass roving. For boat building and commercial applications, heavier grades of cloth may be used than those normally associated with aerospace components. "E" glass is typically used, but the higher-modulus "S" glass is used in aerospace for special items requiring high performance. (For details, see MIL-C-9084 [United States] and British Standard BS 3396 [United Kingdom]. For applications that require the most strength, use unidirectional tape. Next, a woven glass cloth is used, which is a compromise between the strength provided by a close-weave cloth and the encapsulating qualities of a loose-weave cloth. Weave style is important; see later sections of this chapter and Ref. 2.4. Resin properties are also important. Some graphs, plotted from Scott Bader Company Data Sheets for its polyester resins, show that the effect of resin mechanical properties is similar for both woven cloth and chopped strand mat when flexural strength properties are related to resin strength, modulus, and elongation. As expected, the strength of chopped strand beams is lower than woven cloth, but the trends are the same. The same trends were observed by Palmer with epoxies. (See Ref. 2.5.) Note that when chopped strand is sprayed in sheet form, the properties are greatest in the plane of the sheet and are not truly three dimensional. For some non-aeronautical applications in which low cost is more important than high strength, a mat may be used instead of woven cloth. The glass fibers are gathered and pressed together loosely. The random placement of fibers in the mat provides a uniform strength in all directions in the plane of the sheet when bonded with resin.

2.1.2.1 Glass Manufacture

Silica sand is heated until it becomes liquid; however, to reduce operating temperatures, other ingredients are added in carefully weighed proportions. These ingredients are:

- **Limestone:** To add toughness
- **Soda Ash:** Allows silica to melt at a lower temperature
- **Cullet:** Broken glass

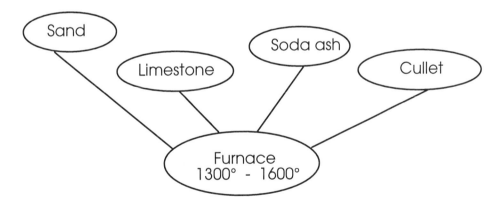

This will produce "A" glass. In the manufacture of "E" glass, borosilicate replaces soda ash and provides greater resistance to heat.

The following describes the various types of glass:

1. **"A" Glass:**

   - Standard soda glass in common use for windows, etc.

   - High alkali content which absorbs water, leading to degradation of material—corrosion

   - Resin adhesion deteriorates with water absorption

2. **"C" Glass:**

   - High resistance to corrosive materials

   - Normally produced and used only as a surface mat to reduce cost

3. **"D" Glass:**

   - An improved electrical grade for modern radomes

   - Lower dielectric constant

4. **"E" Glass:**

   - Low alkali content—hence, minimum water absorption

   - Good resin adhesion properties

   - Good temperature properties

5. **"R" Glass:**

   - American and European version of "S" glass

6. **"S" Glass (Silica-Alumina-Magnesia):**

   - Strength claimed to be 40% in excess of "E" glass

   - Temperature stable in excess of 700°C (1292°F)

   - Used primarily in aerospace, but also in some marine applications in high-performance boats when cost is not as important as performance

2.1.2.2 Glass Fiber Manufacture

"E" glass marbles from a hopper are heated and passed through a bush as a liquid into a setting zone. As solid filaments, the glass is coated with a sizing compound to prevent self-abrasion. The sizing compound remains on the filament until the material is woven. It must then be removed as described later and a finish applied which must be compatible with the resin system to be used.

One glass marble will produce 90 miles of filament. From the sizing ring, the filaments are formed together into a strand and wound on a skein or bobbin. From several bobbins, the strands are formed into a "cheese" of roving or yarn. From rovings, glass is woven into cloth or laid as a tape. For less important applications, it may be made into chopped strand mat.

For sizing and finish, see Section 2.3. See Figures 2.4, 2.5, and 2.6. For the fabrics most commonly used for aircraft repairs, see Table 2.3.

2.1.3 Carbon Fibers

Carbon Fiber Manufacture: Carbon/graphite fiber manufacture begins with a precursor of polyacrylonitrile (PAN), pitch, or rayon. PAN-based fibers are superior to either pitch or rayon. The precursor is pre-oxidized and stabilized at 200 to 300°C (392 to 572°F) for one hour. These pre-oxidized fibers are better able to withstand carbonization/graphitization temperatures and have improved carbon chain orientation. To produce the intermediate strength fiber, the pre-oxidized fiber is then carbonized in a furnace at temperatures greater than 1200°C (2192°F) for one or two minutes. Graphitization takes place at 2000 to 3000°C (3632 to 5432°F) for one to two minutes and produces high-strength, high-modulus fibers. The furnace temperatures remove hydrogen, nitrogen, and oxygen, leaving oriented carbon chains. Carbonization yields a random amorphous structure. Graphitization produces an increased crystalline orientation with higher density and modulus. The new fiber is then surface treated by oxidizing to improve the adhesion to matrix resins.

Epoxy-compatible sizing is applied to aid in handling and processing of the material. The material is then wound and shipped to tape and fabric manufacturers. For further details, see Section 2.2.2.4.

Future Development: The future development of carbon fibers will probably be directed toward the improvement of high-strain fibers in terms of higher tensile strength and correspondingly higher strain to failure because most aerospace applications of carbon fiber involve this high-strain material. Most designs now work to 4000 microstrain, an elongation of 0.004 or 0.4%, depending on the choice of description. This provides a design strain reserve factor of 3.

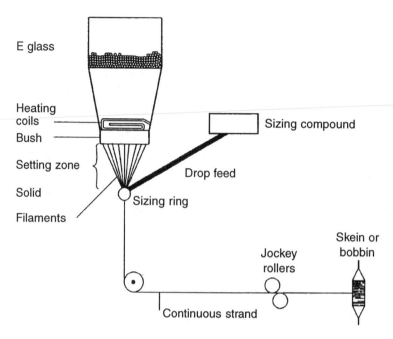

Fig. 2.4 Glass fiber manufacturing. (Courtesy of CS Interglas.)

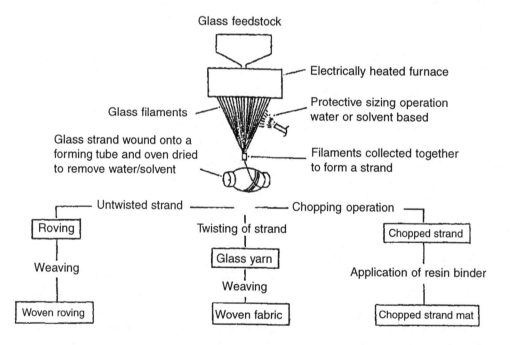

Fig. 2.5 Basic manufacturing process—fiberglass products. (Courtesy of Aero Consultants [United Kingdom] Ltd.)

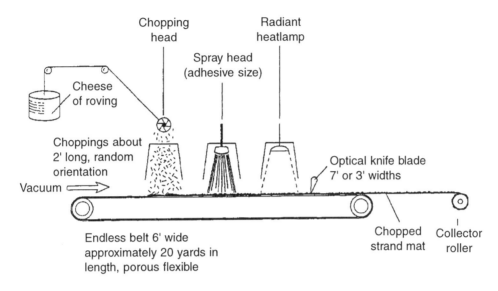

Fig. 2.6 Chopped strand mat machine. (Courtesy of CS Interglas.)

Table 2.3
Fabrics Most Commonly Used in Aircraft Repairs

Glass (White Color); Specific Gravity 2.54 (E Class), 2.49 (S Glass)				
Style	120	181	1581	7781	P6/22
Harness Type	4H	8H	8H	8H	Plain
Name	Crowfoot Satin	Satin	Satin	Satin	Plain
BMS 9-3 Designation	Type D	Type H	Type H-2	Type H-3	
Thickness (in.)	0.004	0.0087	0.0098	0.0098	0.006
Strength (lbf/in. width)[1]	125/120	350/330	350/330	350/340	170/150
Weight/Unit Area (gm/m²)	106	297	306	305	182
Yarn Designation	EC5 11x2	EC27 22x3	EC9 34x2	EC6 68	EC7 22x3

1. The two figures given for strength refer to the warp and waft directions, respectively. Warp is given first.
2. 7781 tends to replace 1581 and 181, but the differences are slight. See MIL-C-9084 and ASTM D-579.

Tensile strength 500 x 10³ psi (E glass), 675 x 10³ (S glass).

Further Details of Carbon and Graphite Fibers: These are made by carbonizing precursor fibers of PAN or pitch. As shown in the diagram, the precursor fiber is heated and carbonized while under tension and is surface treated before being wound on a spool prior to weaving. Further details are shown in Figures 2.7, 2.8, and 2.9. (See Ref. 1.1.) Figure 2.9 shows the important effect of heat treatment. The original treatment proceeded as follows:

- Untwisted tows of continuous PAN fibers were wound on sturdy frames and roasted in an air oven at 220 to 330°C (428 to 626°F). The fiber, which is originally white, slowly turns yellow, then brown, and finally black.

- During this time, oxygen is being absorbed, and, in spite of the loss of volatile material, weight increases. This volatile material comprises some water, a little ammonia, and a fair quantity of hydrogen cyanide. This is a poisonous gas, and appropriate precautions must be taken to neutralize it.

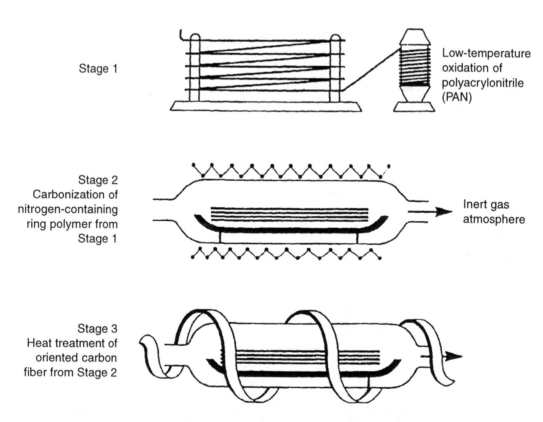

Fig. 2.7 Preparation of carbon fiber from PAN, Royal Aircraft Establishment process. (Courtesy of The Design Council and Springer-Verlag.)

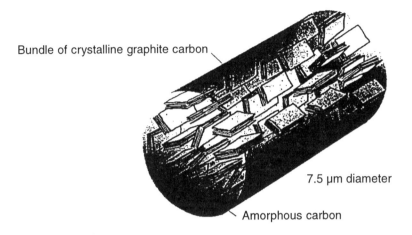

Bundle of crystalline graphite carbon

7.5 µm diameter

Amorphous carbon

Fig. 2.8 Schematic diagram of carbon fiber structure. Bundles of oriented crystalline carbon are held in a matrix of amorphous carbon. (Source: Royal Aerospace Establishment, 1972. Courtesy of The Design Council and Springer-Verlag.)

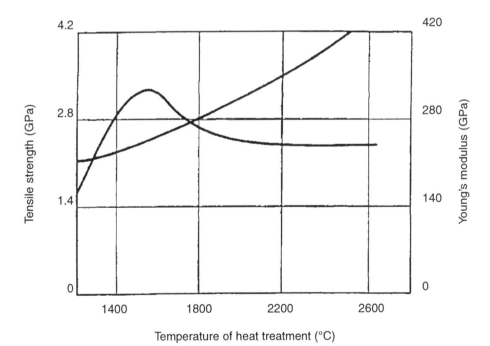

Fig. 2.9 Properties of carbon fiber against temperature of treatment. Fiber modulus rises steadily with increasing temperature of treatment. However, the tensile strength shows an optimum at approximately 1600°C (2900°F). (Source: Watt, 1970. Courtesy of The Design Council and Springer-Verlag.)

- When the process of oxygen absorption is substantially completed, the oxidized fiber can be cut from the frames and packed into graphite boats for carbonization. This comprises heating in an inert atmosphere—"white spot" grade nitrogen is suitable— at temperatures from approximately 400 to 1600°C (750 to 2900°F).

- Some tar and further gases are evolved during carbonization. They include cyanogen (C_2N_2), which is another deadly poison and is treated in the same way as hydrogen cyanide. Also, at the higher end of the temperature scale, some grades of PAN produce a white crystalline deposit around the walls of the vessel, which is sodium cyanide (Ref. 2.6).

- A final, optional stage is heating the boats and their charge of carbon fibers for approximately one hour to even higher temperatures, in furnaces equipped with carbon resistance heating elements, at approximately 2300°C (4200°F).

- The effect of temperature on the mechanical properties of the resultant fiber is shown in Figure 2.9. Note that the modulus rises smoothly throughout the temperature range because the amount of crystallinity in the fiber increases at the expense of the amorphous material. This is shown in Figure 2.9. On the other hand, the tensile strength, after going through a maximum at approximately 1600°C (2900°F), falls to a plateau which is virtually unchanged to the highest temperatures used. From this behavior, we can select three distinct and reproducible types of carbon. Modern developments have resulted in more types than this. See *Carbon and High-Performance Fibers Directory*, Ref. 2.7. The three types described below indicate the range of properties that can be achieved.

Type 1: This is the stiffest fiber and the one discovered first.

Type 2: This is the strongest fiber, carbonized at the temperature at which the best tensile strength was recorded. Because the stiffness is lower and the strength is higher than the first type, the failure strain (or elongation at break) is somewhat increased, a move appreciated by aircraft engineers, the makers of sports equipment, and others. Also at the optimum temperature of 1600°C (2900°F), a more durable platinum-wound furnace can be used, which lowers the production cost.

Type 3: This denotes the cheapest fiber. Following the above theme, an acceptable fiber can be produced from a bank of normal laboratory furnaces equipped with standard Nichrome resistance heaters. The stiffness is lower than the previous grades, but the strength is reasonably good. As Figure 2.9 shows, modulus increases with increasing temperature. Delmonte (Ref. 2.8) describes fibers treated above 2000°C (3600°F) as "graphite." Graphite fibers consist essentially of carbon atoms, which by high-temperature heat treatment have obtained a crystalline state with a high degree of three-dimensional order as determined by x-ray diffraction patterns. Carbon fibers show a lower degree of three-dimensional order or may be almost amorphous. Graphite fibers have higher modulus and strength and are better electrical and thermal conductors. The term carbon/graphite is sometimes used in recognition of the carbon origin of graphite as well as the likelihood of less-than-perfect graphite crystalline structure, with its probable regions of some amorphous carbon. Unfortunately, this leaves the definition less than clear, but it partially explains why both terms are used and why no clear line between the two has been drawn. It is recommended that the term "graphite" should be used only for fibers treated above 2000°C (3600°F) and "carbon" for those treated below that temperature, as Delmonte describes them.

As the diagram from Boeing (Figure 2.10) shows, carbonization occurs at a lower temperature than graphitization. Therefore, it should be possible to describe each type of fiber correctly if its final heat treatment temperature is known. See Figures 2.10 and 2.11.

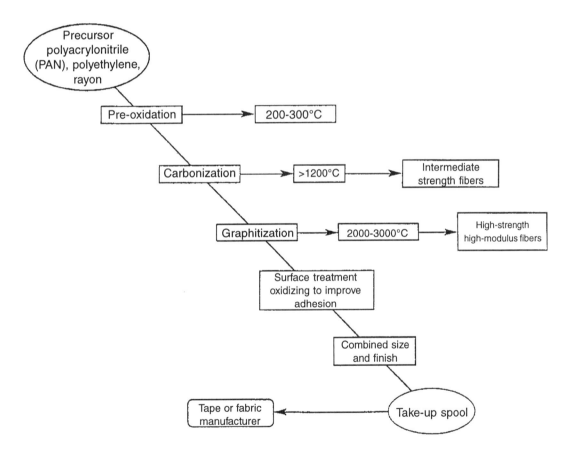

Fig. 2.10 Graphite fiber manufacturing process. (Courtesy of Boeing Commercial Airplane Group.)

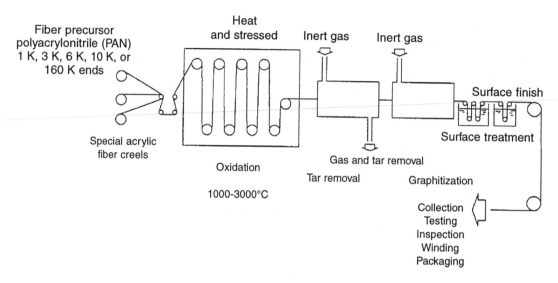

Fig. 2.11 PAN-based carbon fiber production. (Courtesy of Aero Consultants [United Kingdom] Ltd.)

Although three basic types of carbon are described above, carbon fiber is now available in five distinct categories:

1. Ultrahigh Modulus—Tensile modulus greater than 394 GPa
2. High Modulus—Tensile modulus in the range of 317 to 394 GPa
3. Intermediate Modulus—Tensile modulus in the range of 255 to 317 GPa
4. High Strength—Tensile strength up to 3.8 GPa
5. High Strain—Tensile strength greater than 3.8 GPa

Lovell, in the fifth edition of his *Carbon and High-Performance Fibers Directory* published in 1991, states that, "The previous classification of carbon fibers into UHM, HM, IM, VHS, and HS types has been amended since the VHS and IM types are merging and the HS type is beginning to be called standard modulus (SM), so that all reinforcing fibers are now divided into modulus ranges as UHM, HM, IM, and SM. It would be logical to extend this by renaming the fibrous carbon and graphite material as low modulus (LM) since these materials are being increasingly used for reinforcement to provide toughness rather than stiffness, especially for concrete, while retaining their use for furnace insulation and gland packings." This last variety is not of concern for aerospace usage. In a developing subject area such as this, it is difficult to establish any rules for naming because new types may emerge as more specialized applications are found.

Comparison of Carbon with Other Common Fibers: The main advanced composite fibers in current use are listed below, with general notes on their relative performance and cost in order to show their relationship to carbon fibers.

Increasing Cost and Stiffness	Advantages	Disadvantages	Approximate Cost
Glass	Good tensile strength Good electrical insulation Low cost	Low stiffness Low fatigue strength	$0.75–7.00/lb
Aramid	Good tensile strength Medium stiffness Excellent impact resistance	Low compressive strength Difficult to machine High moisture uptake	$12–20/lb
Carbon	High strength or stiffness Good fatigue strength Good damping characteristics	Low impact resistance	$12–100/lb
Carbon Fiber Fabric (Black)	Specific gravity = 1.75–1.76		
Approximate Single-Ply Laminate	Tensile strength = 500 MPa		

Courtaulds Aerospace PLC (Fothergill and Harvey) Designations

Style No.	A–0002	A–0009
Weave	Plain	5H-satin
Thickness (inches)	0.008	0.0126
Weight/Unit Area (gm/m^2)	189	283
Fiber Type	T–3000	T–3000
Number of Filaments	3K	3K

Also available as 8H (satin) and 4H (crowfoot satin) all 3K filaments.

Many suppliers produce material of this type.

Additional terminology may be defined as follows:

- **Tow**: A bundle of fibers with no twist (equivalent to a roving in glass), e.g., 3 K = 3000 filaments/tow.

- **Selvedge**: All carbon cloths are "Jet" loom fabrics; therefore, the loose weft yarns are locked (usually with Kevlar yarns). As with glass, always trim off the selvedge. Do not use the selvedge in the lay-up.

Carbon is available as a fabric (up to 1625 mm [64 in.] wide) and in tape form (approximately 4 in./100 mm).

For sizing and finish, see Section 2.3.2.

2.1.4 Aramid Fibers

Aramid fibers are now made by at least three companies:

- Du Pont (the original inventor)
- Akzo-Enka, whose aramid is called Twaron T-1000
- Teijin Ltd. of Japan, whose fiber is called Technora T-220

Kevlar (Du Pont) is the aromatic form of nylon (i.e., it contains benzene rings in its molecular structure). Its full technical name is poly para phenylene terephthalamide.

This material retains useful properties up to 180°C (350°F) (see Figures 2.1 to 2.3) but carbonizes at 427°C (800°F). It shows little embrittlement at -196°C, the temperature of liquid nitrogen. All aramid fibers absorb moisture from the atmosphere and must be dried before use as detailed in Section 3.2. Aramid composites absorb more moisture than carbon, glass, and most other composites because both the fiber and the resin absorb moisture; in most composites, only the resin absorbs moisture. They are lighter than glass and are of higher strength and modulus. Aramid fibers are highly aligned, high-performance polymer fibers that are capable of high strength and high modulus. The most widely known aramid fiber is Kevlar, which was introduced into the composites market in the 1970s by Du Pont. Kevlar fibers have a tensile strength approximately equivalent to high-strength carbon fibers, but their tensile modulus is significantly lower. Their compression strength and modulus are lower than the tensile values, and this restricts their use in compression-dominated applications. However, the low density of aramid fibers means that, on the basis of strength or stiffness per unit weight, they are superior to high-strength carbon fiber. Aramid fibers are produced by an extrusion and spinning process, followed by a stretching and drawing operation. One unique characteristic of aramid fibers is their ability to absorb ballistic energy. They have excellent impact resistance as a result of the failure mode of the fibers under impact load conditions. Fibers tend to delaminate into numerous fibrils on impact because the structure produced by stretching and drawing is almost completely crystalline. Because of this, aramid fibers are frequently used as the outer ply on airframe structures prone to stone-impact damage. See U.S. military specification MIL-Y-83370 for aramid fibers.

Design Considerations: Aramid fibers have good abrasion resistance. However, on the negative side, they do not resist flame as well as other fibers, and they burn through more quickly. For this reason, their use in aircraft interiors is not as common as it was in the past.

Adhesion to the matrix is also less for aramids than for other fibers. In impact applications such as helmets and body armor, delamination is a useful method of energy absorption. However, these are "one-shot" uses.

Use of Aramids in Repairs: Numerous successful applications indicate that the use of Kevlar for the repair of Kevlar parts should be considered when possible. Airframe manufacturers both allow and recommend the repair of Kevlar parts with fiberglass, but they do not consider the weight penalty to the operating airline. When Kevlar and other aramids are used for flight control surfaces, where weight and balance are important, it is particularly worth considering the use of Kevlar in the repair. Experience has shown that when used in Kevlar/carbon hybrids, delamination may occur between layers of carbon and Kevlar after some time in service in areas subjected to a wide range of cyclic temperatures. This seems to occur only on large aircraft flying at high altitudes and especially on engine cowlings where the range of temperature is high. Commuter aircraft and helicopters operating at lower altitudes have not reported problems. Aramid fibers are degraded by ultraviolet light and, when used for ropes, are usually shielded. In a composite structure, only the surface layers are affected.

Creep Properties of Aramids: In spite of their high inherent tensile strength, and even in unidirectional configurations, aramid fiber composites generally have higher creep rates than those of glass or carbon composites. (Ref. 1.1.) This may be partly due to fiber adhesion problems because glass fiber composites also have significant creep rates if untreated fibers are used. Glass fiber composites have a higher creep rate than carbon fiber composites.

Aramid Fabrics Used on Aircraft: Kevlar fabric is also known as aramid. It is yellow and has a specific gravity of 1.44. For mechanical, physical, and chemical properties, see the data sheet for each specific type. Typical fabrics commonly used for aircraft parts include:

Style	120	281	285
Weave	Plain	Plain	4H-crowfoot satin
Thickness (in.)	0.0043	0.0095	0.0086
Weight/Unit Area (gm/m^2)	60	175	175
Yarn TEX	22	127	127

Several versions of Kevlar are available:

Kevlar	Commercial grade (e.g., ropes)
Kevlar 29	High tensile strength
Kevlar 49	High modulus (main aerospace version)
Kevlar 149	High modulus (low water uptake version)

The selvedge is "cut" weft yarns produced on "Jet" looms; trim this off when using rolls of fabric. Kevlar is available in fabric and tape form. Do not remove the selvedge from tapes. For sizing and finish, see Section. 2.3.3.

Chemical Properties of Aramid Fibers: Aramid fibers of different types have different levels of chemical resistance. Kevlar 49, for example, has a high resistance to neutral chemicals but is susceptible to attack by strong acids and strong bases. (Ref. 1.1.) On the other hand, Technora aramid fiber exhibits high strength retention in both acids and alkalis. Check the data sheet for each type and manufacturer before selecting an aramid for a particular application.

2.1.5 Boron Fibers

Boron fibers are produced in a different way, in that they are formed by chemical vapor deposition of boron gas on a tungsten filament. Because the thermal expansion of cross-plied boron tapes is fairly close to that of aluminum alloys, these materials are being used to make composite repairs to aluminum alloy aircraft parts. Because of the fairly large diameter and stiffness of boron fibers, they are not supplied in the woven form in the same way as most other fibers. Boron fiber repairs (or similar repairs with other types of fiber) avoid the need to drill more holes for rivets or bolts and have been used for many years on the Mirage fighter, F-111 bomber, Hercules transport, Lockheed Orion maritime patrol aircraft, and Macchi trainer by the Royal Australian Air Force. Boeing and Federal Express have conducted a flight evaluation of this type of repair on a Boeing 747 aircraft. See Ref. 2.9 and Boeing report.

Boron fibers are produced by Avco (Lowell, Massachusetts) and supplied as pre-pregs by Textron Inc. (Lowell, Massachusetts). They are produced only in unidirectional tape form because they are large-diameter, stiff fibers that do not readily bend around the tight radii required by the weaving process.

Boron fibers have the following properties:

Tensile Strength	500,000 psi
Tensile Modulus	58,000,000 psi
Specific Gravity	2.57

2.1.6 Other New Fibers

There are a considerable number of new fiber types. The following list is not complete but describes some of these new types of fibers. For more extensive details of these and other new fibers, see the latest edition of *Carbon and High-Performance Fibers Directory* (Ref. 2.7). This directory is updated every few years and contains a wealth of information on fibers, fabrics, and hybrids.

Alumina Fiber: Alpha alumina fiber, FP alumina from Du Pont (Wilmington, Delaware), measures 20 μm in diameter. The fiber's modulus is high: 55 million psi. It can serve continuously to over 982°C (1800°F). However, pure alumina is relatively brittle. Nevertheless, it is well suited to reinforcing aluminum.

A semicontinuous alumina fiber, "Safimax" from ICI (Wilmington, Delaware), has a fiber diameter of only 3 μm. Its strength of 300,000 psi compares with that of FP alumina, but it is not as stiff and has a modulus of 33 to 40 million psi.

Another alumina fiber, from Avco, contains 85% alumina and 15% silica. Its strength of 210,000 psi and modulus of 28 million psi are less than those of pure alumina fibers; however, it is more flexible and can be woven. The fiber remains stable in molten metals and will not corrode. Transparent to radar, it retains 90% of its strength at 1249°C (2280°F).

Alumina/Silica Fibers: Alumina and silica are melted, then poured in a 50:50 ratio, and attenuated into fibers. The fibers, "Fiberfax" from Standard Oil Engineered Materials (Niagara Falls, New York), are comparable to glass, with strength up to 300,000 psi and modulus of 13 to 15 million psi. They can be used continuously to 1260°C (2300°F) and can endure excursions to approximately 1649°C (3000°F). For details of many similar fibers, see Ref. 2.10.

Alumina/Boria/Silica Fibers: These fibers with the trade names Nextel 312, 440, and 480 are monocrystalline oxide fibers from 3M (St. Paul, Minnesota). Nextel 312 is the least crystalline; Nextel 480 is the most crystalline. The higher crystallinity of Nextel 480 enables it to resist high temperatures. All three fibers have a 3:3 mole ratio of alumina to silica. Nextel 312 contains 14% boria; Nextel 440 and 480 each contain only 2% boria. Alumina/silica/boria fibers are available with diameters of 8 to 9 μm or 10 to 12 μm. The fibers resist high temperatures and chemical attack, and are suitable for reinforcing ceramics or metals. They also are compatible with epoxy, silicone, phenolic, and polyimide matrices. All three types can be exposed to a flame at 1093°C (2000°F) for fifteen minutes without penetration. Nextel maintains a low thermal conductivity, allowing it to serve as a heat shield.

Some properties of Nextel are listed below:

Nextel 312	Tensile Strength	250,000 psi
	Tensile Modulus	22,000,000 psi
	Specific Gravity	2.7–2.9
Nextel 440	Tensile Strength	300,000 psi
	Tensile Modulus	28,000,000 psi
	Specific Gravity	3.05
Nextel 480	Tensile Strength	280,000 psi
	Tensile Modulus	33,000,000 psi
	Specific Gravity	3.05

Nextel 312 retains most of its strength up to 982°C (1800°F), and 50% of its strength up to 1204°C (2200°F). Nextel 440 retains most of its strength up to 1093°C (2000°F). Nextel 480 retains 80% of its strength up to 1315°C (2400°F), and 50% up to 1399°C (2550°F).

Polyethylene Fibers: Polyethylene fibers are made by Allied-Signal Inc.—Fibers Division and have the trade name "Spectra." They have a low density, low adhesion to matrix resins, and a low service temperature; however, strength and modulus are good and their dielectric constant is low, which is useful for radomes. Moisture uptake is low. Polyethylene fibers can be treated with gas plasma, which modifies the surface, improving fiber-to-matrix bonding. The plasma removes hydrogen atoms from the polymer backbone, which are subsequently replaced by polar groups. The polar groups on the fiber surface enhance wetting and reactivity with the matrix. Treating with plasma also causes micropitting, enabling the resin to better lock onto the fiber.

Some properties of Spectra 900 are:

Tensile Strength	375,000 psi
Tensile Modulus	17,000,000 psi
Specific Gravity	0.97
Maximum Temperature for Continuous Use	110°C (230°F)
Dielectric Constant, Polyethylene/Epoxy	2.28

Some properties of Spectra 1000 are:

Tensile Strength	408,000 psi
Tensile Modulus	23,800,000 psi

Quartz Fibers: Quartz fibers are produced by J.P. Stevens (Greenville, South Carolina) and FMI (Biddeford, Maine) and have the following properties:

Tensile Strength	500,000 psi
Specific Gravity	2.2
Low Thermal Expansion	0.54×10^{-6} /°C

Norton and other companies are using quartz fibers for radomes because the fibers have a low dielectric constant.

Silicon Carbide Fibers: Silicon carbide fibers have the trade name Nicalon and are produced by Dow-Corning (Midland, Michigan). A polymer precursor is first spun into a fine thread and then is pyrolized to form a silicon carbide (SiC) fiber. The fiber, typically 15 μm in diameter, consists of fine SiC crystallites, with small amounts of free silica and free carbon. As with quartz, SiC is a ceramic and retains significant strength above 983°C (1800°F).

Tensile Strength	Up to 400,000 psi
Tensile Modulus	Up to 28,000,000 psi
Specific Gravity	2.32–2.55

SiC monofilaments from Avco contain a tungsten or carbon core, on which SiC is chemical vapor deposited until it builds up to a diameter of 140 μm. Properties of SiC by this method are:

Tensile Strength	Up to 600,000 psi
Tensile Modulus	Up to 60,000,000 psi

Polybenzimidazole Fibers: Polybenzimidazole (PBI) fibers are produced by Hoechst-Celanese Corporation (Charlotte, North Carolina). They are another organic fiber that maintains strength and modulus to over 349°C (660°F). PBI fibers remain stable in air up to 450°C (840°F) and up to 1000°C (1830°F) in nitrogen. They resist a laser and will not burn in air. The fibers flex easily and can be employed to reinforce and toughen bismaleimides (BMI), thermoplastics, and elastomers. PBI fibers resist most chemicals, solvents, fuels, and steam. They also have a low radar cross section. PBI filaments can be drawn or spun; drawn filaments are stronger.

Tensile Strength (Drawn)	Up to 65,000 psi
Tensile Modulus (Drawn)	Up to 800,000 psi

The filaments have a dog-bone cross-sectional area of 90 to 180 μm^2, equivalent to a diameter of 11 to 21 μm.

Vectran HS Fibers: Vectran HS liquid crystal polymers are being spun into fibers by Hoechst-Celanese (Charlotte, North Carolina). The fibers absorb little moisture—less than 0.1%. With a melting point of 327°C (621°F), they are more stable at high temperatures than polyethylene. Similar to polyethylene, they have a low dielectric constant of 2.6. They resist most chemicals, even at high temperatures, and they inherently retard a flame. Fiber diameter is 23 μm. Some properties of these fibers are:

Tensile Strength	Up to 450,000 psi
Tensile Modulus	Up to 15,000,000 psi
Specific Gravity	1.4

2.2 Forms of Reinforcement

2.2.1 Tapes

Tapes are usually defined as woven fabrics having a width of less than 4 in. (100 mm). Tapes have a woven selvedge which should remain on the tape, whereas it is common practice to cut the selvedge from wider fabrics. Unidirectional materials are described as tapes regardless of width and have occasional transverse threads to hold them together. Their fibers all run in a straight line (unidirectional). Because the fibers are unidirectional, the material will exhibit the maximum possible fiber-dominated characteristics for the fiber system used. The fibers also can be obtained in the form of collimated tow, strand, or roving.

2.2.2 Fabrics

Fabrics are made by weaving fibers or filaments into a cloth. A variety of weave patterns are possible, but in practice they are limited to a few standard types. (See Refs. 2.4 and 2.7). Fabric is the term used for woven fibrous or filamentary materials. The material itself is furnished as a dry cloth or as a pre-impregnated material. Fabrics may be used in the hand lay-up method of fabrication with a wet lay-up resin or in the form of a pre-preg. Pre-pregs also may be laid up by automated methods. In these operations, one or more plies of wet resin-impregnated cloth or pre-preg material are usually laid up on some form of tool and then cured under carefully controlled conditions of heat and pressure. From a structural standpoint, the material has the disadvantage of having mechanical properties lower than those of unidirectional material because the fibers pass over and under each other; therefore, when a tension load is applied, they try to straighten. When a compression load is applied, they are already buckled, and thus their compression strength is reduced compared to that of a similar unidirectional material. When using fabrics, it is important to note the warp and weft directions. Warp direction is defined in AIR 4844 as, "The yarn running lengthwise in a woven fabric. A group of yarns in long lengths and approximately parallel. Fabrics are tensioned in the warp direction during weaving. The weft is not tensioned." Weft direction is defined as, "The transverse threads or fibers in a woven fabric. Those fibers running perpendicular to the warp. Also called fill, filling yarn, or woof." The important point is that the warp direction must be noted when cutting fabric to ensure that the fabric can be aligned correctly in the direction specified on the drawing or in the SRM. The warp direction is along the length of a roll of material and parallel to the selvedge edge and is the direction in which each layer of fabric must be aligned according to the warp clock on the drawing. The number of layers in each direction is chosen by the designer to meet the loads that the part is expected to carry in each direction in service. The reason for emphasizing this is that the warp direction is the direction of greatest strength in a fabric. Several fabrics have significantly fewer yarns in the weft direction and therefore are significantly weaker in the weft direction. Fabrics may be supplied in several forms. See Figs 2.12 to 2.14.

2.2.2.1 Woven Fabric Weave Styles

The mechanical properties of the fiber composite depend on the woven fabric construction as well as on the fiber type and resin matrix. Many fabric weave styles are available, but the more commonly used aerospace weave styles are shown here in diagrams and text.

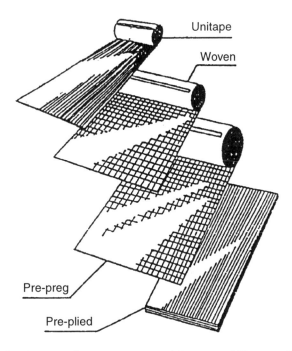

Fig. 2.12 Fiber forms and weave patterns. (Courtesy of Heatcon Composite Systems.)

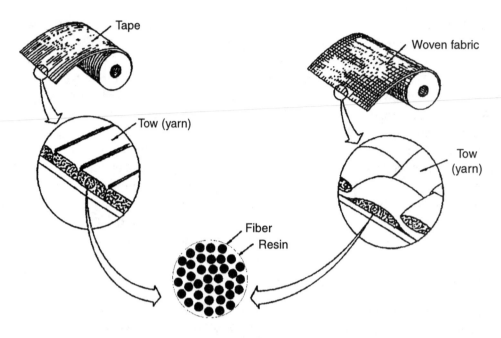

Fig. 2.13 Available forms of composite materials. (Courtesy of Aero Consultants [United Kingdom] Ltd.)

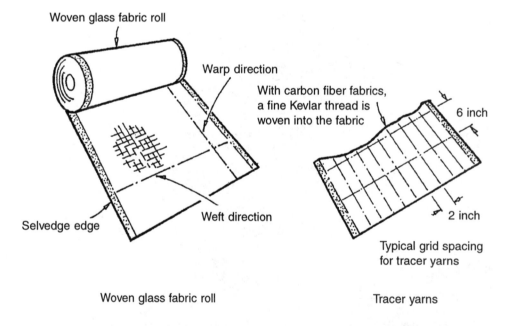

Fig. 2.14 Woven fabrics terminology. (Courtesy of Aero Consultants [United Kingdom] Ltd.)

Weave styles have been standardized to a large degree and are usually defined in aircraft manufacturers' materials specifications and appropriate U.S. military (MIL) specifications and others. Typically, these specifications define the following points:

- **Weave Style Reference:** Usually a three- or four-figure number (i.e., 120 style)

- **Fabric Count:** The number of warp yarns per inch and weft (fill) yarns per inch (i.e., 60 x 58 means 60 warp and 58 weft yarns per inch)

- **Warp Yarn Definition:** A definition of the glass yarn buildup or, in the case of carbon fiber, the number of individual fibers in the yarn: 3,000 (3 K), 6,000 (6 K), or 12,000 (12 K)

- **Weft Yarn Definitions:** Same as Warp Yarn Definitions

- **Weave Style Name:** Defines the weave style (i.e., plain weave, crowfoot weave, 2 x 2 twill weave)

- **Fabric Weight:** Usually expressed in ounces per square yard or grams per square meter and is the weight of the dry, unimpregnated fabric

- **Fabric Thickness:** Usually expressed in inches (i.e., 0.005 in.) or in millimeters

Some weave styles kink or crimp the yarns to a greater or lesser degree than others. This amount of fiber crimp strongly influences the capability of the fabric to be distorted or draped to conform to complex shapes and the resulting mechanical properties of the composite component manufactured from the fabric. The less the fiber crimp, the better the drapeability of the fabric and the better the laminate mechanical properties. Less crimp also means more fraying. For weave patterns, see Figure 2.15. See the selvedge picture in Figure 2.16.

2.2.2.2 Effect of Weave Style

Ideally, the same material, weave, and weight should be used for repairs as was used for the original construction because the fiber tensile properties and modulus in a composite are affected by the weave style. (UD is unidirectional tape; 8HS is 8 harness satin weave; 5HS is 5 harness satin weave; and 4HS is 4 harness satin weave.) In tension, plain, twill, and 4HS satin provide lower values than 5HS, 8HS, and UD. If plies of UD tape are laid at angles to each other, then any such angle ply lay-up will have a lower modulus and strength than pure unidirectional material.

In modulus, all weaves give lower values than UD, with only 8HS approximating to the UD value. However, drapeability increases through the range of weave styles from UD–plain–twill–crowfoot or 4HS–5HS–8HS.

Weave Patterns **Plan View** **Side View**

Plain View: The basic and most common textile weave. One warp end weaves over one weft pick and under the next alternately.
Characteristics: This is the firmest and most stable of industrial weaves, with fair porosity and uniformity of strength in all directions.

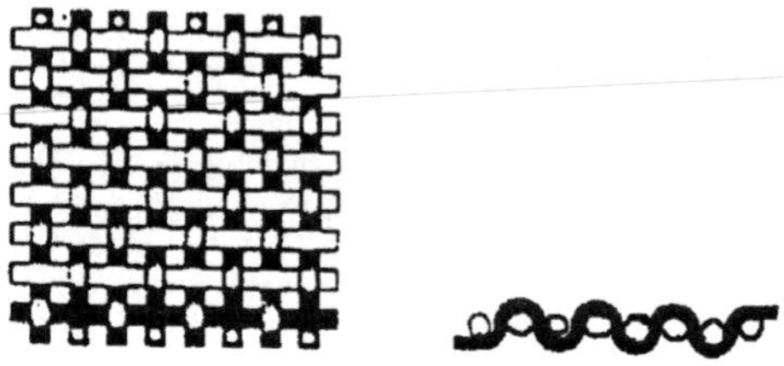

Twill Weave: Constructed with one or more warp ends weaving over and under two or more weft picks in a regular fashion. This produces either a straight or broken diagonal line in the fabric.
Characteristics: More pliable than the plain weave. Folds and hangs better than plain woven fabric and has better sewing characteristics than satin weaves.

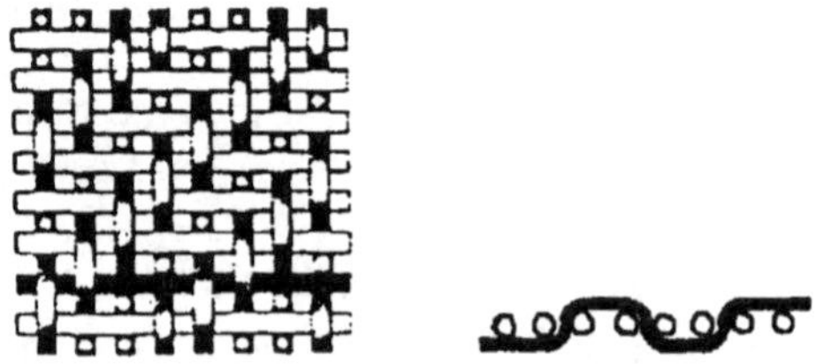

Four Shaft or Crowfoot Satin: Constructed with one warp end weaving over three and under one weft pick.
Characteristics: More pliable than plain woven fabric. Well designed for accurate forming around complex and compound curved surfaces. Allows for high density of yarns to be woven compared with plain weaves.

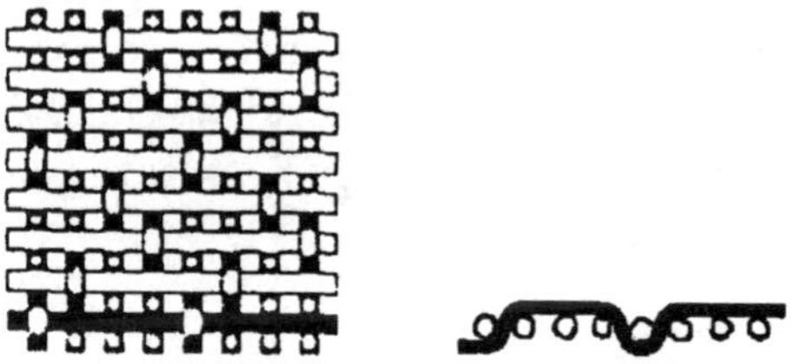

Eight Shaft Satin: This fabric has one warp and weaving over seven and under one weft pick.
Characteristics: Excellent pliability and performance on compound curves. High strength in all directions and can be woven in the highest density. The design allows for close, tightly woven fabrics.

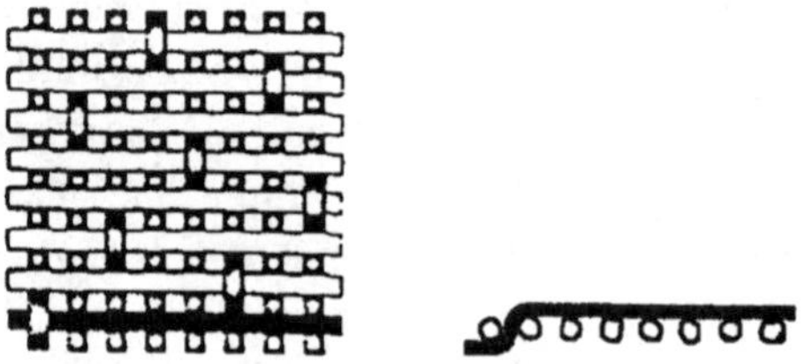

Fig. 2.15 Weave patterns. (Courtesy of CS Interglas Ltd.)

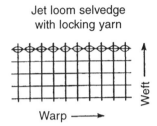

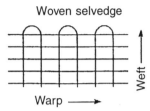

Fig. 2.16 Selvedge. Selvedge is never incororated into the work; it is always cut off (approximately the last 0.5 inch). (Courtesy of CS Interglas Ltd.)

If the identical material is not available, then consideration must be given to the strength and stiffness of the original material and whether additional layers may be required because, for example, plain weave may have to be used when 8HS is unavailable. Property data for both types then becomes necessary.

Fabrics have the advantage over UD composites in drapeability, reduced lay-up time (approximately 2:1), and more consistent laminate properties (3 to 4% variations instead of 5 to 9%), permitting higher design allowables. Also with fabrics, hole bearing load capability is higher (approximately 10 to 15%), and crack propagation is lower than in UD laminates.

2.2.2.3 Noncrimp Fabrics

These are a relatively new development. Before noncrimp fabrics were introduced, composite materials tended to fall into well-defined categories (i.e., unidirectional, woven cloth, and mats). Each offered an advantage in terms of production cost or engineering performance. Selection always involved a tradeoff among cost, performance, and production convenience for each application. Noncrimp fabrics (NCFs) combine the easy deposition of fabrics with the mechanical performance of unidirectional composites. NCFs take the form of multilayered, multi-angular material. The fabric comprises several distinct layers, each of a highly aligned fiber. Several layers are built up in sequence, and then "Z" direction stitching fastens the layers to form a fabric. Within broad limits, the fiber angle of each layer is adjustable, allowing a multilayered fabric to be constructed with each layer at a separate orientation. However, it is important to note that each layer approximates to a unidirectional ply, and the fibers are not woven in the conventional sense. The stitching yarn also may be used to provide some "Z" direction reinforcement and can give some resistance to damage. The fabrics are normally produced as a continuous roll with a width of 1.2 to 1.5 meters. Noncrimp fabrics have many advantages over conventional woven cloths and unidirectional composites. Their versatility offers designers a choice of number of layers, orientations, and fiber types. Their ease of handling is enhanced through high levels of drapeability and stability, although thicker materials and those with many orientations may be more difficult to handle.

NCFs can be supplied as pre-pregs, but a new technique for thermoplastic matrices has been developed. Intraspersed materials, known as Type 1, have filaments of thermoplastic dispersed uniformly among the carbon fibers, as shown in Figure 2.17.

Each NCF layer contains a mix of fibers and thermoplastic filaments at the appropriate volume fraction. Type 1 fabrics have the advantage that minimum matrix flow is required during processing; however, a disadvantage is that the thermoplastic and structural filaments tend to separate during production, giving rise to an uneven distribution of fiber and resin within the composite. Interspersed materials have tows of carbon and thermoplastic filaments placed as separate layers within the NCF (Figure 2.17). This overcomes the problem of uneven fiber distribution and facilitates closer control of the quality of the

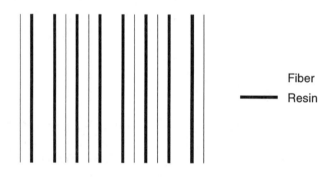

Intraspersed filaments of fiber and resin
Type 1

Interspersed filaments of fiber and resin
(A film of thermoplastic resin may be used instead)

Fig. 2.17 Intraspersed and interspersed noncrimp fabrics (NCFs).

layers. This method also allows the fiber lay-up to be tailored for optimized handling because the thermoplastic filaments completely disperse during processing and thus their initial orientation is unimportant. In both cases, the "Z" direction stitching uses thermoplastic yarn similar to that of the matrix material. "Z" direction stitching can be permanent and can be done with aramid, carbon, or glass fibers. Aramid is tougher and more flexible. See Figure 2.18.

2.2.2.4 Nonwoven Randomly Oriented Mats

These are different from the noncrimp fabrics in that they are randomly oriented mats. They can be supplied in two basic forms made from a range of fiber types.

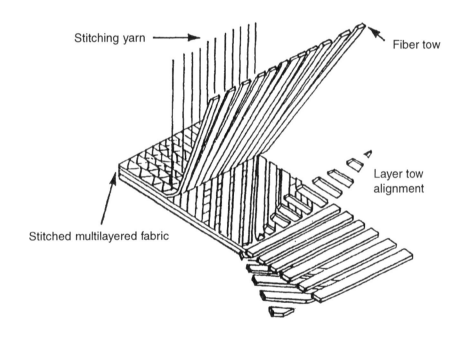

Some of the Fibers and Fabrics Available (1992)

Alumina	Aramid	Boron	Carbon
Cotton	Flax	Glass	Polyethylene
Polybenzimidazole	Quartz	Silicon Carbide	Vectran HS

Fig. 2.18 Noncrimp fabrics (NFCs). The technique of stitching different layers of fibers gives the strength of a unidirectional tape with the ease of use of a woven cloth.
(Courtesy of TechTextiles International Ltd.)

49

Chopped Strand Mats: In chopped strand mats, the fibers are typically 2 in. (50 mm) long, are randomly oriented in the plane of the fabric sheet, and are held together with a binder. The binder must meet several requirements to perform adequately:

- It must hold the strands together strongly enough to permit laminating.

- It must neither inhibit nor catalyze the resin cure.

- It must not cause discoloration of the laminating resin.

- It must not affect the efficiency of any coupling agent applied to the fibers to ensure a good bond to the resin. Chopped strand mat can be supplied in thin or thick sheets. The strength of the final resin-impregnated product is considerably less than that of woven fabric laminates, but it is adequate for many purposes and is considerably cheaper. Chopped strand products respond to different resin mechanical properties in the same way as woven fabrics, but at a lower level of strength.

- It must dissolve in the resin at a rate suitable for the manufacturing process involved. For hand lay-up, mats with a high or medium solubility binder are used. Low solubility mats, formulated so that the binder does not dissolve too quickly in the resin, are used for matched metal die molding.

Continuous Strand (Filament) Mats: These are similar to chopped strand mats, in that they are held together with a binder. However, in this case, continuous fibers of considerable length are employed, and they may be oriented in a swirled or random pattern. Continuous strand mats are available in a range of weights.

2.2.2.5 Fiber and Fabric Glossary

Although SAE AIR 4844A Composite and Metal Bonding Glossary is recommended for reference, it may be useful to extract a limited number of terms at this point for use with this section.

- **Fiber**: Material form showing high length-to-diameter ratio and normally characterized by flexibility and fineness. A general term used to refer to filamentary materials. Often fiber is used synonymously with filament. It is a general term for a filament with a finite length that is at least 100 times its diameter, which is typically 0.10 to 0.13 mm (0.004 to 0.005 in.). In most cases, it is prepared by drawing from a molten bath, spinning, or deposition onto a substrate. See *Whisker*. Fibers can be continuous or specific short lengths (discontinuous), normally no less than 3.2 mm (0.125 in.).

- **Filament**: A single fiber. The smallest unit of a fibrous material. The basic units formed during drawing and spinning, which are gathered into strands of fiber for use in composites. Filaments are usually of extreme length and small diameter, usually less than 25 μm. Typically, filaments are not used individually. Some textile filaments can function as a yarn when they are of sufficient strength and flexibility.

- **Yarn**: An assembly of fibers, of similar or variable length, held together by twisting. An assemblage of twisted filaments, fibers, or strands, either natural or manufactured, to form a continuous length that is suitable for use in weaving or interweaving into textile materials.

- **Tow**: A loose bundle of fibers, having little or no twist, normally associated with carbon fiber materials. An untwisted bundle of continuous filaments. Commonly used in referring to man-made fibers, particularly carbon and graphite but also glass and aramid. A tow designated as 140 K has 140,000 filaments. Equivalent meaning to roving in glass.

- **Roving**: An assembly of compact fiber bundles, or strands, usually with no twist; usually heavier (higher TEX) materials. The term is often used for glass fibers.

- **Denier**: The weight in grams of 9000 meters (29,529 ft) of a yarn, tow, or roving; often used with aramid materials.

- **TEX**: The weight in grams of 1000 meters (3281 ft) of a yarn, tow, or roving. (9 x TEX = Denier.)

- **End**: A thread, yarn, or tow, etc., running along the length of a fabric (the warp).

- **Pick**: A thread, yarn, or tow, etc., running across the width of a fabric (the weft or fill).

- **Warp**: The yarn running lengthwise in a woven fabric.

- **Weft**: The transverse threads or fibers in a woven fabric. Those fibers running perpendicular to the warp. Also called fill, filling yarn, or woof.

- **Hybrid**: Hybrids can take two forms:

  1. Hybrids in which carbon and glass fibers are woven into the same cloth, and aramid and glass fibers are woven in the same way. Any combination of any type of fiber could be woven in this way.

  2. Hybrid constructions in which a laminate is composed of layers of cloth in which some layers are all carbon and other layers are all glass or aramid. Each cloth is made from only one type of fiber, but the construction is a hybrid because not all the layers are made from the same type of fiber.

- **Pre-Preg**: Either ready-to-mold material in sheet form or ready-to-wind material in roving form, which may be cloth, mat, unidirectional fiber, or paper impregnated with resin and stored for use. The resin is partially cured to the "B" stage and supplied to the fabricator, who lays up the finished shape and completes the cure with heat and pressure. Two distinct types of pre-preg are available:

1. Commercial pre-pregs, in which the roving is coated with a hot melt or solvent system to produce a specific product to meet specific customer requirements

2. Wet pre-preg, in which the basic resin is installed without solvents or preservatives but has a limited room-temperature shelf life

Note that "B" stage pre-pregs also have limited shelf lives. The first is the shelf life under refrigerated storage (usually 6 to 12 months at -18°C [0°F]), and the second is the out-time at room temperature (usually 7 to 30 days at 20°C [68°F]). See Chapter 3, Handling and Storage.

- **Pre-Plied Plies**: Multiple plies of pre-impregnated material that have been stacked and compacted together to form a lay-up and packaged or stored prior to being cured individually or in combination with other parts.

Figure 2.15 shows the four most common weave patterns: plain weave, twill weave, four shaft or crowfoot satin, and eight shaft satin. Five shaft satin is also used. All of these materials can be obtained in a range of fiber types, either as dry fabrics with finishes suitable for particular resins or as pre-pregs with a "B" stage thermosetting resin (usually an epoxy) already incorporated in the fabric ready for a subsequent hot cure. They also can be obtained with thermoplastic matrices or thermosets other than epoxy.

See Figures 2.19 to 2.21 for additional details of glass fiber fabrics and Table 2.2 for materials in common use.

2.3 Fiber Sizing and Finish

2.3.1 Glass Fiber Finishes

Oil is used on the mills when the glass cloth is manufactured. However, if cloth containing oil is used with resins, it will resist the resins, and a product of low strength will result. To remove the oil, the rolls are baked in an oven at 600°F (315°C), which burns off all the oil, resulting in a pure glass cloth. For the cloth to better absorb the resin, a pre-wetting treatment is given. This produces a cloth that readily allows the resin to flow completely around, or encapsulate, each fiber to produce a strong lamination. These pre-wetting treatments or finishes are important to good adhesion and especially to durability in a hot/wet environment. It is important to recognize the difference between a *size* and a *finish*. A *size* is applied to glass fibers as they are drawn and before they are twisted into a yarn to provide lubrication and to prevent damage during twisting, weaving, and handling. A fabric supplied without removal of the size is described as "loom state," "greige," or "untreated."

For glass fibers, several surface finishing treatments are available, depending on the resin system to be used. They are used for three basic reasons:

Fabric Design

Properties of fiberglass

The versatility of glass as a fiber makes it a unique textile material. Fiberglass in fabric form offers an excellent combination of properties from high strength to fire resistance. A wide range of yarn sizes and weaves permits the design of an unlimited number of fabrics, enabling the end user to choose the best combination of material performance, economy, and product flexibility.

Dimensional stability

Fiberglass is a dimensional stable engineering material. Basically, it does not sretch or shrink even after exposure to extremely high or low temperature. The maximum elongation for "E" glass at break is 3.5%, and it has a 100% elastic recovery when stretched nearly to its point of rupture.

Moisture resistance

Glass fibers do not absorb moisture. They do not change physically or chemically on exposure to water.

High strength

The high strength-to-weight ratio of fiberglass makes it a superior material in applications where high strength and minimum weight are essential criteria. In textile form, this strength can be undirectional or bidirectional, allowing design as well as cost flexibility.

Fire resistance

Fiberglass is an inorganic material and will not burn or support combustion. It retains approximately 25% of its initial strength at about 500°C.

Chemical resistance

Many chemicals and chemical solutions have little or no effect on glass fiber. The inorganic glass textile fibers will not rot, become mildewed, or deteriorate. They are, however, affected to varying degrees by prolonged exposure to strong mineral acids.

Electrical properties

Fiber glass is an excellent material for electrical insulation. A combination of properties such as low moisture absorption, high strength, heat resistance, and low dielectric constant make fiberglass fabrics ideal as a reinforcement for circuit boards and insulating varnishes.

Fiberglass yarn

How glass fiber is made:

A tightly controlled series of manufacturing processes are used to produce glass fibers and yarns. A specific formulation of ingredients is made into a batch and fed into a high-temperature furnace where it is melted to form glass. Fine, precisely controlled filaments ranging in number from 51 to 1632, are drawn from the molten glass through a platinum bushing, combined to form a strand, and coated with a size to obtain strand integrity.

Twisted single strands are referred to as single yarns. By plying or twisting these single yarns together, a variety of textile yarns are made. These yarns are referred to as plied yarns.

Glass Composition—By Weight		
Composition	**"E" Glass**	**"S" Glass**
Silicon dioxide	52-56%	64-66%
Calcium oxide	16-25%	-
Aluminum oxide	12-16%	24-26%
Boron oxide	8-13%	-
Sodium and potassium oxide	0-1%	-
Magnesium oxide	0-6%	-

Fiberglass yarns are available in different formulations. "E" glass (electrical) is the most common all-purpose glass; "S" (high-strength) glass is made for special applications.

The variety of fiberglass yarns produced requires an exact system of nomenclature for identification. The nomenclature used for fiberglass consists of two basic parts–one alphabetical, the other numerical.

The example below shows that letters describe the basic strand by composition and type, while the numbers identify the filament diameter, the strand weight, and the yarn construction.

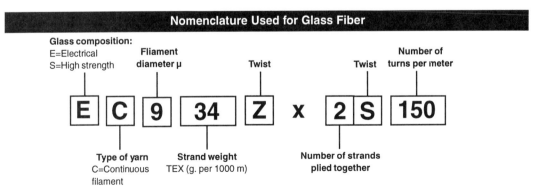

Fig. 2.19 Fabric design and nomenclature for glass fiber. (Courtesy of CS Interglas Ltd.)

Description of Continuous Filament Glass Fibers			
Filament Designation	Strand Weight (TEX)	Nominal Filament Diameter	Number of Filaments
3	34	3.8	1224
4	34	4.5	816
	68	4.5	1632
5	2.8	5.8	51
	5.5	5.8	120
	11	5.8	204
	22	5.8	408
6	34	6.3	408
	68	6.3	816
	136	6.3	1632
7	22	7.3	204
9	34	9.1	204
	68	9.1	408
	136	9.1	816
11	204	10.9	816
13	68	12.9	204
	136	12.9	408
	272	12.9	816

Textile glass formed by filaments

Single filaments   Single yarn

Assembled yarn   Plied yarn

Z-turn single yarn   S-turn plied yarn

Fig. 2.20 Textile glass details. (Courtesy of CS Interglas Ltd.)

1. To improve bond strength
2. To improve moisture and abrasion resistance
3. To stabilize the weave and increase thermal insulation

A *finish* is usually a coupling agent (a double-ended molecule), one end of which bonds to the fiber and the other end to the resin. In some cases, it is more similar to a primer. The wrong finish or the absence of a finish can be disastrous in terms of the composite strength and durability; therefore, it is essential to ensure that the resin and fabric finish are correctly matched. Manufacturers of glass fiber and fabric provide tables to facilitate the choice of the correct finish for each resin system. In many cases, a finish is compatible with more than one resin system.

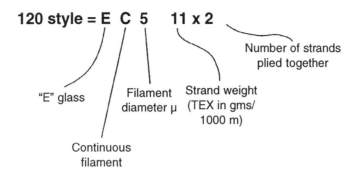

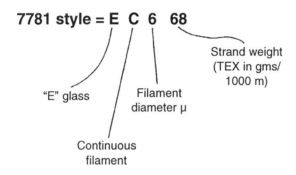

Fig. 2.21 Additional nomenclature for glass fiber. (Courtesy of CS Interglas Ltd.)

Finishes commonly used on glass fabrics are as follows:

- Vinyl silane (methacrylate silane finish) for use with polyester resins
  Clark-Schwebel Code P734, Compliance Standard BS 3396, Part 3, Grade "S"
  Courtaulds (Fothergill Engineered Fabrics) Code 208

- "Volan" (methacrylato chromic chloride) used with epoxy and polyester resins

  Clark-Schwebel Code P703 chrome finish, Compliance Standard BS 3396, Part 3, Grade "C"
  Courtaulds (Fothergill Engineered Fabrics) Code 205

- Amino silane A-1100, may be used with epoxy, phenolic, or melamine resins
  Clark-Schwebel P705

- Epoxy silane Z-6040, used on pre-pregs, suitable for epoxy and phenolic resins

2.3.2 *Carbon Fiber Finishes*

It should be understood that the continuous carbon fiber produced for sale has been sized (i.e., given a protective surface coating) (See Ref. 2.11.) because raw fiber straight from a carbonizing furnace is likely to pick up surface charge in rubbing contact with guides, rollers, etc., and is then awkward to handle. The electrostatic charges are clearly seen on passing a comb through a bundle of the tows, because the fibers fly apart. Winding or unwinding from a spool, passage through guides, weaving, and other textile processes lead to snagging, breakages, and fine floating particles, which can short electrical machinery.

In ascending order of susceptibility to damage, weaving is more abrasive than simple filament winding; braiding machines inflict the most severe abrasion. The fiber intended for braiding needs a heavier coating of size.

Many different polymer solutions have been tried, but the general consensus is that a low-molecular-weight epoxy resin (or a blend of liquid and solid for braiding) is the most satisfactory. This is normally applied without added hardener.

In addition to a size, carbon fiber may require a surface treatment to improve adhesion to the matrix. Surface treatment is essentially an oxidative process. The idea is not simply to clean the surface, but also to produce an etching or pitting of the surface and to provide reactive oxide sites by chemical reaction.

An early wet oxidation, first introduced by AERE Harwell, was sodium hypochlorite solution. The "short-beam" shear test, a good measure of interlaminar adhesion, demonstrated a useful increase; however, the tensile strength and the impact strength both decreased considerably. When continuous carbon fiber became available, the hypochlorite method was abandoned in favor of the electrolytic process (Ref. 2.12).

Because carbon fiber is a good conductor of electricity, a section of tow passing between a pair of separated conducting rollers can be made into the anode of an electrolytic cell. A number of anions are theoretically available (for example, sulfate, phosphate, chloride, and permanganate) which on discharge at the anode produce nascent oxygen, an effective oxidizing agent. In practice, salts are chosen to yield the minimum permanent residue.

By varying temperature, concentrations of electrolyte, current density, and speed through the bath, surface attack on the fiber can be accurately controlled. Moreover, the speed is convenient to interpose the electrolytic process between the last carbonization furnace and the sizing bath. Another method of surface treatment which leaves no residue is thermal oxidation. It is usually carried out at 300 to 450°C (572 to 842°F) in air. Atmospheres of other gases have been used to speed the process.

The reactivity of carbon fiber toward both electrolytic and thermal oxidation varies with the temperature of the original carbonization. The least reactive surfaces are those that are most crystalline (i.e., those that were exposed to the highest temperatures during formation).

2.3.3 Aramid Fiber Finishes

There is a need for a good finish for aramids to improve adhesion to epoxy resins for structural applications and where long-term durability is required. Normally, any sizing used is removed by a scouring process involving passage of the fibers through a bath containing chemical agents, followed by drying, before bonding with epoxy resins (Courtaulds Code 001 finish) or equivalent define the process. In the same way as with other fibers, different surface finishes are recommended to suit the chosen matrix resin, and it is necessary to ensure that the correct one is specified. In the case of polyester and vinyl ester resins, a proprietary finish (Courtaulds Code 286) is recommended for Kevlar. However, the performance of a finish for aramids is limited because, above a certain bond strength to the fibers, they will disbond by fibrillation. For this reason, the bond strength cannot usefully be increased beyond this level, but to reach it would be helpful. A finish that could ensure durability of the resin/fiber bond would be useful.

2.3.4 Sizing and Finish for New Types of Fibers and Fabrics

Each new type of fiber and hence fabric produced may require its own specific surface finish. Manufacturers' data sheets should be consulted to ensure that the correct finish is specified for the resin matrix to be used in each case.

2.4 Matrix (Resin) Systems

Resins fall into two general types: thermoplastics or thermosets. Both types are discussed separately and in more detail in this section.

2.4.1 Thermoplastic Resins

These are already high-molecular-weight, strong solids. They soften on heating, and on cooling they regain their original mechanical properties.

The most common thermoplastics are:

- Nylon
- Polycarbonate (PC)
- Polyimide (Condensation Type) (PI)

- Polyetherimide (PEI)
- Polysulfone (PS)
- Polyethersulfone (PES)
- Polyphenylenesulfide (PPS)
- Polyetheretherketone (PEEK)

Several resins from this list are beginning to be used as composite matrices. These include:

- Polyimide (PI), used as a matrix material for composites and as an adhesive, a fiber, and a film
- Polyethersulfone (PES), made by ICI as Victrex
- Polyphenylenesulfide (PPS), made by Phillips Petroleum as Ryton
- Polyetheretherketone (PEEK), made by ICI

Thermoplastics can be remelted many times and for this reason are said to be easy to repair. However, because they have high glass transition temperatures (T_g) (a desirable property), they also have high melting points. This makes repair a problem because the temperatures involved are dangerous near fuel tanks and would seriously damage the heat treatment condition of aluminum alloys. With suitable "corona discharge" surface treatments, repair with epoxy systems is possible in some cases. Repair of thermoplastic parts requires new techniques specially developed for the purpose. These are needed now for those parts already in service. See Boeing B.747-400 SRM and papers by Kinloch in Ref. 2.13.

2.4.2 Thermosetting Resins

Thermosetting resins harden or cure by a process of chemical cross-linking, whereby resins of low molecular weight and good solubility grow into products of high molecular weight and limited solubility. Cross-linking is an irreversible process, and these resins can never melt. They may become softened to a considerable extent and will eventually char and burn, but they cannot melt. Many systems will cure at, or slightly below, room temperature and emit some exothermic heat. Exothermic heat is a real problem in some cases, which limits the amount that can be mixed at any one time. Other systems require heat to approximately 120 or 180°C (250 or 350°F) to cause a cure. Very low temperatures will slow or prevent chemical reaction and cure. In all cases, the speed of cure increases if heat is applied.

Thermosetting resins are the most commonly used today, although thermoplastics are slowly entering service in aircraft and other applications. The most common thermosetting materials are the following:

- Phenol-formaldehyde
- Melamine-formaldehyde
- Resorcinol-formaldehyde

- Urea-formaldehyde
- Polyester
- Vinyl ester
- Epoxy
- Acrylic
- Polyurethane
- Silicone
- Polyimides (PMR type) (PI)
- Bismaleimides (BMI)
- Cyanate ester

BMI and cyanate ester are more difficult to process than the others and are reserved for high-temperature applications.

Phenolic Resin Systems: This title covers a range of products. Each subtype consists of a range of materials tailored to suit specific applications.

- **Modified Phenolic Resin Pre-Pregs**: Phenolic resins used in pre-pregs provide excellent fire performance properties to composites, which on burning produce low smoke emission and low evolution of toxic gases. Hence, they receive extensive use in interior passenger cabin composite components as a matrix resin. They are used in ships, railway carriages, and aircraft, and they have good heat resistance up to 120°C (250°F). Phenolic pre-preg resins can be used with a range of fibers—primarily carbon, glass, and aramid. Fibredux 917 is an example of this type. It is an ideal facing material for lightweight honeycomb sandwich panels for use in interior furnishings in passenger transport systems.

- **Phenol-Formaldehyde**: Some forms are used as wood adhesives and for plywood, particle board, pulp board, and hardboard manufacture, and for the impregnation of paper to make decorative laminates. Others are used for the bonding of grit to form abrasive grinding wheels, abrasive paper, and cloths, and for the treatment of filter paper for oil filters.

- **Polyvinyl Formal Phenolics**: These were the first metal-to-metal bonding structural adhesives known as Redux 775. Although first developed in 1943, they have been proven to be more durable than epoxies in service and are used today. Some aircraft with Redux 775 bonding continue to be in service after more than 30 years.

- **Resorcinol-Phenol-Formaldehyde**: These are used for bonding a wide range of materials to porous substrates, e.g., brick, concrete, unglazed porcelain, expanded plastics such as ebonite, polystyrene, polyurethane and PVC, industrial and decorative laminates, leather, cork, linoleum, nylon; natural and synthetic rubbers but not silicone, and sheet metals to wood.

Melamine Formaldehyde: These are used as laminating resins for the impregnation of paper for decorative laminates.

Urea-Formaldehyde: These are used for bonding wood, in the manufacture of plywood, and the bonding of veneers. Some versions are made for bonding decorative and nondecorative laminated plastic sheets such as Paxolin, Tufnol, Formica, and others.

Polyester: These two- or three-part systems offer good environmental resistance and are good to 150°C (300°F). They are used as wet resin and pre-preg. Although easily fire retarded, they emit considerable smoke when burning; consequently, their use in aircraft is diminishing. Polyester resins are not as strong as epoxy. They can be dangerous if incorrectly mixed (i.e., risk of explosion); therefore, the manufacturer's instructions must be followed carefully. Polyester resins are widely used in boat building because of their low cost. Two basic types of polyester are orthophthalic and isophthalic. Isophthalic resins have better mechanical properties and chemical resistance and are used as a gel coat or barrier coat because of their greater resistance to water permeation.

Vinyl Ester: These resins are increasing in importance in boat building, and, although more expensive, their high quality has resulted in their consumption being almost equal to polyesters at the time of this writing. They offer good impact and fatigue resistance, and they make a good permeation barrier to resist blistering in marine laminates. Vinyl esters fall between polyesters and epoxies in both cost and performance.

Epoxy: Epoxy resins are very strong and provide good environmental resistance. They can have high-temperature resistance (greater than 200°C [392°F]) and are used as wet resin, pre-preg, or film adhesive layers. Epoxy resins will usually burn readily, but additives can reduce this considerably. They also emit a significant amount of smoke when burning. Epoxy resins are the most commonly used system in aerospace, although their durability has not been developed to the level of the phenolic Redux 775.

Acrylic: These are finding considerable use in automobile applications and many other areas, and are closely related to the anaerobic type commonly used for thread locking and the bonding of bearing housings and bushes. Their low modulus and high elongation to failure render them tough and impact resistant. Two types are made: the resin plus activator type in which the resin is applied to one surface and the activator is applied to the other, and the two-part type in which the activator is mixed with the resin before use. Both types achieve good performance from a room-temperature cure. Another version is the ultraviolet curing acrylic for bonding glass and clear or translucent plastics, which contains photoinitiators in the resin to trigger the curing reaction. A further type is the epoxy cationic version for opaque substrates. In this system, ultraviolet light is used to initiate the cure, which proceeds slowly enough to allow a short assembly time.

Polyurethane: Polyurethane resins can be rigid or flexible, with poor strength and fire performance. They are used infrequently in modern aircraft. Made in a variety of types, they can be thermosetting two-component systems, single component using atmospheric moisture to initiate the cure, thermoplastic, or solvent-based, depending on the formulation. Polyurethane resins are not very water resistant, but this can be improved by the use of primers, especially on metal surfaces. They are flexible and tough with a range of applications in the shoe-making industry, bonding windscreens in motor vehicles, and bonding large panels in coaches, buses, and caravans.

Silicone: Silicone resins are sealants rather than adhesives, but they can be useful when a flexible adhesive of no great strength is required. Cure initiation is by surface moisture for the one-part types. When large-area bonding with a flexible adhesive is necessary, then two-part versions are available which must be mixed in the same way as other two-part systems. They remain flexible down to very low temperatures (-60°C [-76°F]) and can be used up to 200°C (392°F) or more in some cases.

Polyimides: Polymerization of monomer reactants (PMR) types are thermosetting, whereas condensation polyimides are thermoplastic. They are fairly weak as a basic resin, but they offer high-temperature performance (greater than 200°C [392°F]) and good fire-smoke performance. They can be blended with epoxy resins to toughen them.

Bismaleimides: Commonly known as BMIs, these are a special type of polyimide, prepared from maleic anhydride and diamines, preferably aromatic diamines. They are easier to process than polyimides and have processing characteristics similar to epoxies. Post cure for a short time is usually required to achieve full conversion and/or optimum properties. A major advantage of BMIs over other polyimides is that their cure and post-cure reaction mechanisms do not involve condensation products; therefore, no volatiles are produced. BMIs can be problematic because they tend to be brittle as a result of their high cross-link density. This is inevitable because this characteristic enables them to retain high joint strength at high temperatures. However, they can be made less brittle by the incorporation of polymeric modifiers such as rubbers. These modifiers have low glass transition temperatures (T_g) and consequently reduce the T_g of the toughened resin, which in turn may reduce the shear strength of joints at high temperatures. Modifications of BMIs with epoxy resins can improve peel strength without loss of high-temperature shear strength as long as the epoxy is carefully chosen. Fillers can be used to adjust flow characteristics to improve processing. Blowing agents can be incorporated to produce foaming adhesives, and minor changes in co-cure chemistry can be made to achieve compatibility with particular substrates and control speed of cure. Redux 326 from Ciba is an example of a recent development in this field. See Ref. 2.14.

Cyanate Ester: These resins are coming into use on the Eurofighter and in other high-performance applications because of their improved properties at high temperature. Although they can be made with T_g values in the range 200 to 300°C (392 to 570°F), they are expensive. These resins are difficult to use for repair purposes because of their high curing temperatures, but cyanate ester composites can be repaired with epoxies if service temperatures are not too high. Cyanate ester resins form trifunctional highly cross-linked and high-temperature stable polymeric networks (B.2.5). However, to ensure chemical stability in harsh environments, blending with epoxies seems desirable. Unfortunately, this reduces T_g and fracture toughness; therefore, careful compromise is necessary. Cyanate esters offer good adhesion to many substrates including metals, plastics, ceramics, and high-temperature thermoplastics such as PEEK. A significant amount of research remains to be done on these materials, and B.2.5 should be consulted for more detail. If repair is needed, it will be essential to consult SRMs and material data sheets for curing procedures and safety requirements.

Further Details on Thermosetting Resins: Most adhesives and resins used in current composite structure manufacture and repair are thermosetting resins (i.e., they change during the cure stage from a liquid to a hard, brittle solid by a chemical cross-linking process, which irreversibly changes the polymer chains that make up the structure of the resin). After curing, the resin cannot revert to its previous liquid form. The cure stage can be achieved by either application of heat to a one-part system, which already contains a hardener that begins to react only as it approaches the cure temperature, or by a chemical reaction resulting from the mixing of a resin and hardener component (two-part system) to form a room-temperature curing resin. Adhesives are available in the form of two-part resin-hardener pastes and liquids or as pre-made film adhesives in which the resin-hardener mixture has already been cast into a thin film that is ready for curing by the application of heat. The adhesives/resin types most commonly used in aerospace structures manufacture and repair include:

- **Epoxy Resins:** Epoxy resins are the most widely used resin types for aerospace adhesives and composite applications. They range in type from two-part, room-temperature curing pastes to hot-cure film adhesives capable of operating up to 150°C (300°F) and for long periods of time in aero engine applications.

- **Phenolic Resins:** Phenolic resins were the first resin types used for aerospace adhesive and composite applications and continue to be used extensively as adhesives for metal-to-metal bonding and as matrix resins for aircraft passenger-cabin furnishing panels where the low smoke and toxic gas emission characteristics of phenolic resins are advantageous. The key problems with phenolic resins are that they produce water as a product of cure and require cure temperatures of 125°C (257°F) to 150°C (300°F) to achieve a cure.

2.4.3 Properties Required of Matrix Resins and Adhesives

2.4.3.1 Physical and Chemical Properties

The resins used to bind the fibers in composites and structural adhesives have many things in common, although their mechanical properties must be different. See Refs. 2.5 and 2.15. The common characteristics required are:

- Must have good wettability to the fibers or surface to be bonded (substrate) and develop good adhesion on cure

- Should not emit volatiles of any cure products during or after cure if they cause corrosion or other ill effects, such as crazing of plastics

- Should have a simple cure cycle process

- Should have a good ambient-temperature storage life

- Must be tolerant of imperfect processing

- Should be tolerant of small inaccuracies in mix ratio (two-part systems)

- Should not shrink during cure

- Should have excellent retention of room-temperature properties when exposed to extremes of temperature and humidity

- Must have a low water absorption rate and a low saturation water content

- Should have a pH of the resin itself, and of any leachable extracts, that will not encourage corrosion of the metal surface being bonded

- Should not have any toxic hazards in either the uncured form or during decomposition in the event of fire (e.g., in an aircraft passenger cabin)

- Must have a pot life (work life) suitable for the job in hand

- Should have a T_g (dry) and a T_g (wet) as high as possible and high enough, ideally, to allow the repair of significant areas of parts manufactured from hot-cured pre-pregs when cured at least 25°C (45°F) below the original cure temperature of the part and preferably 50°C (90°F) lower

- Should be supplied with a viscosity appropriate for each end use

The last point is important and may help to explain why so many different resins exist. Resin transfer molding (RTM) requires a low-viscosity resin that can flow through the compacted fabric in the mold without either moving the layers out of position or taking an unacceptable length of time to flow through the fabric under a reasonable pressure. Resin viscosity can be reduced by heating in the RTM process. For wet lay-up resins, used at room temperature, experience has shown that the preferred viscosity for impregnation is 10 to 20 poise. Resins are usable for fabric impregnation with difficulty at approximately 150 poise; their use is possible at 200 poise, but the personnel involved complain regularly. One such resin was reformulated to a lower viscosity because of the number of complaints from those who worked with it. The lower values of 10 to 20 poise are preferred. RTM resins are in this band; however, their viscosity can be reduced to 0.5 poise or less when heated. There is a tendency to believe that the highest possible fiber content in a composite is a desirable objective; up to a point, this is the case. However, an adequate resin content is essential for good composite performance, and there is an optimum value below which a resin-starved product is likely to be produced with a lower mechanical performance. Not only is the correct resin content necessary, but the correct distribution of that resin is even more important. Tightly woven fabrics may be difficult for the resin to penetrate, and bundles of dry fibers may exist if the resin is too viscous to penetrate the interior of the bundles. The correct choice of fabric weave style, resin viscosity, pressure, and cure temperature is essential to the manufacture of a good product. Where low-viscosity resins are needed, it is important that they should also be as tough as possible and have a good elongation to failure. For adhesives, a paste viscosity of approximately 200 poise is generally desirable. For gap-filling materials, a higher nonslump viscosity may be required. Fillers may have to be added to achieve thixotropic behavior. No single resin system exists that provides outstanding performance in all of these areas—in effect, this is a "wish list," but a good one at which adhesive development chemists can aim. However, a wide range of matrix resins and adhesives exists that meets many of the above requirements.

2.4.3.2 Mechanical Properties

The work of Palmer (Ref. 2.5) and Armstrong (Ref. 2.15) shows that the mechanical properties required of matrix resins and adhesives are somewhat different. The ideal matrix resin should have the following mechanical properties:

Tensile Strength	8,000–10,000 psi (56–70 MPa)
Tensile Modulus	600,000 psi (4140 MPa)
Elongation to Failure	3–6%
Fracture Energy	>0.5 kJ/m^2

Likewise, Armstrong's work shows that the strength of a lap joint, using the standard test to ASTM D-1002 and aluminum alloy adherends, depends on the fracture energy of the adhesive up to a value of approximately 1.0 kJ/m^2. Above this figure, no increase in lap shear strength was found. An even better correlation with fracture toughness was observed. It was found that most toughened adhesives had a tensile modulus of approximately 350,000 psi (2415 MPa). The highest value of modulus for any adhesive tested was 450,000 psi (3105 MPa) for Redux 308A. It was considered that the reason for this was that it also had a high value of fracture energy. This value of modulus would also allow it to function as a fairly successful composite matrix resin. It was found that, although most of the required properties were common to matrix resins and adhesives, the tensile modulus value (from which the shear modulus can be calculated if the Poisson ratio is known) had to be higher for the matrix resins if good composite compression properties were to be achieved. In contrast, for adhesives' high joint strength could not be achieved with the high-modulus resins because adequate toughness could not be produced at the same time.

2.4.3.3 Epoxy-Based Matrix Resins and Adhesives for Aerospace Use

These systems tend to fall into three categories:

1. **175°C (350°F) Cure**: These systems are used primarily for components that will meet elevated temperature conditions, and they are the most resistant to moisture absorption. Most major structural items such as airframe Class 1 items are based on these resins.

2. **125°C (250°F) Cure**: These systems are used on less highly loaded structures such as fairings and access panels, and they are less resistant to elevated temperatures and moisture absorption.

3. **Room-Temperature (RT) Cure**: These systems are used for repair of composite parts, and recent developments in paste adhesive technology provide systems with a performance close to that of 125°C (250°F) cure resins.

Note that when interpreting the above statements, it is correct to say that the performance of 175°C (350°F) curing systems is better than 125°C (250°F) systems, which in turn are better than RT cure systems in general regarding performance at temperature and moisture absorption. However, also note that the 175°C (350°F) curing systems suffer a greater loss of performance over the dry state for each 1% of water absorbed than found in the 125°C (250°F) curing systems, and they in turn suffer more than the RT cure systems (e.g., the reduction in Tg for each 1% of water absorbed is far less for RT curing systems than for the hot-cure ones). They begin with lower values, and fortunately they lose less; therefore, the final difference in the wet condition is not as great as might be expected (Ref. 2.15).

175°C (350°F) Cure Temperature Epoxy Resins: These systems need a high-temperature cure to be able to develop attractive elevated temperature mechanical properties in structural service. However, the degradation of properties after long-term exposure to humidity means that these systems generally are limited to service temperatures of approximately 135°C (280°F). Matrix systems of this type are good for the fabrication of solid composite laminates, but it is generally impossible to cure these pre-pregs directly on honeycomb core and achieve satisfactory core-to-skin bonds. If these resin systems are to be used for composite sandwich panel skins, it is customary to use an epoxy film adhesive to bond the pre-cured skins to the honeycomb core in a secondary bonding operation. This is not always necessary, but the need for it should be checked by test.

125°C (250°F) Cure Temperature Epoxy Resins: These systems are designed to operate in aerospace applications such as exterior secondary structures for civil and military subsonic aircraft and helicopters. Generally, the long-term continuous operating temperature for structures using these resins does not exceed the 93°C (200°F) level, and their use tends to be limited to secondary structural items. The 125°C (250°F) cure temperature modified epoxy systems can be formulated to have the characteristic of one-shot laminating and bonding to honeycomb core without the need for a separate structural adhesive. The matrix resin flows and fillets around the cell ends of the hexagonal honeycomb core to form a high-strength adhesive bond. This honeycomb bonding behavior considerably simplifies the fabrication of sandwich panels using 125°C (250°F) cure matrix resins.

Room-Temperature Cure Epoxy Resins: The convenience of mixing, applying, and curing a two-part epoxy liquid or paste adhesive at room temperature is attractive in terms of repairs to composite structures. However, until recently, these resin systems had relatively poor elevated temperature strength and poor toughness characteristics. Improvements in room-temperature-curing epoxy resins now allow greater use of these materials for composite structure repair. Development of room-temperature systems continues.

2.4.4 Epoxy and Phenolic Pre-Pregs and Film Adhesives— Curing Stages

Pre-pregs and film adhesives involve resins in the "B" stage. These "B" stage resins may be the matrix resins of pre-pregs or film adhesives which may or may not contain a carrier cloth to improve handling strength, to ensure a controlled glue-line thickness, or both. For other stages and critical points, see the next sections of this chapter.

Resins can exist in one of three stages:

- **"A" Stage**: This is the initial state of the resin as produced by the manufacturer. It is an early stage in the polymerization reaction of certain thermosetting resins (especially phenolics), in which the material is linear in structure, soluble in some liquids, and fusible. The "A" stage usually is considered as a point at which little or no reaction has occurred. Pre-preg in an "A" stage condition would be extremely sticky, lumpy, and have little integrity. This is also called "resole."

- **"B" Stage**: The "B" stage is an intermediate stage in the reaction of certain thermosetting resins in which the material softens when heated and is plastic and fusible, but may not entirely dissolve or fuse. This helps to facilitate handling and processing, and is also called "resitol." The resin in an uncured pre-preg usually is in this stage. These resins require refrigerated storage at -18°C (0°F) to prevent gradual cure to a point at which they become unusable. They have a shelf life that varies from one product to another but is usually one year. They also have a limited out-time at room temperature, which varies from one week to one month.

- **"C" Stage**: This is the final stage in the reaction of certain thermosetting resins, in which the material is practically insoluble and infusible. It is sometimes referred to as "resite." The resin in a fully cured thermoset molding is in this stage.

Gelation: This is an important stage in the cure of any adhesive. It is defined by AIR 4844 as, "The point in a resin cure when the resin viscosity has increased to a point such that it barely moves when probed with a sharp instrument." At this stage, the cure is well on the way, and parts cannot be moved relative to each other. Three other words may be defined at this stage:

- **Gel**: The initial jellylike solid phase that develops during the formation of a resin from a liquid. A semisolid system consisting of a network of solid aggregates in which liquid is held.

- **Gelation Time**: That interval of time, in connection with the use of synthetic thermo-setting resins, extending from the introduction of a catalyst into a liquid adhesive system until the start of gel formation. Also, the time under application of a specified temperature for a resin to reach a solid state.

- **Gel Point**: The stage at which a liquid begins to exhibit pseudoelastic properties. This stage may be conveniently observed from the inflection point on a viscosity time plot. The point in a cure beyond which the material will no longer flow without breaking down the matrix network formed to that point. The point at which the matrix transition from a fluid to a solid occurs.

Glass Transition Temperature: Glass transition temperature (T_g) has several definitions:

- The approximate midpoint of the temperature range over which the glass transition takes place. A wide range of materials exhibits glass transition temperatures, i.e., the temperature at which a material changes from being rigid and brittle to becoming rubbery and flexible with a considerable reduction in mechanical properties. For example, glass and silica fibers exhibit a phase change at approximately 955°C (1750°F) and carbon/graphite fibers at 2205 to 2760°C (4000 to 5000°F).

- The temperature at which increased molecular mobility results in significant changes in the properties of a cured resin system.

- Also the inflection point on a plot of modulus versus temperature.

The measured value of T_g depends, to some extent, on the method of testing. See ASTM D-3418, Transition Temperature T_g Standard Test Method for Transition Temperatures of Polymers by Thermal Analysis, Differential Thermal Analysis (DTA), or Differential Scanning Calorimetry (DSC).

Heat Deflection Temperature: Heat deflection temperature (HDT) was previously known as the heat distortion temperature, and both terms can be found in the literature. Some manufacturers quote HDT in preference or in addition to T_g. The method of test is defined in ASTM D-648 Test Method for Deflection Temperature of Plastics Under Flexural Load. It is widely used and measures the temperature at which a standard bar of the cast resin system, which is under a defined load, undergoes an arbitrary deflection. The test provides a means of comparing the temperature resistance of resin materials. HDT is a useful guide to developing an optimum cure cycle and in general indicates the extent of cross-linking. (See Refs. 2.16 and 2.17.) In general, T_g values are higher temperatures than HDT. Although some companies continue to use HDT, it is becoming the norm in most places to use T_g.

Viscosity Changes During Cure: A pre-preg or film adhesive is supplied in the "B" stage, and the purpose of curing is to bring it to the "C" or fully cured stage. If the material is within its shelf life and is in good condition, it will melt in the early stages of cure and reach a low viscosity. During this time, it can wet the surfaces that it is being used to bond and flow into any crevices or surface variations. In the case of composites, any excess resin can be bled away through perforated release sheets and bleeder cloths. Adhesive also may flow outward from the edges of bonded metal parts and be absorbed into similar bleeder cloths. Flashbreaker tapes may be used to achieve a neat edge and to prevent adhesive bonding to the metal outside the joint area. As the curing process takes place, cross-linking of the resin will occur. At the gel point, the material permanently returns to the solid state. On completion of curing, the material is a fairly hard solid and will never melt again, although it will soften to some extent if heated to the cure temperature. Further temperature increase will cause charring and burning. If the temperature is high enough (e.g., in the case of lightning strikes), then the resin may gasify and disappear completely, leaving the fabric appearing as though it had never been bonded with a resin.

2.4.5 Mixing and Mix Ratios for Epoxy Wet Resins

Wet resins are usually two-part epoxy systems in which the following statements are true:

- Part "A" is a polyepoxide resin (base resin)
- Part "B" is a curing agent (hardener)

See SAE ARP 5256 Resin Mixing for more detail.

2.4.5.1 Weighing

The ratio of Part "A" to Part "B" will vary according to the particular epoxy material, and the manufacturer's instructions regarding mix ratio and other aspects such as health and safety should be strictly followed. Keep both parts of each kit together to prevent mistakes from being made by using the wrong curing agent.

Mix ratios are usually given by weight, and an accuracy of $\pm 1\%$ should be the goal. Mixing by weight is preferred to mixing by volume. Use digital electronic scales with an accuracy of 0.1%. The weights of resin and curing agent must be carefully calculated to determine the mix ratio required for each resin or adhesive (see data sheet). These ratios vary considerably; therefore, it is important to check each time. Most two-part systems will accept an accuracy of $\pm 5\%$; however, following the principle that the accuracy achieved in anything is seldom better than the target for which one has aimed, it is wise to aim for a higher target than the minimum acceptable. There are other reasons for this:

- The material performance improves when the mix is near the correct value.

- Some remaining material always adheres to the side of the container in which mixing occurs. If the viscosities of the two components are very different, then the one with the highest viscosity is most likely to adhere to the container.

- Surplus curing agent that is not reacted can be leached out by water and other fluids and can corrode metal parts.

- Surplus curing agent leads to higher water uptake in the cured resin because the curing agent absorbs more water than the base resin.

2.4.5.2 Mixing

Mixing must be thorough because accurate weighing is not the only important factor. Even if weighing is extremely accurate, the mix ratio will be incorrect at every point in the mix unless thorough mixing occurs. Trials have shown that it is important to mix for at least three minutes and preferably five minutes to achieve a consistent mix with a good, uniform dispersion of the hardener in the resin. Some adhesives and resins are colored to allow an even mix to be easily seen. If longer than five minutes is required to achieve a uniform color, then the extra mixing time will be necessary. Mix slowly, and avoid generating air bubbles as much as possible. One of the advantages of pre-pregs, film adhesives, and one-part paste adhesives is that accurate weighing and mixing has been completed by the manufacturer and only the cure remains to be done by the end user. However, wet lay-up resins and two-part paste adhesives are primarily room-temperature curing systems; therefore, the end user exchanges the responsibility of correct curing for that of correct weighing and mixing. Both mixing and curing must be done accurately; in both cases, the end user controls the quality. Mix in nonmetallic receptacles such as waxless paper cups because some resins and hardeners may dissolve wax and become contaminated. Use flat, not round, nonmetallic stirrers (e.g., wooden or plastic spatulas). If considerable exotherm is likely, it may be better to mix in shallow aluminum trays to allow maximum escape of heat. Alternatively mix only small quantities. The adhesive data sheet usually offers advice on this. If air is introduced into the mix, it may be desirable to vacuum degas or to allow the mix to stand for some time to permit air to reach the surface. Bubbles can be popped with a pin if they can be seen in a clear resin. Opaque resins prevent gas bubbles from being seen. The higher the viscosity of the adhesive or resin, the fewer gas bubbles will reach the surface and escape. Therefore, care should be taken to avoid trapping air during mixing. (See also SAE ARP 5256 Resin Mixing, which is a useful document that is worth reading.)

2.4.5.3 Definitions Related to Mixing and Application

- **Pot Life**: Usable life of mixed resin in the pot (mixing receptacle). Pot life may be extended by decanting mixed resin into smaller units because this allows more cooling and reduces the temperature produced in the curing resin by exothermic heat. Mixing in shallow trays can help to remove heat and reduces the temperature reached as a result of exotherm. Many epoxies exotherm as the cure initiates. In extreme cases, they can ignite and emit a large amount of vapor and smoke. It is important to mix in small quantities and to follow the manufacturer's recommendations. (SAE AIR 4844 definition.) The length of time at some specified temperature that a catalyzed thermosetting resin system retains a viscosity low enough to be used in processing. Also called work life.

- **Work Life**: Another term for pot life. Usable life, after the mixed resin has been wetted into the reinforcement, during which lay-up can be completed. In the case of adhesives, it is the time after mixing during which the viscosity is low enough to allow wetting of the substrate and adequate spreading over the surfaces to be bonded. (SAE AIR 4844 definition.) The period of time during which a liquid resin or adhesive, after mixing with catalyst, solvent, or other compounding ingredients, remains usable. See also gelation time and pot life. It is important to be cautious and to avoid using work life to its limit. The lower the viscosity, the better the wetting of the surfaces to be bonded and the easier the adhesives are to spread. A high-viscosity adhesive is virtually certain to produce an excessive glue-line thickness. It is good policy to work well within any manufacturer's recommended limits. With quick-setting adhesives, the short work life will mean that the mixed adhesive or resin must be used quickly and that the size of the job that can be done will be limited by the work life.

- **Cure**: (SAE AIR 4844 definition.) To irreversibly change the properties of a thermosetting resin by chemical reaction (i.e., condensation, ring opening, or addition). Cure may be accomplished by the addition of curing (cross-linking) agents, with or without heat and pressure. Wet resins usually will achieve a complete cure at ambient temperatures (21 to 24°C [70 to 75°F]) in 16 to 72 hours, although they will harden to an acceptable handling strength in less time. Some require seven days for complete cure. Check the manufacturer's data sheet because the variation among different materials is considerable, and it is important to allow sufficient time to achieve handling strength before moving parts and full cure before using them. For Boeing aircraft, the cure temperature should be below 66°C (151°F) for an adhesive or resin to be classified as room-temperature curing. The manufacturer's recommendations should always be followed for optimum performance because they are based on extensive research.

2.4.6 Polyester Resins

Aerospace usage of polyesters is quite small. They were used for many years for radomes and some interior parts, and the high-grade types were usually employed.

In the marine and automotive trades, the use of cheap, room-temperature curing, polyester resins is common. Polyester resins have three parts: a base resin, a catalyst, and an accelerator. An explosion may occur if the catalyst and accelerator are mixed directly. It is essential for safety to comply with the manufacturer's mixing instructions. The use of excess catalyst may cause fire. Hot-curing polyester resins also are used for RTM applications in the motor industry. Research is ongoing to automate the resin transfer molding process as much as possible to reduce costs and to increase production rates. Some problems have been found with hot-curing systems. They tend to have short gel times, making close process control necessary. Cure starts almost immediately, leaving little time for the resin to dissolve any binders and surface treatments on the pre-form and to wet the fibers completely. In some cases, these hot-curing systems were found to generate considerable exothermic heat which produced temperatures as high as 230°C (446°F) in the molding. This caused the styrene in the resin mixture to boil and thus degraded the moldings. It also demonstrated a need to maintain pressure in the mold cavity during gelation. Except for the most specialized "one-off" cars or home builds, it can be safely assumed that wet lay-up with fiberglass or other fabrics will have little place in the automotive field. RTM is virtually certain to take over the manufacture of medium-sized production runs for composite car parts. Racing cars are made mainly from pre-preg fabrics and tapes.

In marine applications, it seems likely that the use of wet lay-up will continue because the parts are larger and the production runs are smaller than those of cars. Polyester resins are likely to continue to be used because of their low cost. Vinyl ester resins are midway between polyester and epoxy resins in terms of performance and cost, and they are widely used in building boats. However, epoxy technology is finding its way into boats. As the interiors of luxury and other boats and railway carriages move toward the adoption of composite materials, it may also be expected that phenolic resins will be used for interior parts, as they are for aircraft, because of their low smoke emission and toxicity in the event of fire.

2.5 Adhesives

Adhesives generally are available in three basic types: liquid and paste adhesives, foaming adhesives, and film adhesives.

2.5.1 Liquid and Paste Adhesives

As their name implies, these adhesives are in a liquid or paste form when mixed. They may be supplied in two basic forms:

- One-part liquids or pastes containing a latent curing agent that begins to work only at high temperature. These one-part paste adhesives are epoxies and require hot curing in the same way as film adhesives. Similar to film adhesives, they also require refrigerated storage.

- Two-part liquids or pastes of various chemical types that require mixing in the correct proportions and that will cure at room temperature. Their cure time may be reduced by curing at higher temperatures; in some cases, their mechanical properties also can be improved. The manufacturer's recommendations should always be followed. It is possible to cure at too high a temperature and to make these materials more brittle than they would be from a room-temperature cure. The two, or occasionally three, parts involved may vary from stiff pastes for base resins to water-thin curing agents. In the mixed condition, they all have approximately the same viscosity as toothpaste. From a safety viewpoint, it is highly desirable to formulate adhesives so that both components are of a high enough viscosity to prevent splashing. This minimizes the risk of eye or skin contact. Today, some good epoxies have a stiff base resin and a water-thin curing agent. However, this means a high risk of splashing the more toxic component into the eyes or on the skin and thus requires more safety precautions. On the other hand, some other good adhesives have two parts of almost equal and fairly fluid viscosity, which makes them safe to handle and easy to mix. There are several types of liquid and two-part paste adhesives. Paste adhesives are used extensively throughout the home-built aircraft industry for secondary bonding operations. Care must be taken to avoid thick glue lines with the associated risk of high void content in the joint. Thick bond lines can be avoided by rigorous attention to contour profile alignment during dry assembly and by using adequate bonding pressure. The latter is sometimes difficult to achieve for large-area bonds, particularly in assemblies where applied vacuum might distort the structure. Home-built aircraft designers may specify a glass flox/epoxy structural filler to allow gap filling when bonding components together with contour mismatch. This is a mistake if called up over large bond areas because flox may not squeeze out to form a uniform bond thickness under pressure. The resin will squeeze out, but the flox, being composed of short lengths of chopped glass fiber, does not tend to shear over itself and therefore will not "flow." The result is potentially high voidage in all except the high spots.

Epoxies: Many two-part epoxies are available, both as adhesives and matrix resins. The adhesives tend to be in paste form after mixing, whereas the matrix resins are relatively fluid to allow penetration of fabrics. Epoxies are less hazardous than some types, but smoke and fire can occur if quantities in excess of those recommended are mixed in one container. This can result because the chemical reaction involved is exothermic (i.e., heat is emitted during cure). In quantities larger than those recommended, this heat cannot escape at the required rate, and the temperature builds up until smoke and fire occur.

Room-temperature epoxies of the latest type can have good properties, but they are not as good, especially with regard to high-temperature performance and water absorption, as those cured at high temperatures. Nevertheless, they are adequate for many purposes. The usual problem is their glass transition temperature (T_g) in the wet condition. If this is sufficient for the intended use, then they should be accepted more often than they are.

Acrylics: These are now widely used and are supplied in two forms:

- As a pre-mix (i.e., supplied as two parts and mixed before use in the same way as epoxies).

- The original version used an activator applied to one surface, then the resin was applied to the opposite face, and the two were pressed together. This was satisfactory for small areas of a noncritical nature, but the pre-mix version has been found to provide more consistent results. Acrylics have a low modulus compared to epoxies and a high elongation, which makes them tough and useful in impact applications.

Polyurethanes: These generally are used for bonding thermoplastics and, in some cases, metals. They are supplied in two parts in a way similar to epoxies and have good peel strength and toughness. Polyurethanes perform best at or below room temperature. Some have an upper operating temperature as high as 120°C (250°F); others have an upper limit of 82°C (180°F). Other versions are good only from 0°C to 40°C (32 to 104°F). Check the performance over the temperature range required. As with other two-part systems, precise weighing, correct mix ratio, and thorough mixing are critical.

Polyesters: These are used primarily in boat building, where their usage is extensive. They also are used for large chemical tanks; in cars, vans, trucks, buses, and trains; and in building construction and other industrial applications. The top grade versions are used for ships as large as 120-m British Navy Minehunters. The Eurostar high-speed train uses a polyester nose cone. Lower grades are used for automobile and small boat repairs and for home and hobby usage.

Phenolics: Many types of phenolics are mentioned above, and these are used for bonding wood, plywood, and veneers. It is important to use the correct type for boats and outdoor work to ensure durability.

2.5.2 Foaming Adhesives

Foaming adhesives are supplied in two forms: foaming paste adhesives and foaming film adhesives.

Foaming Paste Adhesives: Foaming pastes are one-part epoxy pastes, which expand on heating and are used for honeycomb core jointing, insert potting, and edge filling. Use enough, but avoid using too much; otherwise, the honeycomb may be forced out of shape by the pressure generated as the paste expands during hot curing. The foaming paste is cured at the same time as the part itself.

Foaming Film Adhesives: Foaming film adhesives are rather thick film adhesives, usually 1 to 1.5 mm (0.04 to 0.06 in.) thick, which foam on hot curing and expand by a factor of 1.5 to 3.0 on their original dimensions. As mentioned for foaming pastes, use enough but not too much. These also are cured at the same time as the component.

2.5.3 Film Adhesives

The two main types of film adhesives are the epoxy type and the phenolic type. Both are described in detail below.

Epoxy Film Adhesives: These may be divided into at least four types with different curing temperatures:

- Low-temperature curing, i.e., cure at 95°C (200°F), usually for repair purposes
- 120°C (250°F) cure for lowly stressed parts
- 150°C (300°F) cure for primary structure
- 180°C (350°F) cure for primary structure

Epoxy films are easier to process because the pressure required during cure is lower than for the phenolic type. They are more popular with airframe manufacturers in spite of their lower durability compared to phenolics.

Some typical Boeing specifications are:

BMS 5-129 (5-80)	American Cyanamid FM 123-2
121°C (250°F) cure	3M Company AF 126
BMS 5-101	3M Company AF 163
121°C (250°F) cure	American Cyanamid FM 73
	Hysol EA 9628
	Metlbond 1133

3M-AF 163 superseded 3M-AF 126 in one airline many years ago because it is less sensitive to moisture before, during, and after cure.

Typical AI/BAe/RR epoxy adhesives are:

Ciba Redux 308	170°C (338°F) cure, engine cowls
Ciba Redux 308ANA	150°C (300°F) cure, engine cowls, Concorde undercarriage doors

Ciba Redux 319A   175°C (350°F) cure, carbon fiber composite parts
Ciba Redux 322   175°C (350°F) cure, Concorde parts (replaced 3M-AF 130)

Ciba has amalgamated this part of its work and the Redux trade name with Hexcel. Liquid and paste resins are supplied by Ciba Specialty Chemicals (United Kingdom) Ltd. under the trade name Araldite.

Film adhesives are one-part epoxy films containing hardener. Hence, they must be stored frozen to stop or severely retard cure initiation. See Chapter 3, Handling and Storage.

Epoxy film adhesives typically are used with primers (of the corrosion-inhibiting type) when bonding metal parts, but primer is not often used when bonding composite parts.

Curing must be carried out to the manufacturer's instructions with regard to temperature, pressure, and time of cure. In some cases, dwell periods at a specific temperature may be required. Care must be taken to ensure compatibility of cure requirements between the adhesive film and the parts being joined. For example, do not cure 121°C (250°F) adhesives with 177°C (350°F) pre-pregs. Similarly, parts made with 121°C (250°F) curing pre-pregs must not be repaired with 177°C (350°F) materials. It is always advisable, when permitted, to repair at a temperature 25 to 50°C (45 to 90°F) lower than the temperature at which the part was made, preferably 50°C (90°F) lower. Film adhesives can be supplied as supported or unsupported materials. The support may consist of a nylon woven fabric, a polyester nonwoven fabric, a filler such as glass beads, or short glass or Kevlar fibers to provide glue-line thickness control. Supported film adhesives are easier to handle and are more resistant to tearing. Unsupported film adhesives are used for reticulating around honeycomb cell ends because of their good flow characteristics. The supported type cannot be used for this purpose. Both types must be warmed to room temperature before removal from their sealed bags for two reasons:

- Moisture must not be allowed to condense on the film.

- In the frozen condition, film adhesives will fracture into small pieces in the same way as glass.

See Figures 2.22 and 2.23.

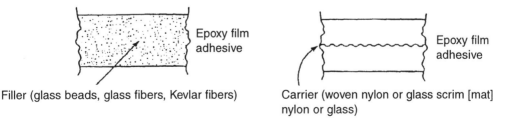

Fig. 2.22 Fillers and carriers for adhesives.

1. **Flow:** The ability of the adhesives to move during heat/pressure cycles and so produce gap-filling properties. High flow is good for honeycomb sandwich bonding because the adhesive moves around honeycomb cell ends. However, excessive flow can cause adhesive wash-out from a joint. This reduces the glue-line thickness below an acceptable minimum, resulting in a weak joint.

2. **Reticulate:** Unsupported film shrinks on heating to form the adhesive around honeycomb cell ends. Used particularly for acoustic panels (because it gives no blocking of perforated liners).

3. **Glue-Line Thickness:** This shall be controlled between 0.003" and 0.006" where possible. Typical film adhesive is 0.01" thick before use, and 0.004" to 0.005" thick after use. Glue-line thicknesses below 0.003" (0.075 mm) and above 0.010" (0.25 mm) are likely to give lower joint strength. The effect varies from one adhesive to another. (See data sheet for Redux 312 and Redux 319.)

4. **Overlap Length for Lap Joints:** The length of overlap in a lap joint should be 50 to 80 times the skin thickness. Overlap lengths in excess of this range add weight but make little difference to the joint strength.

5. **Adherend/Substrate:** This is the material being bonded.

6. **Joint Design:**

   Single Overlap: Joint can fail in peel before shear. A good joint strength is achieved only with a tough adhesive.

   Glue line 0.004"

   50-80t overlap

   Double Overlap: Joint is less likely to fail in peel; higher shear loads are available.

   Flexible substrate

   Peel: Peel performance is improved if tough adhesive is used.

   Rigid substrate

Fig. 2.23 Terms applicable to adhesive films.

Cleavage Peel:

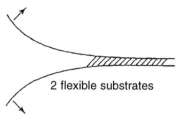

2 flexible substrates

Butt Joint: Weak joint; small surface area and adhesive in tension. Do not use.

Improved by use of butt strap; load is then carried in shear.

7. **Adhesive Failure:** Adhesive joint fails by adhesive film detatching from one or both substrates. This is usually caused by bad surface preparation. The remedy is good chemical pretreatment.

8. **Cohesive Failure:** Adhesive joint fails through the adhesive film. This gives the highest possible joint strength, and good surface treatments must be employed to ensure that cohesive failures can also be achieved after long periods in service.

9. **Fillet: (VERY IMPORTANT)** Bead of adhesive at edge of bond line. Do not remove fillets because they aid environmental resistance and can provide joint strength.

fillet

fillet

Fig. 2.23 (continued)

Vinyl Phenolic Film Adhesives: These were the first to be produced in the 1940s. They cure at approximately 180°C (356°F) and require high pressures during cure because water is a product of the reaction. Redux 775 vinyl phenolic film adhesive has a long history of good durability in service when bonding aluminum alloys. It was used on the De Havilland Hornet, Dove, Comet, and Trident, and is used today on the BAe 146 (now known as the RJ Series). It has also been used and remains in use on aircraft made by Fokker. Epoxies have been improved considerably but do not yet match the durability of vinyl phenolics. Redux 775 may be cured between 150 and 180°C (300 and 356°F).

2.5.4 Glue-Line Thickness Control

This can be achieved in several ways:

- By the use of carrier fabrics in film adhesives. These may be woven nylon or random mat polyester (3M-AF 3306). (See Ref. 2.16.)

- By the use of glass or phenolic microballoons, other fillers, or chopped fibers. The importance of glue-line thickness varies from one adhesive to another and should be kept within the limits specified on the data sheet. For many adhesives, it can vary from 0.05 to 0.25 mm (0.002 to 0.010 in.) without significantly affecting lap shear strength. When the epoxy resins are used for honeycomb bonding, closer regulation of the glue-line thickness is required. This often is accomplished by the use of either supported or unsupported epoxy tapes of a predetermined thickness. In honeycomb sandwich constructions, the major part of the adhesive bond consists of the fillet. The amount of adhesive trapped between the edge of the core and the skin is too small to contribute significant strength to the joint. Therefore, it is necessary to regulate the thickness of the glue line, and hence the thickness of the fillet, to obtain reproducible results. See Figure 2.24.

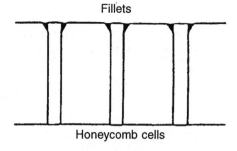

Fillets

Honeycomb cells

Fig. 2.24 Adhesive fillets on honeycomb.

The thicker the fillet, the higher the strength of the adhesive fillet; correspondingly, more adhesive is required and the weight of the adhesive is greater. Optimum values for each specific application can be determined, usually in terms of core strength rather than adhesive failure under test. It is generally found that 3 to 5 mm (1/8 to 3/16 in.) cell size aluminum alloy honeycomb of 0.025 mm (0.001 in.) wall thickness fails by tearing of the metal, and 0.05 mm (0.002 in.) and thicker fails in the adhesive bond. To illustrate the dependence of strength on the thickness of the adhesive film in honeycomb constructions, an unsupported film adhesive having a weight of approximately 0.05 lb/ft^2 and giving a peel strength of the order of 50 lb/in. width will, at a weight of 0.09 lb/ft^2, give a peel strength of the order of 100 lb/in. width (Ref. 2.16). A cohesive failure in a honeycomb construction means that the fillet will remain on the cell walls, and the fillets will tear, leaving a layer of adhesive remaining on the skin. This leaves a honeycomb pattern on the skin but with a thickness equal to the core wall plus two fillets. See Figures 2.25 and 2.26. These figures were obtained during an MSc project to select a suitable fabric that would allow vacuum bonding. See Chapter 10, paragraph 10.4.6.2. The first two illustrations in Figure 2.26 were obtained using unsuitable and fairly thick nylon fabrics. The bottom illustration using a thin nonwoven polyester fabric (3M-AF 3306) shows a cohesive failure in the fillets similar to that obtained with autoclave bonding that does not need a fabric.

For good results, and especially for long-term durability, the aluminum alloy skins must be anodized (unsealed) using chromic or phosphoric acid methods, and the honeycomb must be of the corrosion-treated and resin-dipped type or be anodized and coated with a suitable primer. When clean and dry Nomex aramid paper honeycomb is used, the core always fails

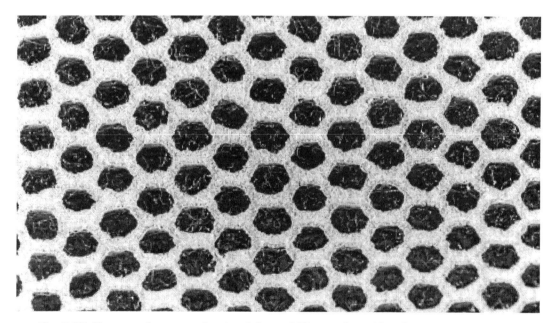

Fig. 2.25 Honeycomb pattern showing failure of fillets and good bond of honeycomb to skin.

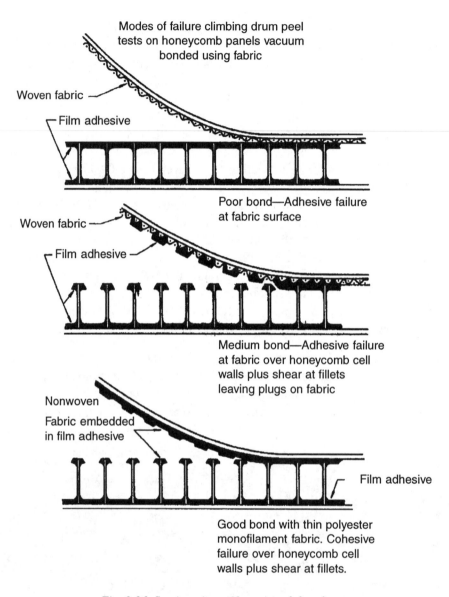

Modes of failure climbing drum peel tests on honeycomb panels vacuum bonded using fabric

Woven fabric

Film adhesive

Poor bond—Adhesive failure at fabric surface

Woven fabric

Film adhesive

Medium bond—Adhesive failure at fabric over honeycomb cell walls plus shear at fillets leaving plugs on fabric

Nonwoven
Fabric embedded in film adhesive

Film adhesive

Good bond with thin polyester monofilament fabric. Cohesive failure over honeycomb cell walls plus shear at fillets.

Fig. 2.26 Section view of honeycomb bond.

before the bond if sufficient adhesive has been used. There are now many types of honeycomb in addition to aluminum alloy and Nomex, and each type and density may require that its failure mode be checked. Honeycombs are available made from fiberglass, carbon fiber, and aramids, in addition to ordinary paper types. See data from Hexcel, Eurocomposites, and others.

In-Service Effects on Resin Systems: Epoxy-based resins and other types tend to gradually absorb moisture and other fluids during their service life until an equilibrium condition is reached. This moisture/fluid uptake seriously affects the glass transition temperature (T_g) of the cured resin, as mentioned previously. Work is ongoing to develop low water uptake systems.

The T_g is a critical point (or narrow band of temperature) at which the cured resin mechanical properties undergo a drastic reduction as the material softens. Hence, if an elevated temperature compression test is conducted on a typical composite panel after it has reached its saturation point of water absorption, this is a good simulation of the worst condition for a composite structure. Another simulation of the in-service impact damage to composite panels is made by impacting a given lay-up of panel at a defined energy level and testing the damaged panel in compression. Because the compression strength of a composite laminate is sensitive to resin mechanical properties, this is a good test of the relative damage resistance of different resins.

2.6 Core Materials

Core materials provide the center section of sandwich panels. A sandwich panel is defined as, "A panel consisting of two thin face sheets bonded to a thick, lightweight honeycomb or foam core." The core can be made of a wide range of materials, provided that they are strong enough to transmit the shear loads involved. In many respects, a sandwich panel is similar to an "I" section beam, where the flanges take the tension and compression loads and the web takes the shear. See Figure 2.27.

In a sandwich panel, the skins take the tension and compression loads and the core takes the shear. The shear normal to the skins is taken mainly by the core and partly by the skin, and the shear parallel to the skins is taken by the core and also by the core-to-skin adhesive bond. If the bond is strongest, the core will fail first. If the core is stronger than the bond, the bond will fail first. If the skins are thin and a weak foam core is used or a honeycomb core which is either weak or of a large cell size is chosen, then the skin on the compression side may buckle either into or away from the core. There are several interesting aspects to sandwich panel design. (See Refs. 2.18, 2.19, and 2.20.) For optimum bending performance, core weight should be approximately equal to skin weight, assuming both skins are equal. If such a design does not provide adequate indentation resistance for a particular application, then this can be improved most weight-effectively by increasing core density rather than by increasing skin thickness. (See the above references.) When repairing sandwich panels, ensure that the correct skin, core, and resin materials are used to restore the original performance. Always use a core material identical to the original or slightly stronger. Never use a core material that is lighter or weaker than the original. Never use a honeycomb of a larger cell size than the original. When repairing radomes, only nonconducting core materials may be used with resins that have no metallic fillers. See the

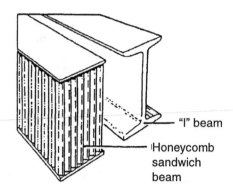

"I" beam

Honeycomb
sandwich
beam

Fig. 2.27 Sandwich versus "I" beam construction. The honeycomb sandwich may be constructed using a variety of metallic or fiber-reinforced plastic facing skins. The structure is analogous to an "I" beam, but with the core extending over the full width, the faces are stabilized and buckling inhibited. (Courtesy of Heatcon Composite Systems.)

SRM for the correct type of core. Also, ensure that the correct skin spacing (i.e., core thickness) is used to avoid additional reflective losses from the radar signal. (See Ref. 2.21.)

Sandwich structures are used where a high stiffness/weight ratio is required. The core material prevents the thin skins from buckling if the panel is correctly designed. Good quality manufacture is also important, and a good bond must be achieved between the skins, core, and edge members. Good sealing is required around the edges, and holes through sandwich panels should be avoided if possible.

A history of sandwich structure use in aerospace follows:

- Wooden sandwich structures—plywood skins/balsa wood core are used on the De Havilland Albatross (1937) and the Mosquito (1942)

- Aluminum alloy skins are bonded to balsa wood core, Chance Vought Cutlass (1948)

- Introduction of aluminum alloy honeycomb on a commercial scale (1954)

- Major use of adhesively bonded aluminum alloy honeycomb components in military aircraft primary structures on the Breguet Atlantic (1961)

- Introduction of nonmetallic Nomex honeycomb (1968), followed by widespread use of sandwich panels with Nomex core and glass/epoxy composite skins in civil aircraft secondary structures

- High-performance polymethacrylimide plastic foam core Rohacell (1985)

- Extensive use of sandwich structures using glass, aramid, and carbon fiber skins in the present generation of civil aircraft (1995)

2.6.1 Wood

Balsa wood has been used in aircraft and model aircraft construction for many years. It also has been used as thermal insulation in ship hulls and for refrigerated road transport vehicles. It was used between plywood skins as the core material for the construction of the fuselage of the Albatross and Mosquito aircraft. Its primary use in modern aircraft has been for the core material of sandwich panels with either aluminum alloy or fiberglass skins. These have mainly been used for floor panels, galleys, and wardrobes. The density of balsa wood varies widely, and only the denser grades are of use in flooring. These vary from 96 kg/m^3 (6 lb/ft^3) to approximately (191 kg/m^3 (12 lb/ft^3). There has been a move away from balsa wood for flooring because of the difficulty of selecting wood within a small band of density. When used in sandwich panels, balsa wood is used end grain (i.e., with the grain normal to the facing skins) because the compression strength is better in this direction and the shear strength is adequate. Panels with balsa wood cores must not be used in wet areas, unless sealing is exceptionally good, because balsa will absorb large amounts of water, causing serious weight increase and corrosion of aluminum alloy skins.

Other woods heavier than balsa wood can be used as core materials for special purposes.

2.6.2 Foam Core Materials

2.6.2.1 Material Types

Several types of foam core are available. They are mostly made in a range of fairly low densities and must be carefully selected according to the importance of their mechanical properties, chemistry, toxic smoke generation in fire, and cost for each particular application.

Polyvinyl Chloride: There are two main types of polyvinyl chloride (PVC): cross-linked and uncross-linked.

Cross-linked materials are available under trade names such as Divinycell. The cross-linked types have a higher shear strength for a given density and are more resistant to high temperatures. The uncross-linked form is made by Airex and others. These uncross-linked materials are more impact resistant and are easier to form around molds.

PVC is a fairly weak material, suitable for low-temperature use (i.e., less than 50°C [122°F]). It is available in densities from 2 lb/ft³ to approximately 25 lb/ft³, and can be heat formed at temperatures above 100°C (212°F).

Permali Ltd. also makes PVC foam under the trade name of Plasticell. Its material is made in a range of densities from 2.5 to 26 lb/ft³. It has a closed cell structure and low water vapor permeability. PVC foam made by Divinycell International AB (Ref. 2.22) has completely closed cells and a low value of water vapor permeability. It is available in densities from 2 to 25 lb/ft³ (30 to 400 kg/m³). It is made in various grades according to end use, and the HT grade is formulated to suit various pre-preg systems and to be compatible with the high process temperatures involved. The FRG grade has low toxic gas emission for applications in which combustion products are restricted, such as aircraft interiors. As with many other materials, it is important to use the correct grade to ensure that design requirements are met.

Polyurethane: Polyurethane (PU) is a fairly weak material, suitable for low-temperature use (i.e., less than 80°C [176°F]). It is more brittle than PVC. The foam surface at the resin core interface tends to deteriorate with age. This material must not be cut with a hot-wire cutter because toxic fumes are emitted. Use sharp knives to obtain a rough shape, and finish to profile by sanding.

Ebonite (Hycar): Hycar is a blown nitrile rubber, fairly weak, used for older radomes. It is suitable for low-temperature use (i.e., less than 80°C [176°F]).

Polymethylmethacrylimide (Rohacell): This is a blown acrylic-based thermoplastic which is increasing in use (e.g., for helicopter blades, interior parts, and some high-temperature parts such as engines). Service temperature is less than 160°C (320°F). It may be heat formed and has a closed cellular structure. This is important for any purposes where moisture ingress could otherwise be a problem. It has been tested for use in boats, and it has been demonstrated that although the foam absorbs water into itself, the water does not fill the cells. It is made by Rohm GmbH under the trade name Rohacell. The material is available in densities from 32 to 190 kg/m³ for commercial use and 51 to 200 kg/m³ for aeronautical use. The aerospace grades are Rohacell "A" and "WF," which have an increased heat distortion resistance compared to the commercial grade. Rohacell is resistant to organic solvents such as benzene, xylene, and monostyrene, and also solvents for paints, fuel constituents, and most technical solvents. It does not withstand alkaline media.

Polyimide Foams (Solimide): These foams, made by Imi-Tech, may be made in a range of densities from as low as 4 kg/m³ (0.25 lb/ft³) for insulation and from 32 to 160 kg/m³ (2.0 to 10.0 lb/ft³) for sandwich panels. The service temperature range is from -184 to 260°C (-300 to 500°F).

Styrofoam: This material normally is used on home-built aircraft, and it is important to buy it to the correct specification. The grade for aircraft and other structural uses is a closed-cell configuration with small cells, and this grade is superior to the type used to make drinking cups. In any honeycomb structure, the core material is used to take shear loads; therefore, it must be of adequate strength. Styrofoam should be used only with epoxy resins because polyester resins will dissolve it. Check your drawings and use only an approved type of resin or adhesive with this material. It may be cut with a hot-wire cutter.

2.6.2.2 Density

These materials are available in a range of densities and can be foamed to almost any density within the available range if required. Mechanical properties are approximately linear with density; thus, it is fairly simple to estimate the density required for a particular purpose when the necessary properties have been calculated or obtained by test.

2.6.2.3 Advantages and Disadvantages

It is fair to say that there are no bad materials—only good and bad applications of them. All core materials have advantages and disadvantages. Foams tend to have lower mechanical properties for a given weight but are sometimes cheaper than honeycombs. If they are closed cell and not naturally water absorbent, then they may be good for boats because a puncture of the outer composite skin will not result in rapid water absorption as would occur with wood, especially balsa wood. It is important for all sandwich structures to balance the factors of cost, weight, mechanical performance, smoke emission, smoke toxicity, flammability, repairability, and durability for each application. The choice of core material will be decided by the most important factors in each case, and data for each candidate material will have to be obtained. Manufacturers' data sheets are usually helpful.

2.6.3 Honeycomb Core Materials

Honeycomb is made primarily by the "expansion" method. The corrugated process is most common for the high-density honeycomb materials.

The honeycomb fabrication process by the expansion method begins with the stacking of sheets of the web material on which adhesive node lines have been printed. The adhesive lines then are cured to form a HOBE® (HOneycomb Before Expansion) block. The HOBE block may be expanded after curing to give an expanded block. Slices of the expanded block to the desired "T" dimension (thickness) may then be taken. Alternatively, HOBE slices can be cut from the HOBE block to the appropriate "T" dimension and then expanded. Each HOBE slice is then expanded to the desired cell shape, giving an expanded panel. The expanded panels are trimmed to the desired "L" dimension (ribbon direction) and "W" dimension (transverse to the ribbon). The "L," "W," and "T" dimensions and the cell size may be expressed in inches or millimeters. See Figures 2.28 to 2.30.

The corrugated process of honeycomb manufacture as illustrated is normally used to produce products in the high-density range. In this process, adhesive is applied to the corrugated nodes, and then the corrugated sheets are stacked into blocks and cured. Panels are cut from these blocks to the required core thickness. The dimensional terminology for blocks made by the corrugated process is identical to that for expanded honeycomb. See Ref. 2.23.

Expansion Process of Honeycomb Manufacture

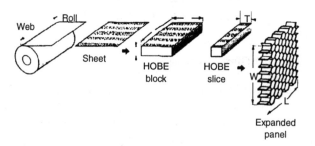

Corrugation Process of Honeycomb Manufacture

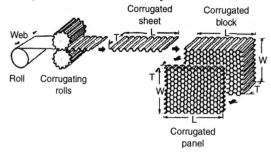

Fig. 2.28 Honeycomb manufacturing processes. (Courtesy of Hexcel International.)

Hexagonal Core

The standard hexagonal honeycomb is the basic and most common cellular honeycomb configuration, and is currently available in all metallic and nonmetallic materials.

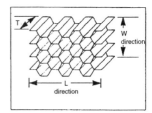

OX-Core®

The "OX" configuration is a hexagonal honeycomb which has been over-expanded in the "W" direction, providing a rectangular cell configuration which facilitates curving or forming in the "L" direction. The OX process increases "W" shear properties and increases "L" shear properties when compared to hexagonal honeycomb.

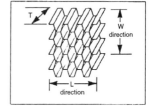

Flex-Core®

The Flex-Core cell configuration provides for exceptional formability in compound curvatures with reduced anticlastic curvature and without buckling the cell walls. Curvatures of very tight radii are easily formed. When formed into a tight radii, Flex-Core seems to provide higher shear strengths than comparable hexagonal core of equivalent density. Flex-Core can be manufactured in most of the materials from which hexagonal honeycomb is made.

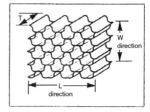

Tube-Core®

Our Tube-Core configuration provides a uniquely designed energy absorption system when the space envelope requires a thin-wall column or small-diameter cylinder. The design eliminates the loss of crush strength that occurs at the unsupported edges of conventional honeycomb. Tube-Core is constructed of alternate sheets of flat aluminum foil and corrugated aluminum foil wrapped around a mandrel and adhesively bonded. Outside diameters can range from 1/2 inch to 30 inches and lengths of 1/2 inch to 62 inches.

Double-Flex™

Double-Flex is a unique large-cell Flex-Core for excellent formability and high specific compression properties. Double-Flex formability is similar to standard Flex-Core.

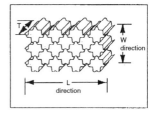

Fig. 2.29 Honeycomb configurations. (Courtesy of Hexcel International.)

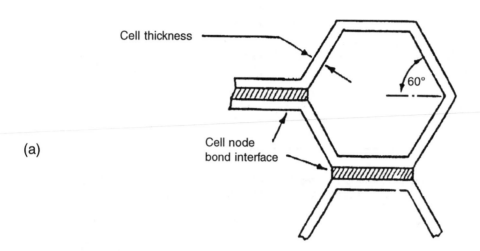

Cell thickness

60°

(a)

Cell node
bond interface

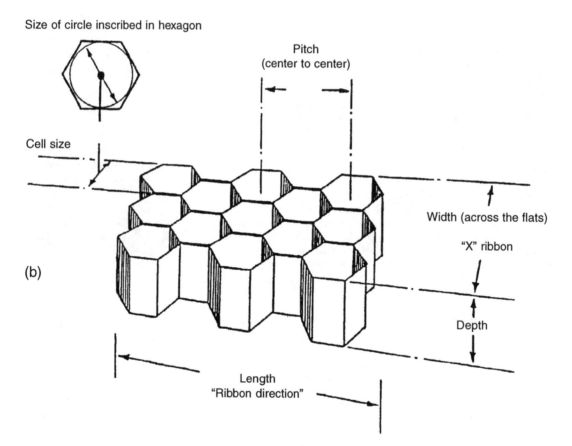

Size of circle inscribed in hexagon

Pitch
(center to center)

Cell size

Width (across the flats)

"X" ribbon

(b)

Depth

Length
"Ribbon direction"

*Fig. 2.30 (a) Standard honeycomb cell; (b) Honeycomb terminology.
(Courtesy of Heatcon Composite Systems.)*

The following defines common honeycomb terminology:

- **"L" Direction**: Length direction or ribbon direction (i.e., the direction of the continuous sheets of honeycomb)

- **"W" Direction**: Width direction

- **Cell Size**: The diameter of an inscribed circle within a cell

- **Node Bond**: That area of the honeycomb core where the cell walls are adhesively bonded

- **Ribbon Direction**: In honeycomb, the direction of the node bonds

- **Foil Thickness**: The thickness of the sheets of metal, paper, or fabric from which the honeycomb is made

2.6.3.1 Honeycomb Material Types

The primary types of honeycomb material in aircraft use are aluminum alloy, Nomex aramid, and fiberglass.

Aluminum Alloy: At least four types of aluminum alloy can be supplied in honeycomb form, and they are discussed as follows in increasing order of strength and cost. These are 3003, 5052, 5056, and 2024. Of these, 5052 is the most commonly used, with 5056 as the next most likely choice; however, for each repair, check the specification of the honeycomb required. It is normally acceptable to use a honeycomb of slightly greater density or strength for repairs, but honeycomb of lower strength or stiffness should not be used. Honeycomb having equivalent plate shear and compressive strengths may be interchanged. It is usually acceptable to use a slightly smaller or slightly larger cell size than the one specified, provided that the alloy and the density are the same. Small cell sizes are preferred because, in the event of disbond, crack propagation resistance is better with the small cells. The simplest arrangement for minimum stock holding is to hold only the 3 mm (0.125 in.) cell size in a range of densities. The mechanical properties of 3003, 5052, 5056, and 2024 differ significantly, and they should not be interchanged unless it is essential and a higher density is used, if necessary, to ensure that the required mechanical properties are met. Hexcel TSB 120 and data sheets from Eurocomposites, and others provide useful comparative data. For aircraft and other structural purposes, only the corrosion-treated and resin-dipped (etched and dipped) or chromic or phosphoric acid anodized honeycomb should be used. This is necessary to ensure the required durability in service under conditions that may be both hot and wet. The corrosion treated form is given an "alodine" conversion coating but not an anodize treatment before receiving a resin coating. Therefore, its preparation for adhesive bonding is not as good as a phosphoric or chromic acid anodized skin. For this reason, it is important that sandwich panels made from aluminum alloy honeycomb faced with aluminum alloy skins should be sealed against moisture or other liquid ingress and preferably have no holes drilled in them. If this is unavoidable, then flanged inserts should be used, made from two halves with an interference fit into each other. They should be adhesively bonded to the skin and core and sealed as effectively as possible to prevent moisture entry that would cause disbond between the skin and honeycomb or corrosion, or both. Any surface treatment of lesser quality than good anodizing will suffer disbond fairly rapidly in the presence of moisture. Contamination of the honeycomb due to poor storage conditions will aggravate the situation and cause early bond failures. (See Chapter 3.) Aluminum alloy honeycomb without the corrosion-treated and resin-dipped surface treatment or better should never be used in structural applications.

Cleaning of aluminum alloy honeycomb should be done by rinsing in cold, clean acetone. However, this should be a spray rinse to waste to avoid contamination and should be used only on a hot-cured primer-coated surface. Cleaning in hot trichloroethylene vapors should not be used because the thin honeycomb warms too quickly for any useful condensation cleansing to occur and the remaining surface is not good enough for bonding. Hot trichloroethylene should not damage the node bonds. A warm industrial detergent solution also is recommended. Household detergents such as "Fairy liquid" should not be used because they contain glycerine to reduce their effect on hands. A detergent wash should be followed by a thorough spray rinse with deionized water. Acids or alkalis should not be used to clean honeycomb. Any residual deposits could quickly lead to corrosion. Some specifications for aluminum alloy honeycomb in common use are Boeing BMS 4-4 and MIL-C-7438.

It is normal to use only aluminum alloy honeycomb with metal skins, but Nomex has sometimes been used. Slotted core is used in acoustic panels on engine cowlings. This type of honeycomb has slots in each cell so that water, which can enter the honeycomb through the perforations in the outer skin required for acoustic attenuation, can drain through the bottom of the core to suitable drain holes. The slots must be correctly oriented to allow the water to drain as completely as possible. When repairing acoustic panels, only the slotted type of honeycomb is acceptable for repairs. The slotted side must be bonded to the bottom skin. If a new section of outer skin is fitted, then any water drain holes used in the original skin must be drilled in the new one.

Nomex Aramid Honeycomb: Nomex honeycomb is made from Nomex aramid paper. Its chemical name is polymetaphenyleneisophthalamide. The paper consists of aramid fibers in a matrix of aramid resin and is available in a range of thicknesses. The paper is dipped in phenolic resin, and tests have shown that, for example, it is better to use three layers of paper and a small amount of resin-dipped coating than to use two layers of paper, resin dipped to the weight of three layers of paper. Some resin coating is necessary to reduce the tendency of Nomex to absorb moisture. The mechanical properties are better when three layers of paper are used.

Nomex honeycomb is available from several sources. Hexcel manufactures HRH-10; Ciba used to manufacture Aeroweb A1 and has now combined this part of its business and the manufacture of film adhesives and pre-pregs with Hexcel. A specification commonly used is Boeing BMS 8-124. HRH-310 aramid fiber paper (Nomex honeycomb) is made from the same aramid paper as HRH-10 but is dipped in a polyimide resin to achieve the final density.

Fiberglass Honeycomb: This can be supplied in fabric form for impregnation with polyester or epoxy resin, or it can be supplied with a variety of resin systems.

- **Hexcel HRP:** This is a glass fabric-reinforced plastic honeycomb dipped in a heat-resistant phenolic resin to achieve the final density. It was developed for use up to 180°C (350°F) or higher for short exposures.

- **Hexcel HFT:** This glass fabric-reinforced plastic incorporates a Fibertruss bias weave dipped in a heat-resistant phenolic resin to achieve the final density. It was developed for use up to 180°C (350°F) or higher for short exposures. HFT has a higher shear modulus than HRP or HRH-10.

- **Hexcel NP:** This is a glass fabric-reinforced plastic honeycomb in which the initial impregnation is a nylon-modified phenolic resin and the final dip coats are polyester resin. This core type is recommended for applications that do not exceed 82°C (180°F) for extended time periods.

- **Hexcel HRH-327:** This is a glass fabric, polyimide node adhesive, bias weave reinforced plastic honeycomb dipped in a polyimide resin to achieve the final density. This material has been developed for extended service temperatures up to 260°C (500°F) with short-range capabilities up to 371°C (700°F).

- **Hexcel TPC:** TPC is the designation used for fiberglass ± 45° bias weave reinforced/thermoplastic honeycomb. This bias weave enhances the shear properties of the honeycomb core. TPC has excellent damage resistance because of the tough thermoplastic resin matrix. This increased damage resistance allows TPC to be formed more readily than thermoset honeycombs, resulting in reduced shop losses and simplified fit-up of parts during assembly. The thermoplastic resin used in TPC allows skin-to-core bonding by the use of thermosetting adhesives.

Other Types of Honeycomb: Honeycomb can be supplied in several other materials, including the following:

- **Graphite Honeycomb Supplied by Hexcel as HFT-G**: This material uses a bias weave fabric (Thornel 300, 1000 tow) reinforcement with a polyimide resin matrix. A thermosetting node adhesive is used with a heat-resistant polyimide resin for web impregnation. This product may be used at service temperatures up to 260°C (500°F) and for short exposures at higher temperatures. HFT-G is also available with a phenolic resin matrix. Special cores with higher densities, other cell sizes, or thermoplastic resin matrix may be supplied on request.

- **Kevlar 49 Honeycomb**: This material was developed by Hexcel as HRH-49 for use as sandwich core material in radomes and antennas because Kevlar has a low dielectric constant. It is made from Kevlar 49 fabric. A thermosetting adhesive is used for the node bonds, and a 180°C (350°F) curing modified epoxy resin is applied for the initial and final impregnations. It is made initially in 1/4-in. cell size and 2.1 lb/ft³ density. Other densities and cell sizes are available on request.

2.6.3.2 Cell Shapes

Four cell shapes are readily available. These are illustrated in Figure 2.28 and are as follows:

- **Hexagonal Cell:** This is the most common cell shape and is relatively simple to produce; therefore, it is also the cheapest. Hexagonal cell shape is available in all metallic and nonmetallic honeycomb types. It is best for flat panels because it cannot take much curvature. In the case of Nomex and other honeycombs, which can be heat formed, the hexagonal cell shape can be formed into curved and even double-curved shapes.

- **Overexpanded Cell (OX-Core):** OX-Core can be easily draped around single curvatures such as wing leading edges and tubes for air conditioning ducts. The "OX" configuration is a hexagonal honeycomb, which has been overexpanded in the "W" direction, giving a rectangular cell shape that facilitates curving or forming in the "L" direction. The "OX" process increases "W" shear properties and decreases "L" shear properties when compared to hexagonal honeycomb.

- **Flex-Core Cell:** This special shape was designed to accept double curvature and is excellent for the nose area of radomes and fairings having significant double curvature. The Flex-Core cell configuration provides for exceptional formability in compound curvatures with reduced anticlastic curvature and without buckling the cell walls. Curvatures of tight radii are easily formed. When formed into tight radii, Flex-Core seems to provide higher shear strengths than comparable hexagonal core of equivalent density. Flex-Core can be manufactured in most of the materials from which hexagonal honeycomb is made.

- **Double-Flex Cell:** This cell shape is a unique large-cell Flex-Core for excellent formability and high specific compression properties. Double-Flex formability is similar to standard Flex-Core.

2.6.3.3 Honeycomb Densities and Cell Sizes

These vary with the material from which the honeycomb is made. Special densities and cell sizes can be made to order. Some readily available densities and cell sizes are as follows:

Aluminum Alloy Honeycomb:

- **5052 Alloy Hexagonal Cell Honeycomb:** Densities can be supplied from 16 to 192 kg/m^3 (1.0 to 12.0 lb/ft^3). Cell size ranges from 1.5 to 10 mm (1/16 to 3/8 in.).

- **5056 Alloy Hexagonal Cell Honeycomb:** Densities can be supplied from 16 to 146 kg/m^3 (1.0 to 9.2 lb/ft^3). Cell size ranges from 1.5 to 10 mm (1/16 to 3/8 in.).

- **2024 Alloy Hexagonal Cell Honeycomb:** Densities can be supplied from 44.6 to 151 kg/m^3 (2.8 to 9.5 lb/ft^3). Cell size ranges from 3 to 6 mm (1/8 to 1/4 in.).

- **5052 Alloy Flex-Core:** Densities can be supplied from 33 to 127 kg/m^3 (2.1 to 8.0 lb/ft^3).

- **5052 Alloy Double-Flex:** Densities can be supplied from 43 to 76 kg/m^3 (2.7 to 4.8 lb/ft^3).

- **5056 Alloy Double-Flex:** Densities can be supplied from 33 to 127 kg/m^3 (2.1 to 8.0 lb/ft^3).

HRP Honeycomb:

- **HRP Hexagonal Cell Glass-Reinforced Phenolic Honeycomb:** Densities can be supplied from 35 to 192 kg/m^3 (2.2 to 12.0 lb/ft^3). Cell size ranges from 4.5 to 10 mm (3/16 to 3/8 in.).

- **HRP OX-Core:** Densities can be supplied from 51 to 111 kg/m^3 (3.2 to 7.0 lb/ft^3).

- **HRP Flex-Core:** Densities can be supplied from 40 to 88 kg/m^3 (2.5 to 5.5 lb/ft^3).

TPC Fiberglass-Reinforced Thermoplastic Honeycomb: Densities can be supplied from 48 to 88 kg/m^3 (3.0 to 5.5 lb/ft^3). Cell size ranges from 3 to 4.5 mm (1/8 to 3/16 in.).

HFT Glass-Reinforced Phenolic Honeycomb (Fibertruss Bias Weave): Densities can be supplied from 32 to 127 kg/m^3 (2.0 to 8.0 lb/ft^3). Cell size ranges from 3 to 10 mm (1/8 to 3/8 in.).

HFT-G Graphite-Reinforced Honeycomb: Densities can be supplied from 32 to 160 kg/m^3 (2.0 to 10.0 lb/ft^3) with one special version at 320 kg/m^3 (20.0 lb/ft^3). Cell sizes range from 4.5 to 10 mm (3/16 to 3/8 in.).

HRH Honeycomb:

- **HRH-327 Glass-Reinforced Polyimide Honeycomb:** Densities can be supplied from 51 to 127 kg/m^3 (3.2 to 8.0 lb/ft^3). Cell sizes range from 3 to 10 mm (1/8 to 3/8 in.).

- **HRH-10 Aramid Fiber-Reinforced Honeycomb:** Densities can be supplied from 29 to 143 kg/m^3 (1.8 to 9.0 lb/ft^3). Cell sizes range from 1.5 to 10 mm (1/16 to 3/8 in.).

HRH-10 OX-Core: Densities can be supplied from 29 to 64 kg/m³ (1.8 to 4.0 lb/ft³). Cell sizes range from 4.5 to 6 mm (3/16 to 1/4 in.).

HRH-10 Flex-Core: Densities can be supplied from 40 to 88 kg/m³ (2.5 to 5.5 lb/ft³).

Note: On 29 February 1996, the Composites Division of Ciba-Geigy merged with the Hexcel Corporation to form Hexcel Composites. All former Ciba Composites and Hexcel product literature is available from Hexcel Composites Ltd., Duxford, Cambridge CB2 4QD, England.

2.6.4 Fluted Core

The simplest type of fluted core, a number of square fiberglass tubes laid side-by-side, is used by several manufacturers for aircraft radomes. Unlike honeycomb, it drains away any liquid water, if oriented correctly, and is preferred by some airlines.

Fluted radomes seem to suffer less damage than the honeycomb type, but they are more difficult and time consuming to repair. Honeycomb radomes suffer from moisture ingress after long periods of time and are more easily damaged; however, they have the advantage of being considerably easier and quicker to repair. For details of various types of fluted core, including complex types, see Ref. 2.24.

2.6.5 Syntactic Core

This material has a higher density than most other core types and is defined later in Section 2.7. For this reason, its use is generally limited to thin sandwich panels. It is supplied as "Syncore" by Hysol. See Ref. 2.24.

2.7 Syntactic Foams and Potting Compounds

Syntactic foams are defined as, "Compounds made by mixing hollow microspheres of glass, epoxy, and phenolic, etc. into fluid resins (with additives and curing agents) to form a lightweight, moldable, fluid mass; as opposed to foamed plastic, in which the cells are formed by gas bubbles released in the liquid plastic by either chemical or mechanical action."

Most manufacturers produce one or more potting compounds for core jointing and edge filling. The terms "potting compound" and "syntactic foam" are sometimes used interchangeably. These are usually epoxies for aircraft use, filled with hollow glass or phenolic microballoons or other materials, and they may be supplied in a range of densities depending on the compression strength required. They are based on epoxy adhesives and matrix resin systems and may be room-temperature curing; high-temperature curing 120°C

(250°F) and 180°C (350°F) systems are available. More recent versions have been developed to meet the latest smoke and toxicity requirements. A surprising number of systems are produced; however, a range of properties may be required, and these help to explain the number of systems on the market.

The following properties are required of potting compounds:

- High compression strength if used to fill honeycomb around bolt holes

- Low density if used for edge filling

- Good fluid resistance if used for sealing

- Flexibility and toughness

- High-temperature resistance for some applications

- Low smoke emission and good fire resistance for interior applications

- Nonsag gap-filling qualities

Several typical products are listed below. They are all described as low-density, epoxy filler pastes.

- Ciba Redux 252 RT cure, two component

- Ciba Redux 253 Hot cure (one hour at 170°C [338°F]), two component, long pot life

- Ciba Redux 255 RT cure, flame retardant, two component

- Ciba Epocast 1610 Hot cure (90 minutes at 120°C [250°F]), one component, long storage

- Ciba Epocast 1614 Hot cure (one hour at 180°C [350°F]), one component, eight-hour work life, good strength up to 180°C (350°F)

- Ciba Epocast 1617 A/B, RT cure, two component, self-extinguishing

- Ciba Epocast 1618 A/B, RT cure, two component, self-extinguishing

- Ciba Epocast 1619 A/B, RT cure, two component, self-extinguishing

- Ciba Epocast 1620 A/B, RT cure, seven days, toughened, self-extinguishing

- Ciba Epocast 1670 A/B, RT cure or five hours at 52°C (125°F), two component, fire resistant

- Ciba Epocast 1671 A/B, RT fast cure, fire resistant, low smoke, two component

- Ciba Epocast 1672 A/B, RT cure or five hours at 50°C (122°F), fire resistant, ultralow density

- Cytec (Cyanamid) Corfil 615, may be cured from RT to 180°C (356°F), two component

- Cytec (Cyanamid) BR 624, one part, may be cured from 107 to 175°C (225 to 347°F)

- Cytec (Cyanamid) BR 626, improved BR624, better hot/wet performance

- Hysol TF3056FR, RT cure, two component, self-extinguishing

- 3M Scotch Weld 3524 FST B/A, RT cure, two component, flame retardant, low smoke, low toxicity, solvent free, designed to meet Airbus Specification ATS 100.001

- 3M Scotch Weld 3500-2 B/A, two part, high-temperature resistant, cure cycle one hour at 175°C (347°F)

- 3M Scotch Weld 3439HS FST, one part, can be cured at 125°C (257°F) or 175°C (347°F) in one hour, self-extinguishing, low smoke, low toxicity; meets all fire, smoke, and toxic gas emission requirements of ATS 1000

- A compound that can be made on site is Shell "Epikote" 815 + RTU + phenolic microballoons in the ratio 100-25-24 by weight; RTU is now supplied as Ancamine 1483 by Pacific Anchor Chemical Corporation

Phenolic or glass microballoons can be added to most resins and paste adhesives to produce a potting compound if necessary in emergency (e.g., Redux 420 or Hysol EA 9321 with 20% microballoons make strong, tough potting compounds).

Hysol produced a film version called "SynCore," which is epoxy resin filled with low-density glass, quartz, or carbon microballoons. It is usually used in thin layers but can be used at thicknesses that will allow it to replace other core materials. Versions curing at 120°C (250°F) and 180°C (350°F) are available. New versions of SynCore using other resin systems are under development. See the Hysol SynCore data sheet.

2.7.1 Filler Materials for Potting Compounds, Resin Systems, and Adhesives

The viscosity of resin systems is increased by the addition of fillers. Particulate fillers cause the least increase; fibrous fillers cause the most increase. Tensile strength may be reduced by particulate fillers but is increased by fibrous fillers if the concentration of filler is not too low. Particulate fillers generally increase thermal conductivity and reduce the coefficient of thermal expansion in proportion to the concentration of filler. Fillers should be of neutral pH or slightly basic and should not react with either the resin or its curing agent. This means that some fillers cannot be used with some classes of curing agents, although most commercially available inorganic fillers are compatible with most epoxy resin systems. The study of the effects of fillers is a big subject in itself. For more details of the effects of various fillers, see Ref. 2.16. All fillers should be dried before incorporation into a resin mix. Some fillers and their effects are listed below.

Milled Fibers: These are used to improve toughness and compressive and tensile strength. Although effective for toughening, the use of asbestos fiber has been progressively eliminated for health reasons. It has been replaced by aramid fibers and possibly others for the toughening of adhesives, which after much work are now proving equally effective. Glass and other milled fibers also may be used in potting compounds. Any fiber normally used in composites can be used in milled form as a filler.

Silica: Silica is used as a thixotrope to provide nonsag and gap-filling properties. It has little effect on resin tensile strength. The thixotropic (nonsag) effect normally results from flocculation of particles when they are wetted thoroughly. Particles are held together at low shear rates by van der Waals forces; at high shear rates (rapid mixing), the particles are forced apart and these forces no longer operate significantly. This convenient characteristic allows for easy mixing but also provides the nonsag effect. (See Ref. 2.16.) Silica is usually described as "fumed silica" and is sold under the trade names Cab-o-Sil or Aerosil. These are the most common names listed in Structural Repair Manuals (SRMs), but there are probably several others. Use only the percentage of silica specified by the SRM, and mix it thoroughly into the resin.

Microballoons: Microballoons are hollow spheres used to reduce density and to provide nonsag and gap-filling properties. They are used in potting compounds. Both glass and phenolic microballoons are available.

2.7.2 Other Fillers

Apart from the most common fillers mentioned above, several hundred commercial fillers are suitable for use with epoxy resins. They range in particle size from approximately 200 mesh to 0.015 micron. Some typical fillers are aluminum silicate, calcium carbonate, ceramic zircon, aluminum powder, kaolin, mica 4X, zirconium silicate, and hydrated alumina. Those most commonly used in adhesives are aluminum, alumina, and silica.

2.8 Protective Coatings

2.8.1 Sealants

Several types of sealants are available. They prevent the passage of fluids in the liquid form, but most allow the penetration of fluids by absorption and permeability. Permeability is defined as, "The passage or diffusion (or rate of passage) of a gas, vapor, liquid, or solid through a barrier without physically or chemically affecting it." Permeability is the product of the diffusion coefficient and the solubility coefficient (or saturation content of the fluid in the material in question) (i.e., $P = D \times S$). Experiments on the strength of adhesive joints after years of exposure have shown that sealants make virtually no difference to joint strength, indicating that they do not keep out moisture effectively. However, sealants often are used to minimize, if not totally eliminate, fluid penetration. They help to provide a smooth finish at joints and give some protection against attack on bond lines by paint strippers, hydraulic fluids, and cleaning materials. They also delay the penetration of moisture into bond lines for some time. In practical applications, they are used to contain fuel, maintain cabin pressure, reduce fire hazards, exclude moisture, prevent corrosion, and to fill gaps and smooth discontinuities on the aircraft exterior to reduce drag and improve performance, economy of operation, and appearance. One definition of sealing describes it as a process that, "Confines gases and liquids in the required area and prevents them from entering areas where they could create a hazard. Sealing is achieved by closing all pathways through which these fluids could escape. The materials used to seal these pathways are applied as wet, easily flowing pastes. After the required period of time for each sealant type, they cure to form a rubbery solid that bonds firmly to the surfaces around the areas to be sealed." Bare honeycomb edges and any water paths to honeycomb structures should be sealed as effectively as possible. Do not apply sealant when the temperature is below 10°C (50°F) or above 49°C (120°F). Reduction in the quality of the sealant will occur. Longer curing times are needed when the air is dry or cold. Air is considered dry when the relative humidity is less than 40% and is considered cold when less than 20°C (68°F).

Sealants also need good surface preparation if they are to remain "stuck." The Boeing 737 dry bay area is an example of the importance of surface preparation before sealing. Many problems have occurred in this area because insufficient time and effort was taken to achieve good surface preparation before applying sealant. It is a classic example of the fact that a job properly done is faster in the long run because it does not have to be done again and again. The Boeing 747-400 SRM 51-20-05 is typical of many good SRM chapters on other types, giving good information on the variety of sealing materials and processes used on modern aircraft. Every type has an SRM chapter on sealing and sealants. They provide guidance on all aspects of sealing, not only sealing of adhesively bonded joints. A few particularly important points are:

- Thorough cleaning is necessary because a sealant will not adhere to a soiled or greasy surface.

- Areas contaminated with hydraulic fluid should be cleaned according to Chapter 20-30-03 of the Boeing Standard Overhaul Practices Manual, D6-51702 or equivalent.

- Sealants can be highly toxic. See Chapter 6, Safety and Environment.

- Do not seal drain holes. Keep drain paths clear. A corrosive environment will develop if drain holes and paths do not remain clear.

- Sealants must be allowed to cure before flight. Study each sealant data sheet for curing times at various temperatures. The lower the temperature, the longer the cure time. Do not confuse work life/pot life with curing time.

Most sealants used on aircraft were of three basic types: polysulfide, silicone, and Viton rubber. The polysulfides are being replaced by Permapol P3 and P5 polythioether polymers, which have considerably better thermal/hydrolytic stability and excellent fuel/solvent resistance. Significant developments have occurred in recent years, and new versions have appeared on the market. These have been driven by the need to remove certain metallic accelerators, used to speed the curing process, because they were found to be toxic and this led to handling and disposal problems. Solvent content regulations have also been imposed on sealant formulation and processing methods. Modern sealant developments are working toward formulations using environmentally safe materials and processes. Critical factors are the temperature range in service, flexibility to accommodate movement due to stress and temperature changes, and the types of fluids they must resist. Another factor is the materials to which they will be applied. Sealants must not cause corrosion or cracking of metals in many applications or crazing of plastics or damage to rubbers in others. Use the specified sealant or one that can safely do the required job. In some applications of silicone sealants, it is necessary to use grades approved for "food contact." Silicones have a low surface energy and thus will bond to many other materials. Conversely, few materials will adhere to silicones—usually only other silicones and even then, only after the application of a suitable primer. This means that seals made with other sealants may be repaired with silicones, but silicone seals cannot be repaired with other sealants.

Silicone and nonsilicone sealants must be stored separately, or they will cross contaminate. Use separate tools and containers.

Sealants are made to a number of specifications, some of which are listed below. Many American military (MIL) specifications are being replaced by SAE AMS specifications. Check SAE lists for AMS replacements for MIL specifications.

- American specifications:
     Boeing BMS 5-26
     Boeing BMS 5-32
     Boeing BMS 5-37
     Boeing BMS 5-63

Boeing BMS 5-95
Lockheed LAC-C40-1279
MIL-S-7502
MIL-S-8802
MIL-S-81733
MIL-S-83318 (AMS 3283)
MIL-S-83430 (AMS 3276A)

- British specifications:
  DTD 900/4079
  DTD 900/4523
  DTD 900/4950
  DTD 900/6021

2.8.1.1 Old-Generation Thiokol Sealants

Old-generation Thiokol sealants include the following:

- Berger Elastomers PR 1301, used for integral fuel tanks and pressure cabins
- Berger Elastomers PR 1440, used for integral fuel tanks and pressure cabins
- 3M-1675 B/A, alternative and equivalent to PR 1440
- Berger Elastomers PR 1436G, pressure, environmental, and fuel cavity sealing
- Essex Chemical Corporation Pro-Seal 870, alternative and equivalent to PR 1436G

2.8.1.2 New-Generation Polythioether Sealants

These have been developed by the Products Research and Chemical Corporation. These new classes of polymers have the designations Permapol P-3 and Permapol P-5. They can be formulated into high-performance aircraft sealants, giving environmentally safe products which include lead-free sealants, solvent-free sealants, chromate-free corrosion inhibitive sealants, rapid-curing sealants, weight-saving sealants, and electrically conductive sealants.

The following shows the new products that replace the previous lead oxide cured materials.

Lead Oxide Cured	Application	Replacement
PR-1301	Low-adhesion access door sealant	PR-1428
PR-1321	Low-adhesion access door sealant	PR-1428
PR-340	Aerodynamic smoothing sealant	PR-1775 or P/S 895
PR-380	Cabin pressure sealant MIL-S-7502	PR-1778
PR-383	Metal/glass bonding sealant MIL-S-11031	PR-1778
PR-1221	Fuel tank and cabin sealant PR-1770	
PR-1223	Low-density sealant	P/S 875
PR-1224	Low-density sealant	P/S 875

New materials showing chemical base include:

Permapol-P3	Silicone	Permapol-P5	Polysulfide
PR-1828	PR-1768	P/S 875	PR-1428
PR-1828G	PR-1990	PR-1775	PR-1448
PR-1829	PR-1991	PR-1776	
PR-1769			
RW-2660-73			
RW-2839-71			

The SRM for each aircraft type lists all the sealants approved for that type and should be consulted before any new task is commenced. Some sealants are not compatible with some primers. Check the type suitable for each application.

2.8.1.3 Silicone Sealants

Silicone sealants are flexible at low temperatures (-60°C) (-76°F) and can operate at temperatures up to 200°C (392°F). Some silicones have a maximum service temperature of 232°C (450°F), others 260°C (500°F), and some 300°C (572°F). They are generally low-modulus, high-elongation materials that allow considerable movement before disbonding or fracture. All silicones have a high resistance to thermal and oxidative degradation. They have a low level of water absorption, but their open molecular structure allows water vapor to pass through more easily than most organic sealants. Silicones show good stability of physical and dielectric properties and thus are frequently used for the encapsulation of electronic components. Most silicone sealants for aircraft use are supplied by Dow-Corning, General Electric, or Loctite. There is a considerable range of color, viscosity, and performance. Most are of the room-temperature vulcanizing (RTV) type and rely on moisture in the atmosphere to initiate cure. During the curing process, they generally emit an acetic acid (vinegar-type) odor. These types of sealants cure properly only if exposed to the atmosphere. Only small beads of sealant should be laid down at any one time because atmospheric moisture cannot penetrate to any great depth. They cannot be used between two sheets of material to bond large areas. For this purpose, two-part versions are made and must be weighed and mixed thoroughly in the same manner as epoxies. These types are effective if tough, impact-resistant bonds are required or if significant amounts of thermal expansion must be accommodated in heat shields. A considerable number of silicones are approved for use in food contact applications. They are available in a range of colors. Silicone primers are often recommended and should always be used if specified in the material data sheet.

Silicones should be stored separately from other sealants and adhesives. See Chapter 3, Handling and Storage. Some specifications are MIL-A-46106 and MIL-A-46146.

2.8.1.4 Viton Rubber Sealants

These are used on Concorde in a number of applications where high-temperature performance and oxidation resistance are required. Viton is the trade name for a copolymer of vinylidene fluoride and hexafluoropropylene. Peroxide catalysts are used to vulcanize or cross-link Viton.

2.8.2 Primers

A primer is defined as, "A coating applied to a surface, before the application of an adhesive, lacquer, or enamel, etc., to improve the adhesion performance or load-carrying ability of the bond. Some primers contain a corrosion inhibitor." Primers are of lower viscosity than adhesives and therefore can wet surfaces more easily. They are more commonly used when bonding metals, and seldom used when bonding composites. When used with metals, they increase the corrosion resistance and durability but generally have a much smaller effect on bond strength. Although not always strictly necessary, if chemical compatibility exists, it is generally recommended to use primers and adhesives from the same company. Most companies recommend specific primers for each adhesive supplied. Where a specific primer is recommended for a particular purpose, it should be used. Primers are useful for the immediate protection of freshly treated metallic surfaces. An immediate coating with a suitable primer protects the surface from atmospheric contamination by reducing the surface energy from a high value, prone to contamination, to a lower value that is less prone to contamination and is highly compatible with the adhesive.

One note of caution should be mentioned concerning the use of primers. An excessive amount of primer is not beneficial. The reverse is actually the case. The correct thickness for a primer coating is specified by the supplier and must be used. If too much primer is used, a much weaker bond will be found on test. The simplest way to use the right thickness of primer is to have the supplier or manufacturer provide a color chart with three zones: too thin, appropriate thickness, and too thick. If an error is to be made, it should be on the side of using a primer coat that is slightly too thin. Tests have shown clearly that too much primer produces weak joints.

Some sealants are not compatible with some primers and other sealants. See each type SRM for details of sealant application. For example, Boeing B.737-300 SRM 51-20-05, Figure 1, Sheet 2, states, "The cure of BMS 5-63 is inhibited by BMS 10-11 and BMS 10-20 primers and by BMS 5-26, BMS 5-79, and BMS 5-95 sealants."

Some silicone sealants are used with special silicone primers. The correct primer specified on the data sheet must be used and applied as directed.

2.8.3 Finishes

Paint finishes serve several purposes:

- They provide an attractive appearance to the aircraft or other vehicle, and, in the form of the company logo, they are an important means of identifying the airline, shipping line, railway, or vehicle operator. They also give an impression, which is hoped to be favorable, to the passenger. This explains why exterior and interior appearance is rated so highly by airlines, in particular, because commercial competition is intense. From a safety viewpoint, a good appearance must be backed by high standards in more critical areas of maintenance, whether of aircraft or other forms of transport.

- They provide a smooth surface to the aircraft and reduce drag, thereby saving fuel.

- For metal airplane parts and automobile, marine, or railway parts, they provide corrosion protection. For composite parts, they minimize the ingress of moisture. Paints are, to some extent, permeable to moisture; that is, moisture is not kept out entirely. For this reason, corrosion-resistant primers should always be used when specified. Paints are, in a sense, weak adhesives. They have a strong similarity to adhesives because the correct surface preparation is required if they are to remain bonded to a surface. Experience has shown, whether working at home or on aircraft or other means of transport, that inadequate cleaning and surface treatment result in paint falling off fairly rapidly.

The correct procedures, primers, and paints for aircraft are always clearly specified in the SRM, and these must be used to achieve adequate corrosion control or composite surface protection. Good paint schemes are not "just a pretty face" for appearance alone; however, from a marketing viewpoint, this is important. Paint schemes are a serious factor in achieving the design life of an airframe. For this reason, paint or primer on the inside of an airplane, automobile, ship, or train and hidden behind the trim is as important as the attractive paint finish on the outside. The protective primer on the inside must be more carefully applied because any deterioration will not be visible until the next inter or even major check. The interval between major checks is usually several years. Great care should be taken to avoid damage to paint schemes when removing parts for repair, especially in less-visible areas.

Paint stripping of composite parts has been a source of concern for many years. Methylene chloride and phenol based paint strippers seriously damage epoxy matrix resins if allowed to remain on the surface for long periods. In one case, on a large transport aircraft some fiberglass panels on the fin were sprayed liberally with paint stripper and left for several days. In this case, two layers of fiberglass could be seen hanging from these panels, leaving the honeycomb core clearly visible. The panel resembled a steamy bathroom with the wallpaper falling off. Even short-term contact with paint strippers (i.e., a few minutes) is not recommended. A variety of methods have been tried for safer paint removal, and the

Society of Automotive Engineers (SAE) has produced a document—MA 4872—on the subject, entitled, "Paint Stripping of Commercial Aircraft—Evaluation of Materials and Processes." IATA has a task force working in this area.

In addition, it may also be fortunate that environmental considerations have forced the study of alternative methods, which in this case may be technically beneficial. Methylene chloride and phenol have been found to be toxic, and the regulations for their usage and disposal are being tightened. The normal alternative method is sanding, but this also requires great care to avoid damaging fibers in the first ply and is too labor intensive for large areas. Check with the SRM for approved methods of paint removal, and use them carefully. These problems have led chemists to study different formulations for paints that can be stripped with less toxic solvents while continuing to perform protective and decorative functions.

British Airways evaluated a new painting and paint removal system for aluminum alloy structure, which seems to show considerable promise. This system leaves the original coat of strontium chromate primer, applied at the manufacturing stage, permanently in place for the life of the aircraft. On top of this, an intercoat is applied before the topcoat of polyurethane. The intercoat and topcoat can be removed together using environmentally acceptable benzyl alcohol stripper, which does not remove the primer. Together with new high-solids/low-solvent paints, containing up to 40% less solvent than in the past, this will help the airline to meet future solvent control rules. The initial trial on Concorde and limited trials on two Boeing 747s have been successful. Further trials are to be stripped on two 757s in August 1997. This paint system will be adopted across the fleet from September 1, 1997.

This paint system is expected to prove suitable for composite materials and to replace the variety of alternative methods that have been studied for some time. Some problems have been found with low VOC (volatile organic compound) paints, as would be expected. They are more difficult to spray because they contain less solvent.

To date, trials seem favorable, and small modifications should overcome these problems. This is encouraging because even aircraft that spend a whole working life with one airline may need to have paint stripped between seven and ten times in their lifetime because of deterioration and company logo changes. It is difficult to estimate how many times an aircraft rented by a leasing company might be repainted, but twenty times is not an impossible figure. When painting, it is important to aim for a total paint thickness of 0.2 to 0.25 mm (0.008 to 0.010 in.) to avoid excess weight, which on control surfaces can put them out of balance. Even large aircraft with powered controls require balanced control surfaces because, in the unlikely event of a jack failure, flutter could occur which would result in serious vibration or structural failure. In the case of radomes, more than the permitted amount of paint will affect radar transmission, and it certainly is not acceptable to use several coats of paint on a radome as a means of changing the color scheme. These special paints are supplied by Courtaulds and Akzo. Courtaulds is advertising this system (as of July 1997) as follows:

- **Desothane® HS High Durability Low VOC Topcoat:** Excellent application properties, UV resistance, and gloss retention. Longer service life, reduced solvent use, and compliance with global regulations. (Desothane is a registered trademark of Courtaulds Aerospace Inc.)

- **F565-4000 Group Intermediate Coat**: Allows the topcoat to be removed with non-halogen paint remover without disturbing detail primer. Saves time and reduces hazardous waste.

This system has been tried by Fokker for several years and has already been adopted by British Aerospace for all aircraft.

2.8.4 Conductive Coatings

Metal Foil: Where a conductive coating is required, a metal (aluminum) foil may be laid into the mold and pre-preg laid up against it to provide a conductive coating to the composite. On some occasions, a layer of Speedtape may be used as a temporary replacement for an area of flame-sprayed aluminum on the surface of a composite part after a repair has removed the original layer.

Flame-Sprayed Aluminum: Composite parts requiring lightning protection often are given a flame-sprayed coating of aluminum for this purpose. After a repair, this will require replacement to restore the original degree of lightning protection. In a few cases, these coatings are used as aerials for radio or navigation aids. If aluminum flame-sprayed coatings are used, usually on fiberglass or aramid components, then the SRM or AMM will specify the area that is permitted to be missing. Small areas sometimes may be repaired without replacing the coating, but limits are provided in the SRM beyond which the coating must be replaced. Electrical continuity must be checked. The Boeing process specification for aluminum flame spray is BAC 5056.

Metal-Coated Fabric: Several metal-coated fabrics are available, some with specific trade names such as Thorstrand, Alumesh, and Alutiss. These fibers are coated with metal and embedded in the outer layer of composite components to provide lightning protection. If damaged, they must be repaired to the SRM and AMM. Electrical continuity must be checked.

Thorstrand, Alumesh, and Alutiss are aluminized glasscloth materials. They can be used either as a single layer or a double layer in areas prone to more severe strikes. Nickel-coated carbon fiber and nickel-coated aramid fibers also have been tried.

Other methods of lightning protection, such as aluminum strips and frames used at the tips of rudders, elevators, and ailerons and diverter strips fitted to radomes, do not fit into the classification of coatings. These also must be fitted and repaired correctly, and they must be checked for electrical continuity as specified in the SRM and AMM. Lightning damage often is more extensive than it first appears. See Chapter 7, Damage and Repair Assessment. Research continues on lightning protection, and new ideas and methods may be expected. Meanwhile, it is essential to carefully maintain the existing methods. See Refs. 2.25 through 2.28, inclusive. In addition, see FAR and JAR 25.581 from the Federal Aviation Regulations (FARs) and Joint Airworthiness Requirements/Joint Aviation Requirements (JARs), respectively.

Wire Mesh: Woven aluminum fabric has been tried, using various wire diameters.

Antistatic Paint: The friction of high-velocity air passing over a large-surface electrical insulator such as a plastic radome or other nonconducting composite or plastic component results in the buildup of static electricity, which periodically discharges by arcing to a metallic portion of the aircraft. This discharging can result in interference with the aircraft's electronic equipment. The use of an antistatic coating having an electrical resistance low enough to allow the static charge to gradually "bleed off" provides a solution to this problem. Antistatic coatings compatible with particular coating systems are formulated by the addition of a small amount of electrically conductive material, generally carbon, to the base coating solution. A 0.025 mm (0.001 in.) thick antistatic coating applied over the primary coating will normally fulfill the antistatic requirements. Typical requirements for an antistatic coating are given in U.S. military specification MIL-C-7439B, which covers elastomeric-type rain erosion resistant coatings. These are special coatings with controlled levels of conductivity for particular purposes.

The Boeing specifications are as follows:

- BMS 10-21, Type I, a conductive coating for use on all fiberglass areas other than antennae or radomes

- BMS 10-21, Type II, antistatic coating, De Soto Super Koropon

Type I is a more conductive material than Type II; therefore, use the correct type. The lower conductivity antistatic coating is used to minimize interference on aircraft radio communications systems. Any higher level of conductivity would reduce radar transmission efficiency. The SRM and Maintenance Manual (MM) should be consulted for details of the type of coating to be used and its precise location on the radome or other component. Other aircraft manufacturers use similar materials from other suppliers in their own geographical regions.

2.8.5 Erosion-Resistant Coatings

These are used on radomes, leading edges of wings, tailplanes, and (in some cases) fins and on the leading edges of helicopter rotor blades to reduce erosion wear caused by rain, hail, dust, or sand. If specified in aircraft manuals, they should be inspected, repaired, or fitted as directed. This is especially important for radomes to ensure that the correct type of erosion-resistant paint is used and that the coating thickness does not exceed the specified value. All painting or bonding of erosion-resistant coatings should be carried out on clean, dry surfaces. It is helpful to have a resin-rich outer surface finely sanded to a good finish as a base, and a good final surface finish has been found to improve erosion resistance. Over the years, two methods have been used for radomes:

1. Neoprene or polyurethane overshoes or "boots" bonded in place with solvent-based neoprene-type adhesives. These normally last one year or longer, but this depends on the thickness of the rubber overshoe. Experience has shown that overshoes of 0.25 mm (0.010 in.) thickness have a life of approximately one year. For a good service life, tests at the Defence Evaluation and Research Agency (DERA) Farnborough, England, have shown that a minimum thickness of 0.5 mm (0.020 in.) should be used. The latest overshoes from 3M are now made 0.5 mm (0.020 in.) thick and should provide a better life. The rubber must not contain any additives that would make it electrically conductive. Work at the Cornell Laboratory in the United States showed that the time to erode through a neoprene coating was proportional to the square root of the coating thickness.

2. Erosion-resistant paint coatings tend to be the most popular method at the present time, but the debate continues as to whether overshoes or paint schemes offer the most cost-effective erosion protection. Several paint schemes are available:

   • A commonly used paint system known as Astrocoat is a polyurethane system developed for the U.S. Air Force by Olin Chemical. Because urethane coatings cure by reaction with atmospheric moisture, control of the spraying and curing environment is mandatory. Urethanes are hazardous to health, and all the required safety precautions must be strictly observed. Urethanes, in the same way as neoprenes, require lengthy spray application and cure procedures. Cure times approximating one week are typical for room-temperature curing. Urethane coatings require primer coatings to ensure a good bond. Urethanes are approximately equivalent to neoprenes in their effect on radar transmission, and their rain erosion resistance has been proven to be superior.

   • Prestec enamel polyester paint used on the Boeing 737.

   • Anti-erosion paint with primer from Bollore S.A. or Celomer is used by Airbus Industrie. Permitted thickness is defined in the Component Maintenance Manual (CMM).

   • Maranyl (methylmethoxy nylon) has been used successfully on helicopter rotor blades. Polyurethane sheets also have been used on helicopter blades.

In some cases, formed sheet titanium is used for leading edge capping to provide resistance to erosion and hail and other impact damage. Some applications use aluminum or stainless steel. Helicopter blades and engine fan blades often have metal leading edge capping. In these cases, the correct surface treatment must be used before bonding. Details of anti-erosion coatings for each component requiring them can be found in the SRM; details of their fitment and repair are usually found in the Aircraft Maintenance Manual (AMM). Special care is required when fitting these coatings because they are fitted either to radomes where signal transmission is important or to leading edges where airflow is important. In the case of helicopter blades, balance is a critical factor to avoid vibration; therefore, SRM procedures must be carefully followed.

2.8.6 Other Protective Coatings

Some other protective coatings may be used to seal the inside of lightweight sandwich panels made from aramid or glass fabrics. They are necessary because only one layer of fabric often is used on these parts, and this is not sufficient to ensure that fluids such as Skydrol hydraulic fluid and water cannot penetrate through pinholes in the skin. Some materials that can be used to seal the inside surface of fairings and falsework panels are:

- Tedlar polyvinyl fluoride plastic (PVF)
- Speedtape thin aluminum foil with an adhesive backing
- A thin coat of polysulfide or polythioether sealant

2.9 References

2.1   Hogg, P.J., "Factors Affecting the Stress Corrosion of GRP in Acid Environments," *Composites,* Vol. 14, No. 3, July 1983, pp. 254–261.

2.2   Jones, F.R, Rock, J.W., and Wheatley, A.R., "Stress Corrosion Cracking and Its Implications for the Long-Term Durability of E-Glass Fibre Composites," *Composites*, Vol, 14, No. 3, July 1983, pp. 262–269.

2.3   Morgan, Phillip [ed.], *Glass Reinforced Plastics*, 3rd edition, Iliffe Books Ltd., London, UK, 1961.

2.4   Starr, Trevor F., *Glass-Fibre Data Book*, Edition 1, Chapman and Hall, London, UK, ISBN 0-412-46280 X, 1993.

2.5   Palmer, R.J., "Investigation of the Effect of Resin Material on Impact Damage to Graphite/Epoxy Composites," McDonnell Douglas Aircraft Corporation, Douglas Aircraft Company, Long Beach, CA, NASA-CR-165677, 1981.

2.6   Watt, W. and Johnson, W., "The Effect of Length Changes During the Oxidation of Polyacrylonitrile on the Young's Modulus of Carbon Fibers," Applied Polymer Symposia No. 9, London, 1969, pp. 215–227.

2.7   *Carbon and High-Performance Fibers Directory*, Chapman and Hall, London, UK, see latest edition.

2.8   Delmonte, J., *Technology of Carbon and Graphite Fiber Composites,* Van Nostrand Reinhold, New York, 1981.

2.9   Belason, E.B. and Shahood, T.W., "Bonded Boron/Epoxy Composite Doublers for Reinforcement of Metallic Aircraft Structures," *Canadian Aeronautics and Space Journal*, Vol. 42, No. 2, June 1996, pp. 88–92.

2.10   Klein, A.J., "Specialty Reinforcing Fibers," *Advanced Composites*, May/June 1988.

2.11   Welling, M.S., *Processing and Uses of CFRP* (translation), VDI Verlag, Dusseldorf, 1981.

2.12   Fitzer, E., *Carbon Fibers and Their Composites*, Springer-Verlag, Berlin, 1986.

2.13   Kinloch, A.J. [ed.], *Durability of Structural Adhesives*, Applied Science Publishers Ltd., London, 1983.

2.14   Irvine, Tania J., "Modified BMI Adhesives for Bonding High Temperature Structural Carbon Composites," Bonding of Advanced Composites Symposium, 4 March 1993, The Institute of Materials Adhesives Group, London, UK.

2.15   Armstrong, K.B., "Selection of Adhesives and Composite Matrix Resins for Aircraft Repairs," Ph.D thesis, City University, London, UK, 1990.

2.16   Lee, Henry and Neville, Kris, *Handbook of Epoxy Resins*, McGraw-Hill, New York, 1967.

2.17   Houwink, R. and Salomon, G. [eds.], *Adhesion and Adhesives, Vol. 1, Adhesives*, 2nd edition, Elsevier, Amsterdam, The Netherlands, 1965.

2.18   Allen, H.G., *Analysis and Design of Structural Sandwich Panels*, Pergamon, Oxford, UK, 1969.

2.19   Armstrong, K.B., "Aircraft Floor Panel Developments at British Airways (1967–1973)," *Composites,* Vol. 5, No. 4, July 1974.

2.20   Armstrong, K.B., Stevens, D.W., and Alet, J., "25 Years of Use for Nomex Honeycomb in Floor Panels and Sandwich Structures," published in *50 Years of Advanced Materials, or Back to the Future*, Proceedings of the 15th International European Chapter Conference of the Society for the Advancement of Material and Process Engineering, 8–10 June 1994, Toulouse, France, ISBN 3-9520477-1-6.

2.21   Armstrong, K.B., "The Repair of Aircraft Radomes," *Advanced Materials and Structures from Research to Application*, Proceedings of the 13th International European Chapter Conference of the Society for the Advancement of Material and Process Engineering, 11–13 May 1992, Hamburg, Germany, pp. 469–473.

2.22   Divinycell International AB, Data Sheet HT.

2.23   Hexcel TSB 120, "Mechanical Properties of Hexcel Honeycomb Materials."

2.24   Niu, Michael C.Y., *Composite Airframe Structures—Practical Design Information and Data,* Conmilit Press Ltd., P.O. Box 38251, Hing Fat Street Post Office, Hong Kong and also Technical Book Company, 2056 Westwood Blvd., Los Angeles, CA 90025, ISBN 962-7128-06-6.

2.25   Payne, K.G., Burrows, B.J., and Jones, C.R., "Practical Aspects of Applying Lightning Protection to Aircraft and Space Vehicles," Proceedings of 8th International Conference, La Baule, 1987, Society for the Advancement of Material and Process Engineering (SAMPE), Elsevier Science Publishers, Amsterdam, The Netherlands, 1987.

2.26   Philpott, J., "Recommended Practice for Lightning Simulation and Testing Techniques for Aircraft," Culham Laboratory Report CLM-R-163, Her Majesty's Stationery Office (HMSO), 1977.

2.27   Burrows, B.J.C., "Designer's Guide to the Installation of Electrical Wiring and Equipment in Aircraft to Minimize Lightning Effects," Culham Laboratory Report CLM -R- 212, Her Majesty's Stationery Office (HMSO), 1981.

2.28   Little, P.F., "A Survey of High Frequency Effects Due to Lightning," Culham Laboratory Report CLM-R-201, Her Majesty's Stationery Office (HMSO), 1979.

2.10 Bibliography

B.2.1   Ciba Data Sheet, Aeroweb Type A1, "High Strength Polyamide Honeycomb."

B.2.2   Ciba Data Sheet, Aeroweb Type 5052, "High Strength Aluminum Honeycomb."

B.2.3   Murphy, John, *The Reinforced Plastics Handbook,* Elsevier Advanced Technology, ISBN 1-85617-217-1, 1994.

B.2.4   Uhlig, C. and Bauer, M., "Modified Cyanate Ester Resins for High Temperature (Aerospace) Applications," presented at the Structural Adhesives in Engineering Conference, Bristol, UK, 1995.

B.2.5   Hamerton, I., [ed.], *The Chemistry and Technology of Cyanate Ester Resins*, Chapman and Hall, Glasgow, Scotland, 1994.

Chapter 3

Handling and Storage

3.1 Shipping and Receiving

All materials must be shipped in adequate packaging to ensure that they are in a serviceable condition when they arrive for use. Honeycomb, in particular, must be strongly packaged and marked "FRAGILE." They must be inspected first by the supplier before dispatch; then again on arrival at the aircraft manufacturer, airline, or maintenance base; and again before being prepared for use. The degree of care required is equally applicable to automobile, marine, railway, and other usage of these materials, although it is a legal requirement for aerospace applications. The required documentation must be provided with each item. (See Section 3.3.5.) Storage must then be provided in accordance with data sheet requirements until the item is either issued for use or disposed of as "time expired."

3.2 Temperature Requirements

The specified storage requirements for each material can be found on data sheets or obtained on request from the manufacturer or supplier. In particular, the storage temperature and humidity should be checked and followed.

Many adhesive materials and pre-pregs require refrigerated storage and have a limited shelf life and out-time. Out-time is the maximum time the material can remain at room temperature before it becomes unusable because it has begun to cure, has lost tack, and will not flow adequately to wet the bonding surfaces during cure. Some materials should not be stored below 0°C (32°F). These materials must not be stored as cold as others that are stored at the specified low temperature. Check the data sheet, and store each material at the recommended temperature.

The temperature and humidity of the stores building should be recorded continuously on a chart recorder. Alternatively, a record book should be kept with readings taken at least twice daily (i.e., at 0800 and 1800 hours each day). A continuous record allows the significance of any variations to be assessed more accurately. Freezers should have a continuous chart record to ensure that any periods of power failure or other failure can be recorded as out-time if the temperature rises above 0°C (32°F). The chart record should be monitored twice daily as suggested above to facilitate prompt corrective action.

3.3 Storage Practices

3.3.1 Temperature Requirements

3.3.1.1 Dry Fabrics

In this context, a dry fabric is a fabric that has not been impregnated with a resin, although it may be coated with a size or finish.

- **Dry Glass Fabrics:** Dry glass fabrics should be stored at a temperature not exceeding 20°C (68°F). The storage environment should be temperature controlled, and the temperature should be monitored twice daily.

- **Dry Carbon Fiber:** Store dry carbon fiber in the same way as dry glass fabrics.

- **Dry Aramid Fiber:** Store dry aramid fiber in the same way as dry glass fabrics, except that Kevlar and other aramids should be stored in clean, sealed black polyethylene bags to minimize water uptake and degradation by ultraviolet light. Aramids should not be stored where sunlight can reach them.

3.3.1.2 Storage of Pre-Preg, Film, and Paste Adhesives, Potting Compounds, and Primers

Pre-Pregs, Films, and One-Part Paste Adhesives, Resins, and Potting Compounds:
Pre-pregs of the above listed types and any other fabrics or tapes, film adhesives, and one-part paste adhesives, resins, primers, and potting compounds must be stored as directed by the manufacturer. In most cases, this means storage in a freezer at -18°C (0°F). For special materials such as pre-pregs, film adhesives, and some one-part paste adhesives and potting compounds, suppliers must provide packaging in dry ice during transportation and labels to show that delivery must be expedited. These items must then be shipped immediately and quickly transferred to a suitable refrigerator on arrival (i.e., within one hour). These items should also carry a temperature monitoring device, preferably giving a continuous record of

time and temperature, to confirm that storage temperatures have not been exceeded in transit. The recorder chart should be retained with the receiving records for the material. Two types are normally used: an electrical type and a temperature recording strip.

In an electrical type, a battery drives a motor to operate a sleeve carrying a chart, and a bimetallic strip operates a pen which marks the chart. This gives a continuous time-versus-temperature record and also the time for which the critical temperature has been exceeded, if this occurs. Date of dispatch is recorded.

A temperature recording strip carries several small areas of salt compounds that turn black when a certain temperature is exceeded. Each section uses a slightly different formulation to change color in intervals of a few degrees. This type can record only the maximum temperature reached; it cannot indicate when that temperature was reached.

The importance of recording time and temperatures cannot be overemphasized because these materials are expensive. If the materials must be scrapped on arrival because their out-time is unknown, further delays and costs will be incurred while the material is reordered. Prior arrangements should be made for the arrival of these special materials to ensure that they do not arrive on a Friday afternoon to await cold storage until Monday or even later during a holiday. On normal working days, some companies are not open 24 hours a day; therefore, it is essential to refrigerate pre-pregs and film adhesives as soon as possible after delivery, even if this means retaining a storekeeper on overtime pay. Stores staff should to be trained to give these materials the special treatment they need.

Two-Part Paste Adhesives, Resins, and Potting Compounds: Some materials (e.g., acrylics) should not be stored below 0°C (32°F) or above 25°C (77°F). Therefore, consult the data sheet for each material to obtain the correct storage conditions. Experience with acrylic adhesives has shown that short excursions of a few degrees above and below these limits have not caused significant problems. Acrylics and some other adhesives can be highly flammable, and storage conditions should consider this. Store on a "first in—first out" basis to minimize waste.

For two-part materials such as epoxies, note that the base resin sometimes has a different storage life from that of the curing agent. Many two-part epoxy systems may be stored at room temperature or below with varying shelf lives. They do not all require low-temperature storage.

Polyesters are usually three-part systems having a base resin, a catalyst, and an accelerator. Storage temperature for the resin should not exceed 20°C (68°F). For other resin and adhesive systems, see the material safety data sheets (MSDS).

Primers: Primers also may require refrigerated storage, and each should be stored according to its own data sheet requirements. When refrigerated storage is specified, the containers must be allowed to warm to room temperature before they are opened. This will prevent condensation from occurring on the surface of the primer. Most primers are hygroscopic and can absorb water or moisture from the air. Lids should be replaced on the containers as soon as the required amount of primer has been removed, and the containers should be returned to the refrigerator immediately.

3.3.1.3 Storage of Sealants for Aircraft Use

Several types of sealants may be called up in Structural Repair Manuals (SRMs). These include polysulfides, polythioethers, and silicones.

Polysulfides: Polysulfides have a shelf life of at least six months when stored between 4 and 27°C (40 and 80°F) in the original unopened containers. They should be stored in sealed containers in dry conditions. The quoted shelf life varies, and some data sheets quote nine months if stored below 25°C (77°F); therefore, check each individual data sheet for instructions specific to each material. Re-lifing for a short time may be possible. For unimportant work, they may be used beyond their specified shelf life if they are tested and found to mix, spread, and cure properly. For important work such as fuel tank sealing, time-expired materials should be submitted to an approved laboratory for testing before a life extension is given. Some polysulfides are flammable and contain a volatile solvent. These should be held in flammable material stores.

Polythioethers: These were developed by Products Research and Chemical Corporation (now Courtaulds Aerospace Inc.) as Permapol P3 and Permapol P5. They tend to replace the earlier polysulfides because of their lower toxicity. As an example, PR 1828 Class B has a storage life of at least nine months when stored at temperatures between 4 and 25°C (40 and 77°F) in the original unopened containers. An unusual requirement that should be noted reads as follows: "The base material should have the mylar disc replaced, and the can should be **flushed with nitrogen** before resealing the can."

Silicones: Storage life for silicones is approximately six months for PR 1991 when stored at temperatures below 27°C (80°F) in original unopened containers. It may be worth checking the data sheets of other silicone materials in case they are different. Silicones have limited shelf lives and become unusable after long periods of storage.

3.3.1.4 Storage of Consumable Items for Composite and Bonded Metal Repairs

Although these commercial items do not become part of an aircraft, they are involved in repairs and should be stored between 10 and 30°C (50 to 86°F) and in a clean, dry environment to ensure that they do not cause contamination when in contact with the parts under repair. The items concerned are vacuum bagging materials, sealants, release sheets, and breather cloths.

3.3.2 *Cleanliness and Damage Prevention*

Rolls of Dry Fabric: Rolls of dry glass fiber fabric must be stored horizontally and suitably protected, preferably in individual cradles. If the fabric is required for issue in small quantities from the store, the rolls must be handled only on benches covered with clean kraft paper to avoid abrasion of the strands and consequent reduction in strength. Fabric must not be stored flat, folded, or on end because the glass fibers are easily fractured. The fiber must be handled as little as possible before use and always with clean gloves. After issue from stores, it should be kept flat or on a roll of large diameter and not folded or bent during cutting and lay-up operations because this is likely to fracture the fibers. It must be kept in a clean, dry environment prior to use, and the roll of fabric and its supporting frame should be protected by a plastic cover at all times except when fabric is being cut from the roll. (See Figure 3.1.)

Rolls of dry carbon fiber fabric and aramid and other types of fabric material should be protected and stored as detailed for glass above.

Rolls of Pre-Preg and Film Adhesive: Rolls of pre-preg or film adhesive material, in their sealed containers, should be stored horizontally while fully supported (i.e., on a shelf or the floor of a freezer and not "on end" or supported on two bars). This will prevent damage to the ends in the former case and creep distortion resulting from partial support in the latter. Film adhesives, if badly handled in the frozen condition, can crack into small pieces and become unusable. Materials should be handled with care at all times to avoid impact with the rolls. They must be lowered carefully into freezers and must not be dropped either

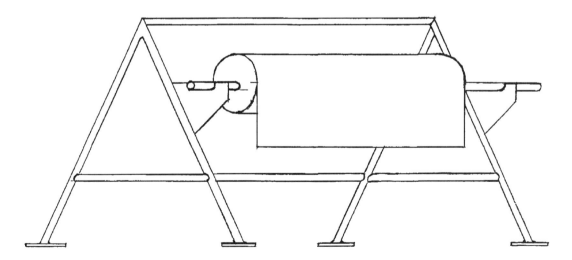

Fig. 3.1 A-frame holder for rolls of dry fabric (plastic cover not shown).

horizontally or "on end." The latter could buckle and crack the edges of a roll. These materials are expensive, and damaged materials (e.g., folded or creased fabrics in which fibers may have been broken or cracked film adhesive) must be scrapped.

Honeycomb: Honeycomb of any type should be kept in a sealed polyethylene sleeve for cleanliness and in its original cardboard box to prevent mechanical damage until it is required for use. Paper or cardboard interleaves between each sheet of honeycomb should be retained to avoid one layer damaging the next during removal from the box or in transit. During transportation, ensure that heavy objects are not placed on top of the boxes containing honeycomb. The boxes should be stored flat on a shelf (i.e., the honeycomb cell walls are vertical). It is good practice to reseal the polyethylene bag to avoid contamination after some of the honeycomb has been used. Store in a dust-free area, away from solvents, oils, greases, exhaust fumes, and other forms of possible contamination. It is better to keep honeycomb clean than to have to clean it. Honeycomb also should be stored in a dry environment to minimize moisture absorption in nonmetallics or corrosion of aluminum alloys.

Nomex honeycomb should be dried before use for sixteen hours at 40°C (104°F), three hours at 100°C (212°F), or one hour at 160°C (320°F). Phenolic and fiberglass honeycombs should be dried similarly. They should be bonded within one hour of drying or dried again. This is required because some atmospheric moisture is absorbed and must be removed to achieve a good bond. Honeycomb of any type should be handled only with clean, white cotton gloves and as little as possible. It should be laid on clean tables covered with clean kraft paper during cutting, and it should be covered with clean kraft paper until required for use. Every effort should be made to avoid contamination at all times to ensure a good bond.

Consumable Items: Consumable items such as vacuum bagging materials, sealants, release sheets, and breather cloths also should be kept clean and dry before use to ensure that they do not contaminate the repair area.

Sheet Metals: Sheet metals should be stored in a clean, dry environment. If covered with protective paper with a lanolin or other treatment, this should be retained as protection against damage and corrosion until the metal is cut for use. If the metal is not supplied in a protected form in this way, it would be helpful to apply such protection on receipt for storage. Sheet metals should be transported in padded racks and strapped sufficiently to prevent damage resulting from high winds or bumpy roads during transit. It is amazing how much scratching and other damage is found, even on aerospace metal sheets, due to lack of protection and care in handling and transportation. A little more care could prevent this type of damage.

3.3.3 Moisture Damage Prevention

Dry Fabrics: Dry glass fabrics should be stored at an ambient relative humidity not exceeding 65%. Dry carbon fiber should also be stored as above for glass. Dry glass fiber and carbon fiber fabrics (i.e., those without pre-preg resin) should be dried under heater lamps at 50°C (122°F) for two hours immediately before use.

Dry Kevlar and other aramids should be stored in the same way as for glass, except that it should be stored in clean, sealed black polyethylene bags. Kevlar fibers and other aramid fibers absorb moisture, and their fabrics should be dried before use at 120°C (248°F) for sixteen hours. The amount of moisture (at equilibrium) at 22°C (72°F) and 65% relative humidity can vary from 1.2% by weight for Kevlar 149 to 4.3% for Kevlar 49 (both from Du Pont). Twaron T-1000 from Akzo-Enka absorbs 7% and Technora T-220 from Teijin Ltd. absorbs 3%.

Pre-Pregs and Film Adhesives: Pre-pregs and film adhesives should be stored in a dry environment. This means that they must be sealed in a polyethylene bag before being placed carefully in a freezer. On removal from the freezer, they must be allowed to warm to room temperature before the bag is opened; otherwise, condensation will occur on the surface of the pre-preg or film adhesive and will be absorbed to some extent by the resin. Humidity of the room should be as low as possible when the bag is opened and when the bag is resealed before returning the roll to the freezer. Room humidity should be low from the time the material is removed from the bag to the time it is cured. Absorbed moisture can affect the cure of the matrix resin or adhesive and may also cause porosity during the cure as it tries to escape.

Honeycomb: Nomex and other aramid honeycomb should stored in a clean environment at approximately 20°C (68°F) and in a relative humidity not exceeding 65% and preferably lower. It should be dried for sixteen hours at 40°C (104°F) immediately prior to use. Alternative drying cycles are one hour at 160°C (320°F) or three hours at 100°C (212°F). Phenolic/fiberglass honeycomb should be dried in a similar way. Dried materials should be used within one hour of drying, or a further drying process immediately before use will be necessary.

Aluminum honeycomb should be stored in a dry environment to prevent degradation of any surface treatment.

3.3.4 Identification

All aerospace materials must be clearly and accurately identified in transit, storage, and before use. Where special disposal requirements exist (e.g., pre-pregs, adhesives, and many chemicals), identification must be maintained up to the time of destruction or final disposal.

Dry Fabrics: Dry glass fabrics must be segregated in weave and thickness in accordance with the requirements of specifications BS 3396: Part 3 (Woven Glass Fiber Fabric for Plastics Reinforcement) and DTD 5546 (Woven Glass Fiber Tape and Webbing for Use as Resin Reinforcement). In the United States, the appropriate specification is MIL-C-9084C or later issue. See your national or international standard. Chemical treatment in the form of finishes must be carefully noted. It is possible to identify the weave and thickness visually, but various chemical treatments cannot be positively identified if the rolls become mixed in storage. Specific finishes are required as coupling agents to assist bonding to particular resins. The requirements of other national and international specifications regarding storage must be observed. Glass fabrics designation and the applied chemical finish shall be marked on both ends of the cardboard roll on which it is supplied, on the release notes and approved certificate, and on labels showing manufacturing batch numbers, quantity, and specification reference (e.g., BS 3396/3/P6/225/E/Grade S/2-10-94/ Courtaulds Aerospace). Labels must be supplied with any parts of rolls issued from stores to ensure clear identification of fabric type and finish. Dry carbon fiber, aramid, boron, and other aerospace fabrics and tapes must be identified clearly to ensure that the correct type is used. As with glass fabrics and tapes, these materials come in a wide range of fiber specifications, weave patterns, and thicknesses; some have surface finishes specific to particular resin types. To ensure that the correct material, or a suitable approved alternative, is used for each repair job, each roll or piece of material must be clearly identified with the supplier's name and the manufacturer's designation number, type and grade of fiber, surface finish, weave pattern, and fabric weight per unit area. If small quantities are issued, it is a requirement that a label identifying the issued piece must be provided and that the original label must remain with the roll in the stores until the last piece is issued. This is essential to ensure sound structural repairs with the correct material.

Pre-Preg and Film Adhesives: As with dry fabrics, the correct pre-preg or film adhesive must be used where specified. In these cases, they must be cured at the correct temperature and pressure for the specified period of time. Correct labeling is essential, and a label should be provided whenever a length is cut from a roll to ensure identification up to the moment of use. This length should be kept in its own polyethylene bag complete with the label. The bag in which the main quantity of material is sealed should retain complete identification at all times. It should also carry details of shelf-life expiration date and a log of out-time so that the true condition of the material can be assessed. If the material has been given a life extension test, then details of the new shelf life should be checked.

One- and Two-Part Paste Adhesives, Primers, and Potting Compounds: In all cases, tins or other containers should be correctly labeled with material identification number, shelf-life details, and release notes and test certificates, where appropriate, until the material is completely used or scrapped.

Honeycomb: Honeycomb materials can be fairly easily identified as to general type of material, but it is virtually impossible to be sure of the density without correct labeling. In the case of aluminum alloys, the grade of alloy cannot be confirmed visually. Therefore, honeycomb should always be correctly labeled with material type, density, cell size, and surface finish (if any). Any pieces cut for a specific job should be labeled with full details so that no doubt exists as to type, grade, or finish.

Sheet Metals: Sheet metals should be clearly identified by printed numbers at intervals along each sheet. After heat treatment, additional identification should be provided so that the precise condition of the material is known to the end user. Aluminum alloys are supplied mainly to the 2000, 5000, 6000, or 7000 series in the United States and in the United Kingdom to British Standards BS L.156 and L.158 (equivalent to 2014A-T4 and A-U4SG), L.157 and L.159 (equivalent to 2014A-T6 and A-U4SG), L.163 (equivalent to 2014A Clad 1050A-T3), L.164 and L.166 (equivalent to 2014A Clad 1050A-T4), and L.165 and L.167 (equivalent to 2014A Clad 1050A-T6). BS L.109 is equivalent to 2024 Clad 1070A-TB; BS L.110 is equivalent to 2024 Clad 1070A-M. The British DTD 5070B is Concorde material CM001and is equivalent to the French A-U2GN/AZ-Z1 and 2618A Clad 7072-T6. BS L.113 is equivalent to 6082-TF. BS 2L.88 is equivalent to 7075 Clad 7072-TF and A-Z5GU/A-Z1. 2024-T3 is equivalent to A-U4G1, 2024 Clad 1070A-T3 is equivalent to A-U4G1/A5, and 7075-T6 is equivalent to A-Z5GU. Some American materials have direct British and French equivalents. In other cases, a British or a French equivalent may exist but not both.

Titanium alloys are supplied to MIL-T-9046 in the United States and to DTD 5023 and DTD 5163 or BS TA-10 in the United Kingdom.

Stainless steels are supplied to MIL-S-5059 in the United States and BS S.524, S.525, S.526, and S.527 in the United Kingdom.

3.3.5 Release Notes and Approved Certificates

All end users of aerospace materials are responsible for ensuring that only the correct materials are purchased from approved suppliers. End users must audit their suppliers either by mail with a suitable questionnaire or by visiting and inspecting each company to establish that adequate procedures, equipment, and trained staff are in place to ensure that production, storage, and distribution meet FAA, JAA, or other national standards. Release notes and approved certificates must be supplied with all aerospace material deliveries and must be retained until the material is used or scrapped and for a longer period if required.

Dry Glass Fiber Materials: Release notes and approved certificates, which must be supplied with all deliveries of glass fabric, must state that the material meets the residual size content, fabric breaking strength, and general requirements of specifications BS 3396:

Part 3: 1961: Grade S or C, DTD 5546, 900/4521, and BS 3496: Part 2: 1962 as applicable, or other national or international standards as required in the country of use. In the United States, see MIL-C-9084 for fiberglass.

Dry Carbon Fiber Fabrics: These should also be carefully identified in storage because many different types now exist. The correct type, weight, and weave must be used because these materials are often used for primary structure parts. Five basic grades of fiber now exist in regular use, and others may come into use.

Dry Aramid Fabrics: Kevlar and other aramid fabrics should also be carefully identified because some have finishes specific to the resin systems to be used. Note that there are several types of Kevlar (i.e., Kevlar 29, 49, and 149). These have different mechanical properties and water absorption levels. Kevlar 49 is the one most used in aerospace; Kevlar 149 will be supplied if sufficient demand exists. Other types of aramid fiber are made by Teijin in Japan (Twaron T-1000) and Akzo in Europe under the trade name of Technora T-220. In the United States, aramid fibers are covered by MIL-Y-83370.

New Fabrics: New fabrics developed in the future must be identified in a similar fashion.

Pre-Pregs and Film Adhesives: Release notes must be provided with all aerospace pre-pregs and film adhesives showing that they meet the specification to which they are sold and that batch tests have achieved acceptable results.

Adhesives, Resins, Primers, Potting Compounds, and Sealants: These materials must also be supplied with release notes and certificates of compliance with the data sheet performance they are required to meet. These certificates require batch testing by the manufacturers, which have the necessary facilities to perform the tests. Airlines and other end users are responsible for checking that these certificates are provided and that the shelf life of the remaining material is sufficient to ensure that the job can be done within the time limits established by the manufacturer.

Honeycomb Materials: These materials must have release notes confirming that they meet the requirements of the specification to which they are supplied.

Sheet Metals: Sheet metals require release notes and approved certificates to confirm that they are to the correct specification and have been tested to confirm conformance.

3.4 Shelf Life/Out-Time

Dry Glass Fabrics: Where prolonged storage either in stores or in production shops is necessary, glass fabrics must be resealed in their original polyethylene containers. Identification labels must be provided for each roll and for any material issued to workshops. Identification must be maintained until the last piece has been used and must include the finish and style of the fabric. This is important because the finish on a fabric cannot be visually identified and this information is necessary to confirm compatibility of the fabric with the resin systems to be used. The maximum storage period of all glass fiber materials under these conditions is one year. Fabrics exceeding this period shall be quarantined and samples from each container and/or batch submitted to laboratories for assessment regarding their suitability for issue and use. Time-expired material usually can be used for training purposes because only the finish deteriorates, not the glass.

Dry Carbon Fabrics: No shelf life is normally quoted for dry carbon fabric. In cases where it is stored for some time before use, clean and dry conditions must be maintained.

Dry Aramid Fabrics: When used with epoxy resins, there is no finish and therefore no limit on storage life for dry aramid fabrics. Refer to drying requirements, and dry before use.

Pre-Impregnated Fabrics and Tapes (Pre-Pregs) and Film Adhesives: When any fabric or tape has been made into a pre-preg by the application of a resin system, shelf life and out-time become critical. Pre-pregs and film adhesives have resin systems that are "B" staged or very partially cured. When heated, they melt, flow, and wet the surfaces to be bonded, whether these are adjacent plies of the same material or metallic or cured composite surfaces. Although they require heat and pressure to melt, flow, and cure properly, they will slowly harden and lose the property of tack or stickiness to a surface on contact. They also will cease to be able to flow and wet surfaces properly after a period of time at room temperature. Eventually, they will not flow at all and will have to be scrapped. For this reason, shelf life and out-time are two factors that become important.

- **Shelf Life:** This is the period of time that a material should remain usable if stored under the prescribed conditions. These conditions are usually storage in a refrigerator at -18°C (0°F). Under these conditions, the shelf life is usually between six months and one year. The data sheet for each material should be checked because requirements may differ considerably. Some may be as little as three months.

- **Out-Time:** Out-time is the time that a material can safely spend outside a refrigerator at room temperature before becoming unusable. A record of the total out-time of each roll must be kept to ensure that limits are not exceeded. Some large jobs may require three days to lay-up and possibly a little more on the production line for large parts. Clearly, such parts need at least three days open time remaining before lay-up starts and preferably more in case problems arise. It is good practice to use fresh material for large jobs to avoid any problems. To minimize out-time, a useful technique is to cut a roll of film adhesive or pre-preg into 5-m (16-ft) lengths and then to wrap and seal each one in a separate polyethylene bag. Longer lengths can be used if this would generate waste. In this way, the large original roll has to be removed from the freezer and allowed to warm before cutting on only one occasion. The roll must be allowed to warm before it is opened, or moisture will condense on the film or pre-preg, which does not help good bonding. Note that film adhesive will crack if cutting is attempted before it has warmed. The smaller 5-m (16-ft) lengths will warm more quickly and, even if not used all at once, will need little out-time before they are fully used.

Two-Part Paste Adhesives, Resins, Primers, and Potting Compounds: Shelf life for the variety of items under this heading varies, as does the recommended temperature. It is important to check each data sheet for details. In some cases, the base resin and curing agent may have different shelf lives. Some adhesive materials (e.g., acrylics) must not be stored below 0°C (32°F); therefore, it is vital to check the requirements for each material. Others, such as polyesters, should be stored at temperatures not exceeding 20°C (68°F). Shelf life may depend much on the temperature of storage.

3.5 Kitting

For parts that are made frequently or for standard-sized repairs, it is common practice to cut patches, skins, or doublers to shape or to approximate size and shape, to wrap them and seal them in polyethylene bags, and to return them to a freezer to save time when an urgent job is required. If this is done, then the out-time already used and the remaining shelf life should be recorded and supplied with the parts. It follows that precise identification of the material part number, type, and weight on a note in each bag should be provided. The shelf life of all such kits should be recorded, and a system should be developed to ensure that they are scrapped when they become time expired. In addition, the orientation of the warp direction for each piece should be marked on the backing sheet so that it can be assembled correctly. For satin weaves, mark also the warp face of the ply.

3.6 Recertification

Dry Fabrics: The finish on glass fiber fabrics may degrade after approximately 12 months, especially if the storage conditions are not good. Many manufacturers will perform recerti-fication tests. If stored as recommended, glass fabrics should be good for several years. Manufacturers will also recertify other fabrics if requested, although they often quote no shelf life for them. Airlines and repair stations are most likely to keep fabrics for a long time because of the low rate of usage. Correct storage then becomes vital. In production units, materials are usually used soon after receipt. Repair stations should buy in small quantities, store carefully, and ensure that the material is used on a "first in—first out" basis.

Pre-Pregs and Film Adhesives: Most film adhesives and pre-pregs can be recertified if they become shelf life expired. This can be done either by a tack test or by making test pieces for destructive testing. Manufacturers of these materials have been asked to specify recertification requirements and test methods on their data sheets, but at the present time this is not always done. Film adhesives can be recertified by lap shear testing to ASTM

D-1002 and if honeycomb core is involved by climbing drum peel testing to ASTM D-1781. Pre-pregs are more likely to use a tack test to determine if the resin matrix remains acceptable. It is possible to recertify for half the original shelf life until this becomes less than three months, in which case recertification is required monthly. If in doubt, consult the material manufacturer. The interlaminar shear test to ASTM D-2344 could be used to recertify pre-preg for important parts. If any material has reached its limit of out-time, it is much less likely to pass a recertification test.

Two-Part Paste Adhesives and Potting Compounds: Adhesives may be recertified by lap shear testing, and potting compounds may be recertified by lap shear or compression testing. Material that fails recertification testing may be used only for training purposes and in so doing may usefully demonstrate the reasons for shelf life and out-time limitations.

3.7 Care of Materials in the Hangar or Workshop

All the precautions mentioned above are as necessary when materials are received in the hangar or workshop prior to use as they are in the stores. If materials are badly stored or transported, they will be useless for the real job. It is equally true that even after storekeepers have done their best, it is possible to damage or destroy the material in the workshop before it can be used. All the care and refrigerated storage mentioned earlier must be continued until the job is complete. The following is only a simple checklist.

Dry Fabrics: Lay out dry fabric on clean kraft paper, and cover it with the same material. Take care not to bend or fold fabrics before use or during cutting, because this will fracture some fibers and reduce the strength of the repair.

Pre-Pregs and Film Adhesives: Store pre-pregs and film adhesives in the freezer until they are required. Ensure that they are correctly identified up to the moment of use. The protective coverings on each side of these materials should not be removed until lay-up. Contact between the protective layer and the pre-preg maintains a surplus of resin at the surface, retaining tack for lay-up. With the coverings removed, the resin will cold flow under surface tension into the pre-preg and away from the surfaces, with a resulting loss of tack (although without any deterioration of the material properties). In the case of film adhesives and pre-pregs, the coverings minimize moisture absorption. It is absolutely essential to remember to remove these protective layers before use. A poor bond and a scrapped part result if the protective layers are left on at this stage. Most suppliers use protective films of a different color from that of the materials to reduce the risk of leaving on the protective layers.

Record the time when the roll is removed from the freezer. Allow the roll to warm before opening it to prevent condensation of moisture on the roll. Check for condensation on the bag every half hour. When there is no condensation, the bag may be opened. As soon as the required amount of film adhesive or pre-preg has been cut from a roll, return the roll to the freezer after resealing it. Do this immediately, and record the out-time in the log provided. Record to the nearest fifteen minutes above the time out.

Two-Part Paste Adhesives, Primers, and Potting Compounds: If two-part paste adhesives and potting compounds are used, weigh the amount needed and return the remainder to the freezer immediately. Replace the lids on the tins immediately. In particular, the curing agent absorbs moisture which will degrade it.

Mix two-part materials thoroughly (i.e., for at least three minutes). If the two parts are of different colors, mix until an even color is achieved, even if this takes longer than three minutes.

Primers are usually one-part materials, but the lids should be replaced as soon as the required amount has been removed and the tins returned to refrigerated storage immediately.

Honeycomb and Other Core Materials: Allow honeycomb or foam or other core material to remain in its box or protective covering until needed. If a small piece has been supplied from the stores, wrap it in clean kraft paper, place it in a polyethylene bag, and seal the bag. Ensure that the material is correctly labeled, whether it is a large or small piece. Sometimes a job is delayed, and it is easy to be unsure of the identity of a piece of material, which could result in good material having to be scrapped.

Dry Nomex honeycomb, other nonmetallic honeycombs, and dry fabrics before use.

Handling: Handle all materials that are to be bonded with clean, white cotton gloves. Handle all materials as little as possible.

Cutting: Cut all materials carefully using approved techniques. This is especially important for honeycomb materials, which require special high-speed bandsaws and special blades. See Chapter 13, Section 13.2.1.

Chapter 4

Manufacturing Techniques

4.1 Filament Winding

Filament winding may be simply described as being similar to winding a bobbin of cotton or thread on a sewing machine. The Glossary AIR 4844 has a more complex definition: "A process for fabricating a composite structure in which continuous reinforcements (filament, wire, yarn, tape, or other), either previously impregnated with a matrix material or impregnated during the winding, are placed over a rotating and removable form or mandrel in a prescribed way to meet certain stress conditions. Generally the shape is a surface of revolution and may or may not include end closures. When the required number of layers is applied, the wound form is cured and the mandrel removed." This is an all-embracing definition and covers the variety of parts that are made by this process. The filament winding technique can be applied to any continuous fiber bonded with a cold- or hot-setting thermoset resin or thermoplastic resin. The angle at which each layer of fiber is wound, and the number of layers at each angle, depends on the stress pattern to be met in the end product. The parts most commonly made by this method are pipes, automobile and helicopter drive shafts, rocket motor nozzles, potable water tanks for aircraft, sewage storage tanks for aircraft, masts for sailboats, and spherical pressure vessels. Circumferential or polar winding may be used, depending on the part. See Figure 4.1.

Circumferential winding is used for shafts and tubes or pipes, but polar winding is needed for spherical pressure vessels or the ends of parallel section pressure vessels. Ref. 4.1 gives a good insight into the filament winding process. The filament winding process has now been extended to the manufacture of golf club shafts and compressed natural gas cylinders. The production rate of these items may soon exceed that of those listed above. With such a variety of parts, the stress patterns and the fiber angles in the lay-ups designed to meet them vary considerably. In most companies, the angle of fiber lay-up is computer controlled to

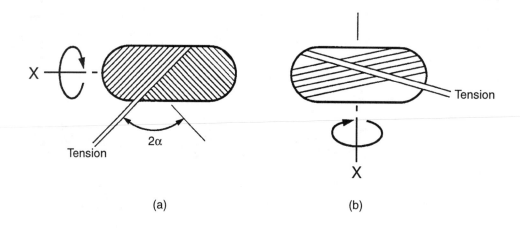

*Fig. 4.1 Filament winding. (a) Circumferential; (b) Polar.
(Courtesy of The Design Council and Springer-Verlag).*

tight limits and can be designed to suit the stress pattern applied to the part in service. In the filament winding process, the original method of compaction was tension in the fiber tow as it was wound onto the mandrel. This method is used when the fiber tow is pulled through a bath of wet resin before being wound on to the mandrel and can be varied as required. The process must be modified when thermoplastics are wound and heated at the point of application. In this case, a compaction pressure, using a roller, must be applied at the same time. One unique advantage of filament winding, compared with other automated processes, is that it is possible to watch each piece of tow being laid on the mandrel. An on-line sensor can be used to inspect each layer as it goes down. Work is being done to enable the process to be stopped if anything goes wrong so that corrections can be made before completing the part. An advantage of computer control and the availability of a printout of the entire process is that a record can be kept for quality assurance purposes and for providing traceability in a recognized BS 5750 or ISO 9000 approved quality control system.

In the case of potable water tanks, the chemistry of the resin matrix is important because any leached products must not be toxic. Whether for manufacture or repair, the epoxies used for potable water tanks must be selected from the limited number that are nontoxic. The Structural Repair Manual (SRM) must be strictly followed in this case, and no alternative materials must be permitted unless they can be demonstrated to meet the health requirements of this special application. These items could also be used in marine and static applications where the degree of control is less than that for aircraft and personnel may not be made clearly aware of the safety aspects. Repair personnel should make inquiries to ensure that they use only the correct resin before repairing these items, regardless of whether the items are used on aircraft, ships, small boats, or trains, or for water supplies in buildings. A variation on filament winding carried out similarly is automated tape laying for the manufacture of industrial shafting. Automated Dynamics Corporation has developed a tape laying machine to make carbon fiber shafts that are stiffer than the metal parts they replace and thus can rotate at higher speeds. The original work used tapes of 12.5 mm (0.5 in.) width, but work is proceeding on 25 mm (1 in.) and 75 mm (3 in.) wide tapes to increase production rates. This particular system uses carbon fiber with a polyphenylene-sulfide (PPS) thermoplastic resin matrix. The tape is heated and the matrix melted by a patented Hot Gas Torch located on the robotic head and consolidated immediately by a compaction roller producing pressures of 100 psi (7 bar) or more. Unlike filament winding, the tape is not under tension during lay-up but relies on the roller for compaction. The process is fully computer controlled and provides consistent quality. Tape can be laid at angles from 0 to 90°. The shafts are precisely machined after laminating, and reprocessing is possible if needed to correct out-of-tolerance parts or to repair surface damage. In both cases, the surface is machined down, the shaft is replaced on the spindle, and new layers of tape are laid as required to restore the part to specification. Other thermoplastic matrix systems can also be used in this type of process.

Another variation on filament winding is the recent production of thermoplastic composite braided tubing (TCBT). At least one company produces thermoplastic braided tubing. The braided reinforcement is interleaved to provide closed layers. Presumably, this is made by producing the braided tube of fiber in a process similar to filament winding and then adding the thermoplastic resin. Alternatively, fibers of thermoplastic resin could be woven into the braiding process with the reinforcing fibers and the complete assembly co-cured. This particular tubing is used for air conditioning ducts on the British Aerospace ATP (Jetstream 61). Although initially more expensive than aluminum tubing, its installed cost is said to be lower. It is approximately 20 to 30% lighter and easier to fit. The material used for the ATP applications is glass fiber with PPS resin. Other applications use a nylon resin, and more advanced aerospace items use Kevlar as the reinforcing fiber. Surface transport applications are being sought in ships, trains, buses, and trucks.

4.2 Lay-Up Methods for Fabrics and Tapes

4.2.1 Hand Lay-Up (Wet and Pre-Preg Laminating)

Wet Lay-Up: This remains the most common method today for aircraft repairs because each repair is unique and each is unsuitable for automated processes. In boat building, it is the most common manufacturing process because the product is large and the number produced is relatively small. First the orientation of each layer should be checked with the SRM or part drawing. Also, check that each layer is of the same material type and thickness. Although uncommon, it is possible for different layers in a lay-up to be of different thicknesses or styles of the same type (e.g., carbon fiber fabric for one layer and tape for another). Not all layers are necessarily of the same weight of fabric or tape. In the case of hybrids, some layers may be carbon and others may be glass or aramid. Also, check the resin type and part number to be used. Each fabric has several possible finishes, which may include no finish at all in the case of aramids. For glass fabrics, a number of finishes are available. The choice depends on the resin type to be used (i.e., epoxy, polyester, or phenolic). The correct finish for the resin type must be ordered using the supplier's code number for that particular finish. These finishes have a large effect on bond strength of the resin to fiber and on durability of the bond under wet conditions. For boat building, it is important to use the correct fabric finish. After the resin and fiber type, grade, weight, and finish have been established, the next step is to impregnate the fabric with the resin. This involves cutting suitably sized sheets of Mylar, Melinex, or other release material which must be clean and dry. Then the resin and curing agent (hardener) must be weighed accurately, using electronic scales with an accuracy of 0.1 g to a mix ratio accuracy of $\pm 1\%$ or better, and then thoroughly mixed for at least three minutes. The amount of resin mixed should be slightly more than the weight of the fabric to be impregnated to give a resin volume fraction of approximately 50%. A little excess is required to allow for the resin that remains on the release sheets. Half of the resin should be spread on the lower release sheet over an area the same as the fabric, and the fabric laid on this. The other half of the resin should be poured on top of the fabric and lightly spread so as not to distort or damage the fabric. Another release sheet should be placed on top and rolled down with a medium-hardness rubber roller to thoroughly impregnate the fabric. Rolling should be done from the center outward in each direction to remove air bubbles as much as possible. The next step is to mark the orientation of the fabric on the release sheet and to cut the release sheets with the fabric layer in the middle to the shape required for the repair. Any other sheets should be prepared similarly as quickly as possible, and then marked, cut, and numbered to ensure correct positioning in the repair lay-up. After checking that the repair area is dry, clean, abraded, and ready for bonding, a thin layer of resin should be carefully applied to the skin and the ends of the honeycomb cells (if applicable). The repair patch layers should then be applied in the correct order and rolled down with a rubber roller. First, the bottom layer of release sheet should be removed and the layer correctly located and rolled down. Then the top release layer must be removed carefully. Each subsequent layer should be applied in the same way. When all the layers are in place, a perforated release sheet should be placed over the whole repair and a suitable arrangement of bleeder and breather cloths applied before the vacuum bag is sealed. (See the repair techniques discussed in Chapter 10.) This method is both the most common and the most suitable for repairs because repairs usually differ slightly and are "one-off" jobs for which automation is seldom appropriate.

Pre-Preg Hand Lay-Up: Most pre-preg lay-ups today are done by hand, as are all pre-preg lay-ups for repairs. Procedures are similar to those for wet lay-up except that the resin is already applied to the fabric or tape. Clean, white cotton gloves should be worn because it is possible to contaminate pre-pregs with skin oils, thereby seriously reducing bond strengths. Pre-preg repairs require vacuum pressure or a higher pressure and hot bonding controllers with heater mats, heater lamps, or hot air heater guns to cure the pre-preg. A layer of film adhesive between the pre-preg patch and the repair surface is typically used to give a tough adhesive bond to the skin and/or honeycomb core. Pre-preg repairs should be made using pre-pregs that cure 25 to 50°C (45 to 90°F) below the original cure temperature of the part to avoid blowing the skins away from the honeycomb around the repair area. This is a common problem that should not be underestimated. The only alternative is to apply additional pressure above vacuum to give approximately 30 psi total pressure over the repair area and the entire heated surrounding area. Again, see the repair techniques discussed in Chapter 10.

4.2.2 Automated Lay-Up

This technique is used for production purposes and is not applicable to repairs.

To hasten the production of composite parts and to improve the accuracy of fiber alignment, several automated tape and fabric lay-up machines have been developed which are computer controlled. After a trial has been run and the program has been debugged, every subsequent lay-up should be correct with a repeatability much better than with hand lay-up. The pre-preg materials used in automated lay-up systems come with the resin already applied, thus preventing mixing problems. However, they require storage in a freezer, and time must be allowed for a roll to warm to room temperature before lay-up can commence. It is essential to check that the correct pre-preg has been drawn from stores and that its shelf life and out-time at room temperature have not expired. Also, a satin weave will have a warp face and a fill face and therefore must also be laid with the correct side facing upward.

4.3 Pultrusion

Pultrusion is a production process similar to metal extrusion. It produces standard sections in composite materials and does not produce parts or components. It is defined in AIR 4844 as: "A continuous process for manufacturing composites that have a constant cross-sectional shape. The process consists of pulling a fiber-reinforcing material through a resin impregnation bath and through a shaping die, where the resin is subsequently cured." The process is performed with glass, carbon, and aramid fibers and several hot-setting resins. Other fibers also could be used in this process. The name "pultrusion" was partly invented to differentiate the process from metallic extrusion, which occurs under compression loading as the metal is forced through the die, and partly from necessity because fibers impregnated with a wet resin are unstable in compression. Pulling or tension loading the fibers is the only practical way to force them through a die. A variety of sections can be produced, including round and square sections, "I" beam sections, angles, and channel sections. Many variations on these shapes are possible. See Figure 4.2.

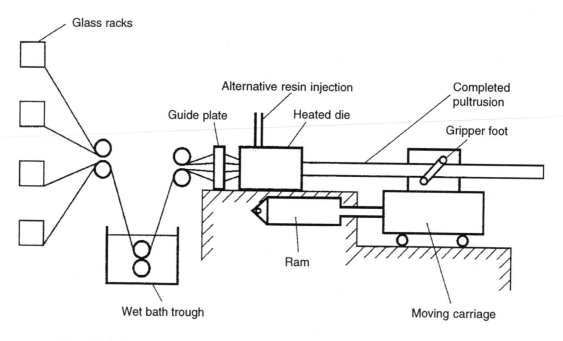

Fig. 4.2 Pultrusion process. (Courtesy of The Design Council and Springer-Verlag).

Matched lengths of fiber tows from a creel are fed, together with enough mat or biaxial fabric to give the pultruded section sufficient strength and stiffness, through a resin bath and saturated with a mixed resin system capable of hot curing. From the bath, the impregnated reinforcement is passed through several wiper rings, which remove excess polymer, and then through a spider which distributes and collimates the fibers before they enter the die. The die may be heated in one or two zones, depending on the complexity of the section. If two zones are used, the first zone is at a lower temperature and serves to reduce resin viscosity and to assist impregnation and compaction. The second zone gels and cures the matrix at a higher temperature. After leaving the die and being allowed to cool, the pultruded section is gripped by rubber pads and pulled to a cut-off wheel. Pultrusion is theoretically a continuous process. If matched lengths of continuous fiber are supplied to the resin bath and the bath is kept topped up with resin, then considerable lengths of a section can be produced. It does not require much imagination to realize that modern demands, for lengths of up to 2 km (1.25 miles) for optical and electrical cable, require some organization, ingenuity, and quality control effort to produce. Pultrusion can provide a high throughput with a high volume fraction of fiber. Designers now have available many standard rods and tubes which can be used in a range of applications.

4.4 Resin Transfer Molding

Resin transfer molding (RTM), sometimes known as resin injection molding, was invented as long ago as the late 1940s by Dr. Muskat, who produced small boats in glass/polyester by this process. Recently, the process has become more popular and involves holding preformed dry reinforcements in a closed mold and then impregnating them under pressure with a liquid thermosetting resin of low viscosity. The resin is pumped in under pressure, and resin coming out can be collected and recycled until the resin being expelled is clear and free from bubbles. At this point, the resin flow is cut off and the temperature and pressure are increased until gelation and cure occur. RTM is a useful process for producing void-free laminates. The molds must be of high quality and have a good surface finish. RTM enables complex parts to be produced, and core materials, inserts, and fittings can be included. Potentially dangerous emissions are reduced, and the process is clean because it is sealed from the atmosphere. Storage problems with pre-pregs are eliminated. Tooling costs are higher than for other composite molding methods, but they are much less than those for steel press tools. The process allows the accurate placement of a range of reinforcing materials, foam cores, and woven, knitted, or unidirectional fibers. One sports car company produces car bodies in two pieces instead of the large number of steel pressings previously used. Some of the present disadvantages of RTM are that pre-form preparation and lay-up is labor intensive and involves considerable waste. Another problem is that if fibers must be oversized to allow air to escape at the edges, then trimming becomes necessary, which is costly. Fibers at the edge allow moisture penetration, which can lead to degradation of the composite. The process also is rather slow, and the time taken from injection to cure and removal from the tool can vary from 20 minutes to several hours, depending on the size and complexity of the part. The Concorde radome was made by this process and has served well. Dowty is using it for a new family of composite propellers. A few years ago, BP Advanced Materials (BPAM) (now GKN/Westland Aerospace) applied the RTM process to the manufacture of blocker doors for aero engines. Blocker doors are an important part of an aero engine thrust reversal system. This is used to supplement the aircraft brakes and to reduce the length of runway required. In the normal position, the blocker doors form part of the acoustic attenuation area of the engine cowling. On landing, they are opened into the bypass airstream to block the flow and to divert the air to flow forward through a series of guide vanes to provide reverse thrust to slow the aircraft. One large aero engine has 12 blocker doors per nacelle, making a total of 48 per aircraft if the aircraft has four engines. This means that the numbers required are reasonably large; thus, tooling and quantity production methods become appropriate for this part.

BP found several factors that demonstrated that RTM was the best process for these items. These were:

- Use of dry, unimpregnated fiber eliminates shelf-life concerns
- Moldings are geometrically repeatable (parts are interchangeable)
- Cure cycles are simple
- Resin is mixed before use, reducing shelf-life problems
- Fiber pre-forms can be produced in batches and held as a buffer stock
- In-process quality control is possible

The perforations in the front panel, required for acoustic reasons, are created within the mold. This enables fiber continuity to be retained, compared to drilling the holes after molding which would break some fibers, reduce strength, increase the chances of edge delamination, and create a pathway for moisture ingress. The back panel has two hinge lugs integrally molded into it. The front and back panels encase the honeycomb core and are bonded in the normal way, using conventional film and foaming adhesives. A weight saving of approximately 25% and a useful cost reduction are claimed.

For vehicle production, the time taken to make each part is critical, but the quantity required is much larger, which helps to justify further work. Research is under way to resolve the problems mentioned above and involves major automobile and aerospace companies. Significant improvements in this process can be expected in the next few years.

4.5 Injection Molding

This process is different from RTM (which, as previously stated, is sometimes described as resin injection molding) because in RTM the reinforcing fibers are laid up first and held in place while the resin is injected under pressure. Only low-viscosity thermosetting resins can be used for this process. Injection molding is quite different because the molten thermoplastic or fiber-filled thermoplastic resin mix is injected into an empty mold until it is completely filled and then is allowed to cool sufficiently before the shaped part is ejected from the mold cavity. The process is defined in two ways by AIR 4844 as follows:

1. "A method of forming a plastic to the desired shape by forcing the heat-softened plastic into a relatively cool cavity under pressure."

2. "The production of a composite component by the injection of resin or a fiber/resin mix into a closed mold."

In the first case, a heated plastic is injected under pressure and cooled. This leads to residual stress problems if cooling rates are increased to give a high production rate. The residual stress can be so high that crazing, cracking, or fracture can occur from the residual stress alone. Sometimes a solvent test is used. In this case, a solvent known to cause crazing above a certain stress level is employed. The injection-molded parts are placed in the solvent, and the time to craze is recorded. The results may lead to a reduction in cooling rates to reduce the residual stress. Another problem may be one of orientation in the molecules because of the speed of flow and friction on the mold surface. Oriented plastics, with molecules aligned preferentially in one direction, behave in a manner similar to wood and have good properties along the molecular chains and lower transverse properties. If the parts split "across the grain," then the injection speed may have to be reduced, the temperature increased, the mold redesigned, or a compromise of all these. Ref. 4.2 states: "In injection molding, the requirements of the material, mold design, and operating conditions are highly conflicting and therefore difficult to optimize."

On the one hand, a low viscosity is desirable for easy mold filling because:

- The use of low temperatures would minimize degradation (or, conversely, because lower pressures could be used while keeping the temperature as it was)

- Weld lines and consequent weaknesses in the moldings are minimized

- Intricate moldings, possibly including inserts, can be more readily achieved

On the other hand, high molecular weight polymer will give a product with the best mechanical properties, and faster cycle times are possible because the difference in temperature between the melt in the barrel and in the mold is greater, and the molding can be ejected after a shorter mold-closed period, although this will also cause residual stresses in the molding which may lead to ultimate distortion. Turbulent flow into the mold is not, of itself, undesirable because it promotes randomness and isotropy. Nevertheless, injection-molded articles are almost inevitably highly anisotropic. See Ref. 4.3. For an introductory treatment of polymer properties, see Ref. 4.4.

In the second case, the resin is injected after mixing into it the required amount of short fibers. These give a considerable improvement in the mechanical properties of the basic resin, but long fibers cannot be used and thus the strength of a composite material with carefully oriented long fibers cannot be achieved. It is a type of halfway house between the properties of the resin alone and the properties of a long-fiber composite material. Only thermoplastic resins are normally used for injection molding. Glass-filled nylon is a common example of a filled thermoplastic resin. Some of those used in addition to nylon are polymethylmethacrylate, polycarbonate, polysulfone, polyethersulfone, or polyether-etherketone (PEEK). Polysulfone is more readily attacked by substances commonly used around aircraft, such as trichloroethane (Genklene). Polyethersulfone, polyphenylene-sulfide, and polyimides are more suitable. These resins may be filled with glass, carbon, or other fibers, depending on the needs of the part. The fibers are compounded into the resin at the granule production stage up to a maximum of 40% fiber content. The fibers have the usual surface treatment appropriate to the resin system to ensure good adhesion and maximum reinforcement. The granules are progressively heated as they approach the point of injection and are forced into the mold under high pressure. Molding pressures may be as high as 200 MPa (29,000 psi) for filled materials. As strength and stiffness are improved by fiber addition, it may be important to consider which of these is most required. If Young's modulus is sufficient and a significant increase in strength is required, then glass reinforcement may be adequate. If a considerable increase in stiffness is required, then carbon fibers are preferred and their cost can be justified. Other advantages of carbon are that it offers lower shrinkage and better conductivity of electricity and heat. It also reduces the rate of wear and lowers the coefficient of friction. Aramid fibers are less successful for reinforcing thermoplastics. They do not chop cleanly, as do glass and carbon, and they tend to separate at runners and gates. Aramids are being used to toughen epoxy resins and others as a replacement for asbestos, which is being eliminated because of health concerns.

The subject of injection molding is vast because it has been a common production process for many years, and considerable literature is available to anyone wishing to pursue the subject. In addition to the more academic references listed, which will give an understanding of the principles involved, a large body of literature also exists on the more practical aspects and the design of the machinery required. Injection molding is a mass production process using a vast range of materials if the full range of polymers, polymer blends, and molecular weights available for each polymer is included. In addition, the extensive range of coloring materials and fillers (other than fibers) can be used to reduce cost and modify properties. Where necessary, ultraviolet stabilizers are used to improve outdoor weathering properties. The products involved vary from household articles and children's toys to fiber-filled automobile and aerospace parts. For some of these products, cost will be the overriding consideration; in others, quality, performance, reliability, and life-cycle cost will be foremost. Parts for aircraft, spacecraft, submarine, and underground railway interiors and public buildings will have the issue of flammability and smoke and toxic fume generation added to the vital specification requirements. For some applications, resistance to particular solvents and chemicals will be required, and this will decide the choice of plastic and possibly also the molecular weight that must be achieved. This raises the important point that, when cleaning plastic parts, the SRM should be consulted for the correct choice of cleaning material. This is especially important for acrylic cockpit and passenger windows on aircraft, which should be cleaned only with a solution of soap and water. Although of lesser structural importance, plastic interior parts also must be cleaned with approved materials to avoid crazing, cracking, or loss of surface appearance.

4.6 Tow Placement

The word "tow" usually describes a narrow flat tape of untwisted fibers of carbon or aramid fiber; the word "roving" is used for the same form of material made from glass fibers. These tows or rovings are often used in the process of filament winding when several tows or rovings may be laid at the same time. The number of tows and their width are chosen to suit the lay-up required, and the angle of lay-up is normally computer controlled to give accurate placement and angle of lay. The total number of lays also is computer controlled, and the angle of each lay can be varied to provide the amount of strength required in each direction.

This method can be used to manufacture tubes, drive shafts, and pressure vessels which may be either spherical or spherical ended in shape. Ref. 4.5 shows a fiber placement system for large composite engine inlet cowls. The system used is a new seven-axis, twenty-four tow Viper CNC Fiber Placement System (FPS) from Cincinnati Milacron, which claims to meet four objectives:

1. Reducing costs
2. Reducing cycle times
3. Creating more damage-resistant structure
4. Achieving consistent part quality

There is unlikely to be any doubt about objectives 1, 2, and 4; however, the means of achieving objective 3 is not mentioned, nor is there mention of the topics of concern to airlines (i.e., repairability and life-cycle cost). Chapter 16 records complaints about the difficulty of repairing filament-wound items, and this appears to have been neglected.

4.7 Press Molding

Press molding is commonly used for the production of aircraft floor panels and other flat honeycomb panels with either composite or metal skins and cores of honeycomb, balsa wood, or foam of various types. For these thin, flat panels, multidaylight presses with steam heating of the platens are often used and can simultaneously press five or more panels. These presses have several platens, one on top of the other with spaces between them (hence multidaylight), and the platens are kept at a constant temperature to cure the prepreg skins or film adhesive in the case of metal skins, while the required pressure is applied by the press. The pressure must be carefully selected to suit the strength of the core material to avoid crushing it, but it must be sufficient to ensure good contact during the curing operation. Mechanical presses also are used for metal bonding using the phenolic resin system Redux 775 and equivalents because of the high pressure, approximately 100 psi (6.8 bar), required to prevent voids from developing in this resin system which emits water vapor during the cure.

4.8 Vacuum Bonding

Vacuum bonding is commonly used for repair work on composites and for the adhesive bonding of metal to metal for the repair of bonded metal parts, although it is often used in production for the vacuum forming of plastic sheets. For this reason, the vacuum bonding process is fully described in Chapter 10.

4.9 Autoclave Bonding

This topic will be discussed briefly here because a separate course covering this specialized activity has been recommended. See Figure 4.3.

In autoclave bonding, the hot, compressed air (or other gas) provides both the heat to achieve the curing temperature and the bonding pressure. It is necessary to have thermo-couples at the glue line or composite surface to enable the cure cycle to be correctly moni-tored. Sometimes nitrogen is used as the heating and pressurizing gas to avoid fire risk. The flexible blanket is forced against the assembly by the pressure resulting from the com-pressed gas in the autoclave and the vacuum pressure under the blanket. Autoclaves are used extensively in the manufacture of composite and adhesively bonded metal parts. They require high-quality tooling to provide each part with the specified shape and to withstand

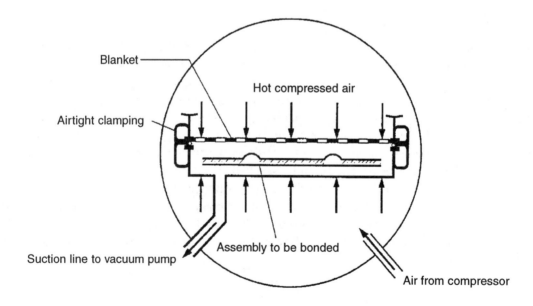

Fig. 4.3 The autoclave bonding principle.

the temperatures involved in curing resins and adhesives. Good tooling also is necessary for large repairs made in autoclaves because the entire part must be heated. Although adhesives that cross-link during cure cannot melt a second time, they soften at high temperatures. Consequently, complete tooling is necessary for large autoclave repairs, whereas cold-bonded repairs and hot-bonded repairs made with heater blankets can be made without tooling unless the damage is extensive. A technique now being adopted allows an autoclave to be used for small repairs with little or no tooling. In this method, the autoclave is simply used as the means of applying pressure (i.e., as a rather expensive vacuum bag). However, the advantage is that pre-preg repairs can be made at the original cure temperature and pressure, and manufacturers allow larger and sometimes unlimited repairs if this is done. A heater blanket and hot-bonding control unit are used to provide the local heating around the repair. By this means, the advantages of hot-bonded repairs to the original manufacturing standard are achieved for small repairs without the expense of tooling but with the added expense of an autoclave. The risk of the skins being blown from the honeycomb in heated areas around a vacuum bonded repair is eliminated. For those airlines and repair stations with an autoclave, the method is a useful but costly option because only repairs of this type can be done at any one time. In a normal autoclave run, several parts can be repaired or made simultaneously. For structurally important parts, the method may be worth the cost, especially as airlines seldom use an autoclave as often as a manufacturer. Consequently, this type of repair may be able to make additional productive use of what would otherwise be idle time for the autoclave.

Parts for autoclave bonding are first sealed in a vacuum bag, and the bag is tested for leaks before bonding commences. This is essential because unless the vacuum bag provides a seal throughout the bonding process, the autoclave pressure may penetrate the part. In this case, although the pressure may be high, if the pressure reaches all areas of the part, there will be no pressure difference to hold the parts together. Therefore, autoclave bonding is similar to vacuum bonding except that the pressure is higher. It is common practice to stop the vacuum pump after autoclave pressure is applied and the vacuum bag has been shown to be sealed. Autoclaves tend to be used for major repairs and complete rebuilds. Smaller repairs can be made more economically with heater blankets or cold-setting adhesives or resins if permitted and if the correct techniques are employed. However, as detailed above, an autoclave can be used for small repairs if a hot bonder also is available and the cost is considered to be justified to meet technical or economic objectives. Advertisements for new autoclaves offer a wide range of options, and prices will vary according to the chosen specification. Sizes are increasing and may exceed the following dimensions:

Diameter	Up to 25 ft
Length	Up to 90 ft
Pressure	Up to 10,000 psi (small autoclaves)
Temperature	Up to 815°C (1500°F) (special orders)
Heating Systems	Electricity, gas, fuel oil, or steam

For aerospace purposes, gas or electric heating are the best options, with gas heating being shown to be more economical. Both can be controlled to maintain accurate temperatures, and the gas-fired system is considered more reliable in service because, in the unlikely event of failure, the burner unit could be changed in five minutes without affecting the internal pressure in the autoclave.

Figure 4.4 shows an autoclave recently supplied to British Airways. It can accommodate parts up to 7.3 m (24 ft) long and 3 m (10 ft) wide. It can operate at temperatures up to 204°C (400°F) and pressures up to 100 psi (6.8 bar). A similar autoclave supplied to BP has a length of 15 m (49 ft) and a diameter of greater than 4 m (13 ft). It can operate at 200 psi (13.6 bar) and 200°C (392°F) but is normally used at 30 psi (2 bar) and 150°C (302°F). The largest autoclave in use at British Aerospace at the time of this writing is 5 m (16 ft) in diameter and 14 m (46 ft) long.

Figures 4.5 to 4.7 illustrate several factors that must be considered in any form of hot bonding. They should be noted carefully and used as a checklist each time.

Fig. 4.4 Autoclave.

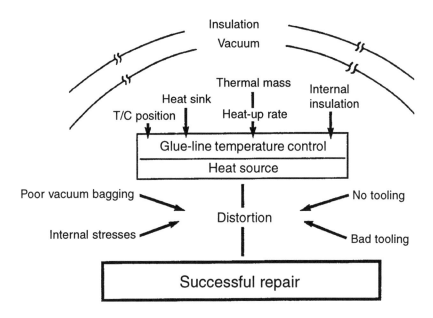

Fig. 4.5 Hot-bonding considerations—The factors involved.

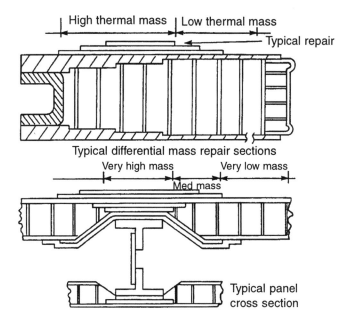

Fig. 4.6 Hot-bonding considerations—Thermal mass.

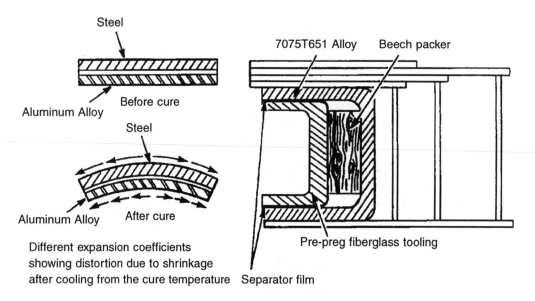

Fig. 4.7 Hot-bonding considerations—packing. Using an interference fit (i.e., wedging the channel open during build by 0.006 in.), the different materials, tool, and component can be adhesively hot bonded with no spring-in after-effects.

Table 4.1 shows the coefficient of thermal expansion (CTE) of various tooling materials.

Table 4.1
Coefficient of Thermal Expansion of Tooling Materials

Material	Coefficient $10^{-6}/°K$
Graphite/Epoxy	3.6
Glass/Epoxy	11.7–13.1
Steel	12.1
Nickel Electroform	13.3
Fiberglass Wet Lay-Up	14.4–18.0
Aluminum	22.5

4.10 Oven Curing

This can be used where vacuum bonding is sufficient, e.g., bonding doublers to skins or with honeycomb if additional film adhesive and 3M-AF 3306 positioning cloth is used between the film adhesive and the honeycomb. (See previous references to positioning cloth in Section 4.6). It also may be used for warm-curing two-part adhesives and resins at temperatures to 80°C (176°F). Fan-assisted ovens use recirculating fans to achieve a uniform temperature throughout the oven. With microprocessor control, these can give a temperature uniformity of ± 3°C (± 5°F). When curing film adhesives or pre-pregs in an oven, it is essential to have thermocouples at the glue line because the temperature that matters is not the oven temperature but the temperature at the glue line or composite surface. Curing time begins when this temperature reaches the required value.

4.11 References

4.1   Stover, D., "What's New in Filament Winding," *High Performance Composites,* May/June 1994, pp. 26–31.

4.2   Lenk, R.S., *Polymer Rheology*, Applied Science Publishers, London, ISBN 0-85334-765-4, 1978.

4.3   Ferry, J.D., *Visco-Elastic Properties of Polymers*, John Wiley and Sons Inc., New York, 1961.

4.4   Treloar, L.R.G., *Introduction to Polymer Science*, Wykeham Publications Ltd., London, 1970.

4.5   "Fiber Placement for Composite Engine Cowls," *Aerospace Engineering,* January/February 1995, p. 42.

4.6   Murphy, John, *The Reinforced Plastics Handbook*, Elsevier Advanced Technology, ISBN 1-85617-217-1, 1994, London.

Chapter 5

Original Design Criteria

5.1 Primary, Secondary, and Tertiary Structures

For design purposes, each part of an aircraft is placed in one of the categories of primary, secondary, or tertiary structures, in accordance with its importance to the safety of the aircraft. See FAR/JAR 25.571 from the Federal Aviation Regulations (FARs) and the Joint Airworthiness Requirements/Joint Aviation Requirements (JARs), respectively. The definition of primary structure is not a simple structural definition but includes all other related events that could result in the loss of an aircraft. For example, failure of leading edge flap structure would not result in catastrophic structural failure. However, loss of the leading edge flap structure during takeoff or landing could result in loss of lift at a critical phase in the flight. This could be considered to be primary aerodynamic structure. In contrast, major cracking in the wing spar structure could result in disintegration of the wing, which would be a more specifically structural event. Tertiary structure is not always recognized as a classification. The structural significance of a given level of damage to a composite part, or any other part, will depend on the location of the damaged area within the specific component. To allow for the different strength, stiffness, and damage tolerance features of the various components in an airframe, the manufacturer will classify his components by one of the following two methods or a similar one. Two systems are used to define the importance of parts and areas of parts. These are the class/zone system and the mechanical grade system. The following has been taken from IATA Document DOC:GEN:3043.

Definition of the Class/Zone System:

- **Class I Primary Structure:** Failure of the component could cause loss of the aircraft.

- **Class II Secondary Structure:** Failure of the component will not cause loss of the aircraft, but disturbances to aircraft control and behavior may occur and/or detached parts could damage primary structure or engines.

- **Class III Tertiary Structure:** Failure of the component will not disturb a safe continuation of the flight.

- **Zone:** To account for some areas being more critical than others, each class shall be divided into a maximum of three zones. They should cover critical, important, and the least important areas. There shall be no critical areas in Class III components.

Although the airframe manufacturer's damage classification systems may not precisely follow the IATA class/zone recommendations, each manufacturer provides in its Structural Repair Manual (SRM) specific damage evaluation guidance for each composite part.

The relation between class and zone is shown in Table 5.1.

Table 5.1
Relation Between Class and Zone

Class	Zone		
	A	**B**	**C**
	Critical	Important	Least Important
I	I-A	I-B	I-C
II	II-A	II-B	II-C
III	–	III-B	III-C

For each of the eight class/zone combinations (I-A, II-A, etc.), allowable damage limits, repairable damage limits, and other related repair information shall be provided.

Definition of the Mechanical Grade System: Strength, stiffness, and damage tolerance features of composite and metal structures can be characterized by allocating mechanical grades which account for certain areas of a component being more critical to safety and performance than others. The reason for this system is that, even on primary structures, noncritical areas exist which are, from a strength and damage tolerance viewpoint, equal to secondary structure.

Five mechanical grades are used. Mechanical Grade 1 is given to the most critical areas on primary structures. These areas have small allowable damage limits and limited repairable damage. Restoration of strength, stiffness, and damage tolerance is critical for the choice of repair methods and materials. Quality control plays a vital role to ensure that damage is properly repaired within the limits defined. Mechanical Grade 5 is given to the least important areas, which have large allowable damage limits and unlimited repairability with relatively simple methods and materials.

The mechanical grade can be considered as a relation between the class and zone combinations. The OEM's in-house classification system has no influence on the mechanical grade system; therefore, any classification system can function as a basis from which a mechanical grade system can be derived. The relation between the mechanical grade system and the in-house system of the OEM is to be determined by the OEM.

Relation Between the Systems: An example is given to explain the relation between an OEM in-house component classification system and the mechanical grade system. The following definitions of classes and zones serve as examples of in-house classification only. Other classification systems should result in similar usage of the mechanical grades.

- **SSI:** A structurally significant item (SSI) is a structural detail/element/assembly which is judged significant because of the reduction in aircraft residual strength or loss of structural function, which are consequences of its failure. (Ref. MSG-3. MSG stands for Maintenance Steering Group, an organization set up by airlines and OEMs to develop aircraft maintenance schedules on the basis of logical analysis of the need for work to be done.) Note that a principal structural element (PSE), according to FAR/JAR 25.571, is always an SSI. However, a SSI need not automatically be a PSE.

- **Class I:** SSIs with either a single load path or poor damage tolerance properties, whose failure has catastrophic consequences. Also SSIs whose failure has a very serious economic impact.

- **Class II:** SSIs with multiple load path and good damage tolerance properties, whose failure has catastrophic consequences or serious economic consequences.

- **Class III:** Non-SSIs.

- **Zone:** To account for some areas being more critical than others, each class is divided into four zones: A, B, C, and D. Zone A is the most critical area in the considered class; Zone D is the least important. However, Zone A of Class I is not comparable with Zone A in Class II with respect to strength or damage tolerance. Zoning determines only the relative importance of areas within one class.

Table 5.2 provides an example of the relation between class/zone combinations and the mechanical grade.

Table 5.2
Mechanical Grade as a Function of Class and Zone

Class	Zone			
	A	B	C	D
I	1	2	3	4
II	2	3	4	5
III	4	5	5	5

The basis of this is that an important area on a Class II component may be as repairable as a less important area on a Class I component. In other words, an equal level of safety (and/or economic consequences) is obtained by each mechanical grade, idealized in the table diagonals where approximately equal mechanical grades are obtained. However, in this example, a big difference in structural importance exists between Classes II and III, which results in the absence of Mechanical Grade 3 within Class III. Because the lowest mechanical grade is 5, the class/zone combinations of III-C and III-D also result in Mechanical Grade 5.

As a rough guide, parts would be classified as follows:

Fuselage Skin and Doublers	Primary
B-747 Flaps, Including Honeycomb Panels	Primary/Secondary
Elevators, Ailerons, and Rudders	Primary/Secondary
Undercarriage Doors with Ribs/Skins Riveted and Bonded	Secondary
Undercarriage Doors, Composite or Metal Skin, and Honeycomb Sandwich Construction	Secondary
Fairings	Tertiary

5.1.1 Sources of Damage to Composite Airframe Components

Service experience from a wide range of carriers indicates that damage occurs from numerous sources. Each of these possibilities (or sadly probabilities) should be considered at the design stage to minimize the damage caused and the consequences that could follow, which are mentioned in the next section and in Chapter 16, Design Guide for Composite Parts.

Mechanical Impact Damage: This can result from several causes:

- Foreign object damage in flight, which could occur from hail impact, bird strike, engine disintegration, or tire protector disintegration. Less violent possibilities are rain erosion and sand erosion (during takeoff or landing) and erosion resulting from the dust of volcanic eruptions.

- Foreign object damage on the ground, which could result from hail impact, runway stones, or debris; collision with ground handling and servicing vehicles; mishandling by maintenance staff; power system overload; tow-bar breakage; or an occasional collision with other aircraft during taxiing or towing.

- Dropped tools or impact with mobile ladders and other equipment during hangar maintenance, or inadvertent control surface operation during hangar maintenance, causing control surface impact with docking structure.

Lightning Damage: Lightning damage occurs at aircraft extremities (i.e., wing tips, fin tips, elevator tips, aileron tips, engine nacelles, and radomes).

Overheat Damage: Overheat damage can be found in engine cowl and wheel bay areas and anywhere a lightning strike occurs.

Erosion Damage: Erosion damage is particularly common in radome, engine cowl, and leading edge areas.

Delamination Damage: This could result from manufacturing deficiencies but also can occur in service due to impact of dropped tools or the use of percussion riveting on composite parts.

Moisture and Aging: This is the ingress of moisture into composite parts in service.

5.1.2 Consequences of Failure

Possible consequences of failure are considered in the allocation of classes, zones, or mechanical grades, and when deciding the size of allowable damage, repairable damage, and repair methods to be used. In the broader context, consequences of failure are carefully considered at the design stage when, for example, the size of the rudder is chosen to be sufficient to cope with engine failures on one side or the size of a flap section or spoiler is kept down to a size that will allow the loss of one section without the loss of aircraft control.

5.2 Types of Composite Structures

The two basic types of composite structures are monolithic laminated structures and sandwich structures. Some parts may be a combination of both of these.

5.2.1 Monolithic Laminated Structures (Solid Laminates)

Monolithic laminated structures consist of a multilayer stack of composite pre-preg plies with appropriate orientation of each ply to provide the best balance of mechanical properties from the finished laminate. The name "monolithic" was originally derived from "monolith," meaning a single block of stone. In composite terminology, it means "made from fiber and resin only," i.e., solid composite as opposed to thin composite skins in a sandwich panel which are bonded to a lightweight core of honeycomb or other material. The usual areas of application for solid composite laminates are wing skins, fin box skins, and horizontal stabilizer skins for military aircraft, and fin box and horizontal stabilizer skins for civil aircraft. These are in widespread use today by many manufacturers. Progress into wing skins is being made on some small civil aircraft (e.g., ATR 42 and ATR 72). Small executive aircraft have been made with all-composite fuselages and wings—almost everything except the engines, undercarriage, and electronics. The Russians are building a filament-wound fuselage approximately the size of a Jetstream 61, and the large aircraft companies are studying the possibility of this method for aircraft of the size of the Boeing 767. This will arouse considerable interest in their repairability.

The complexity of the ply lay-up that can be used is indicated by the fact that, on the horizontal stabilizer skin for the F-16 aircraft, the skin is composed of a multi-ply assembly ranging from 10-ply thickness at the trailing edge to 56-ply thickness at the root section. The Eurofighter and JAS 39 wings are almost certain to be even more complex. These types of structures are usually made from unidirectional pre-pregs, in which all the fibers in the pre-preg layer are aligned in one direction. To achieve the desired balance of strength and stiffness in the component, it is necessary to follow a detailed ply orientation schedule when building the part. This schedule defines the angle at which each ply must be laid to form the finished laminate. The laminate structure for other parts may consist of layers of pre-preg fabric of either a plain or satin weave, with or without layers of unidirectional material. Both types are formed into structures by curing on a suitably shaped tool at the required temperature and pressure for a specified period of time to ensure full curing of the resin matrix. After a good tool has been made, it is an advantage of composite materials that complex shapes can be produced easily. It is important in repair situations to determine the ply orientation pattern for the item to be repaired and to strictly follow the ply orientation instructions. This applies regardless of whether the lay-up is of unidirectional tape, woven fabric, or a mixture. Note that it is becoming common practice to specify the orientation of glass fiber fabrics. Years ago, these materials were used exclusively for lightly loaded parts, and the orientation was seldom, if ever, noted. Today, it is important to check the orientation of glass plies in addition to carbon, aramid, and other fibers. Engineering drawings and SRMs define the angle of each ply. Solid laminate structures often are fairly thick, and a particular rule of thumb for repair design should be noted. The best bonded joints are scarf joints, preferably of 1 in 50 or at least 1 in 30 taper. A taper of 1 in 20 is permitted at panel edges, which are made thicker to allow for the heads of countersunk fasteners. Overlap joints in composites are usually 50 times skin thickness at full patch thickness with the ply drop-offs carried beyond this. Above a skin thickness of 2 to 3 mm (3/32 to 1/8 in.), it is better to bolt or rivet a repair than to bond it. This has been determined because above this thickness, a bonded joint becomes too large and a bolted joint can be made at equal or less weight. A bolted or riveted joint also can be bonded, and this improves the fatigue life. It is to some extent a relief to know that the heavier and more structural composite parts will be repaired by bolting. This saves significant worry about the dryness, cleanliness, and surface preparation quality required to ensure the performance of a bonded joint. To the aircraft manufacturer, it means that the original design must allow for the inefficiency of bolted joints and the structure must have sufficient spare capacity to tolerate repair joints with an efficiency of significantly less than 100%.

5.2.2 Sandwich Structures

Sandwich structures consist of thin composite skins bonded to a low-density honeycomb or other type of core material (see Chapter 2, Section 2.6). The lightweight core acts as a support for the thin composite skins and prevents them from buckling when the component is subjected to bending and twisting loads. These are an efficient type of structure in which the bending and buckling loads are carried almost entirely by the thin skins and only the shear loading is carried by the low-density core. However, note that the core does real work and that it must be strong enough to perform its function. When a sandwich part requires repair, the core must be replaced with the correct type and density or one that is slightly heavier and stronger. Sandwich panels are commonly used as floor panels. In this application in which point loading from stiletto heels is high, the density of the core has a marked effect on panel life. Because the loading is less severe, floor panels used under the seats in passenger aircraft have a lower-density core than those used in gangways and doorways. It is important to ensure that the correct grade is used in gangway areas. It may seem odd to refer to "structural carpet," but it was found, during some impact tests, that the resistance of lightweight sandwich-type floor panels to impact was greatly increased by the use of carpet and underlay. The energy required to damage the floor panel skin was eight times as high in the presence of the carpet and underlay as it was on the bare floor. Clearly, the correct grade of carpet and underlay must be used if the intended life of the floor is to be achieved. This method of construction allows us to obtain the highest efficiency possible for structural members that are required to carry bending or column loads. Structural efficiency is defined (in this case) as the lightest and most rigid method of meeting this type of load-carrying requirement. Sandwich structures typically are used in aircraft design for components such as flaps, ailerons, rudders, elevators, spoilers, fairings, and access panels, as well as for floor panels, galleys, wardrobes, and most of the cabin interior trim areas in large civil aircraft. Because the skins are thin and the mechanical loading is usually low, a sandwich structure is most easily repaired by adhesive bonding, regardless of whether it has metal or composite skins. To achieve good results, this requires either chromic or phosphoric acid anodized skins for aluminum alloys and dry, clean, and lightly abraded surfaces for composites. See Chapters 10 and 15.

Sandwich structures can be made with a wide range of combinations of lightweight core material and high-strength and high-stiffness skins. However, the most commonly used combinations found in aerospace structures are listed below.

Core Material	Skin Material
Polyvinyl Chloride (PVC) Foam	Glass/Epoxy Composite
Polymethacrylimide Foam (Rohacell)	Kevlar/Epoxy Composite
End Grain Balsa Wood	Carbon/Epoxy Composite
Phenolic Coated Kraft Paper Honeycomb	
Aramid Honeycomb (Nomex)	
Aluminum Honeycomb (Various Foil Grades 3003, 5052, 5056, and 2024)	

Normally, adhesive film is used to bond the skins to the sandwich core. With some glass, Kevlar, and carbon pre-pregs, the resin in the uncured pre-preg can flow and bond well to the honeycomb core during the cure cycle, thus eliminating the need for a separate adhesive. This is known as a co-cured component. The alternative technique is to pre-cure the sandwich skins and to bond them to the core in a second operation, often referred to as secondary bonding. Co-curing also can be done with a film adhesive. The film adhesive virtually guarantees a good fillet and a good bond to the honeycomb. The film adhesive must be only approximately 150 gm/m^2 (0.03 lb/ft^2) weight in this case, as it is additional to the resin from the pre-preg.

A typical breakdown of a honeycomb sandwich panel is shown in Figure 5.1.

In addition to the adhesive bond joining the core to the skin, it is equally important to bond sections of the honeycomb core to each other and to the various inserts and edge members that may be built into sandwich panels. This adhesive bond transfers shear between the honeycomb core and edge members or other pieces of honeycomb.

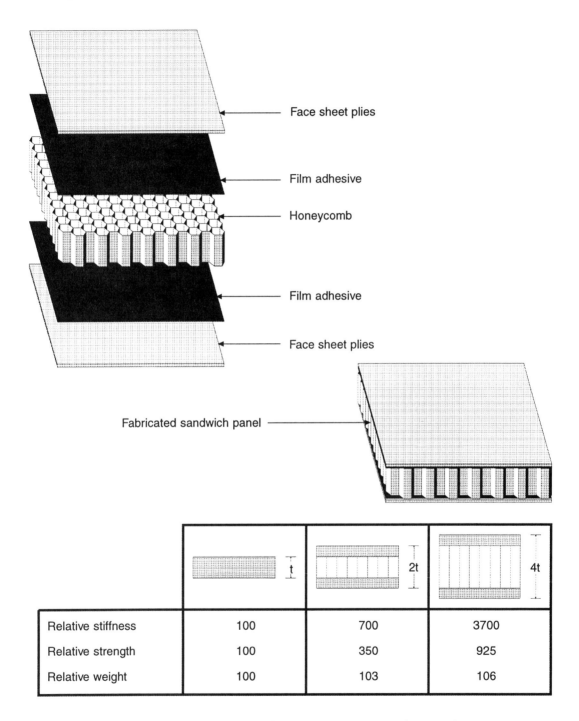

	t	2t	4t
Relative stiffness	100	700	3700
Relative strength	100	350	925
Relative weight	100	103	106

*Fig. 5.1 Typical construction of a honeycomb sandwich panel. The table provides an example of how honeycomb stiffens a structure without materially increasing its weight.
(Courtesy of Abaris Training Resources Inc.)*

The adhesive or potting compound used for this purpose is often referred to as core splice adhesive. It can take several forms:

- A potting compound or two-part paste that cures at room temperature

- A one-part foaming paste that can be hot cured and that expands during cure to fill the void between two sections of core or between the core and an edge member

- A foaming film adhesive that expands during cure to fill the void

These core jointing compounds, especially the foaming ones, must not allow moisture penetration. The foaming types must generate a closed-cell foam to prevent liquid water penetration. Severe water ingress problems have occurred with porous types of foaming adhesive.

Sandwich Panel Inserts: A special situation occurs with sandwich structures when a need exists to fasten other components to a sandwich panel. With a solid metal sheet, it is only necessary to drill a hole and attach the component with a bolt or rivet. The procedure is not as simple with sandwich structures because if a hole is drilled through a sandwich panel and a bolt is tightened, the core will be crushed because it is not strong in compression. Several methods can be used to avoid this type of damage:

- Local inserts can be bonded into the panel to act as hard points. These can be metal, Tufnol, wood, or solid composite inserts, and they must be connected to the surrounding core with a core splice adhesive.

- A hole can be drilled in one face of the sandwich, and a bobbin insert can be bonded in with epoxy two-part paste adhesive. See Shur-Lok and similar types of fasteners.

- Two-part metal fasteners are available that can be fitted to the panel. These two halves are assembled by tapping them together with a small hammer to lock them by interference fit while bonding them with a tough adhesive to the skins. In the case of composite aircraft floor panels, special inserts with large-diameter flanges are used. These are purchased already anodized and primed, and they must be kept completely clean before use because they are required to carry shear loads of 454 kg (1000 lb) or more, according to the airframe manufacturer's specification. They must be carefully handled with clean cotton gloves before and during installation to avoid contamination. These inserts must carry shear loads in the crash case and are structural items.

- Threaded inserts can be bonded into the panel with paste adhesive. They must have bonded flanges and be potted in place to enable them to resist the torque loading from screws or bolts as they are tightened. The part carrying the screw thread must be fitted from the far side of the panel relative to the bolt to prevent the insert from being pulled out by the bolt.

- For lightly loaded attachments, a paste epoxy adhesive can be injected into the panel, and this insert can be drilled to take screws.

Note that it is highly undesirable to use bolt holes through any part of a sandwich structure where a honeycomb or other core is present. If it is unavoidable, then the third type mentioned above (two-part metal fasteners) should be employed, and adequate adhesive should be used both in the hole and to bond the flanges to the skin. This is essential to prevent or minimize water penetration, which leads to disbond and corrosion in the case of metal skins and considerable weight gain resulting from water uptake in composites.

Typical insert designs are shown in Figure 5.2.

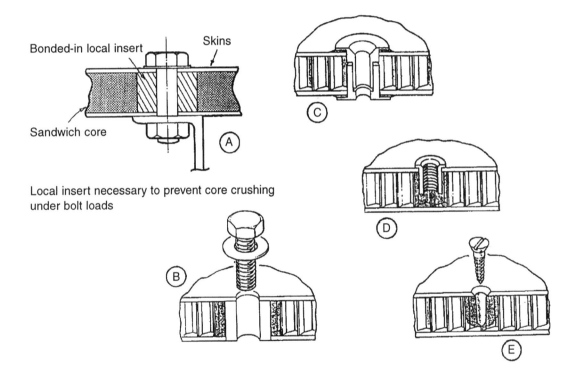

Fig. 5.2 Typical fastening/insert techniques and insert designs.
(Courtesy of Aero Consultants (United Kingdom) Ltd. and Shur-Lok.)

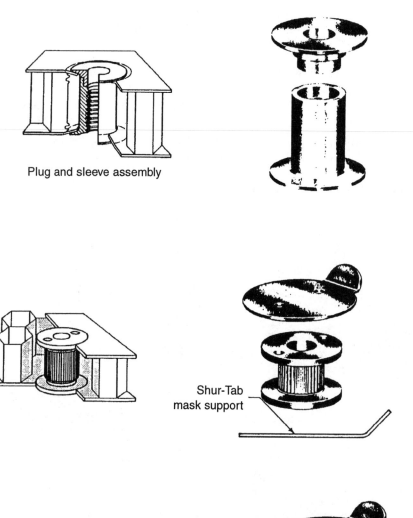

Plug and sleeve assembly

Shur-Tab
mask support

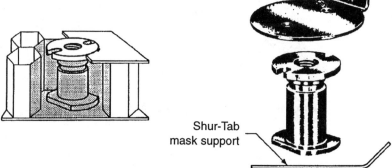

Shur-Tab
mask support

Fig. 5.2 (continued)

5.3 Ply Orientation

As previously mentioned, the orientation of each ply in a composite lay-up is carefully chosen to meet the loading and stiffness requirements of each individual part. Therefore, it is vital that the ply orientation in repairs is correct. By comparison, metals are relatively simple because they are isotropic (i.e., they have similar properties in all directions). Rolled bars and sheets show some strength reduction at 90° to the plane of the sheet, but the difference is small compared to composites. Unidirectional composite tapes have a maximum strength and stiffness when the load is applied parallel to the fibers, but these properties drop off sharply as the loading angle moves away from the fiber axis. At 90° to the fibers, the strength falls to approximately 1/30 of the fiber strength because, at this angle, the load depends on the lesser of either the strength of the matrix resin or the strength of the fiber-to-matrix bond. In the case of woven fabrics, with equal numbers of fibers in both the warp and weft, minimum properties are found at 45° to the warp direction because beyond this angle the weft fibers take up the load. See Figures 5.3, 5.4, and 5.5.

5.3.1 Warp Clock

To achieve correct orientation in manufacture and later in repair, a diagram known as a warp clock is used on the manufacturing drawings and also in the SRM and in certain Component Overhaul Manuals. This warp clock has not yet been made into an international standard. Although the systems used vary from one manufacturer to another, they are largely similar in concept. A datum is chosen, and the orientation of each ply is measured from this. See Figure 5.6.

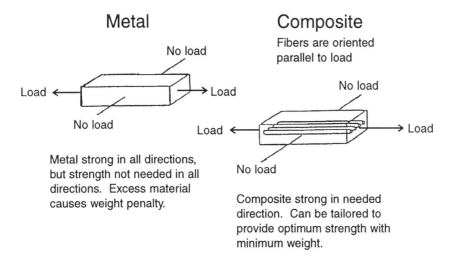

Fig. 5.3 Tailored design of composites. (Courtesy of Heatcon Composite Systems.)

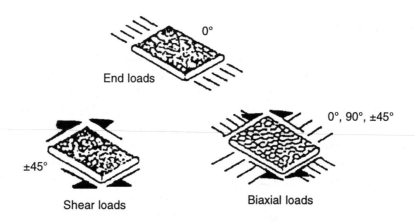

Fig. 5.4 Composites require definition of fiber patterns to suit applied loads.
(Courtesy of Heatcon Composite Systems.)

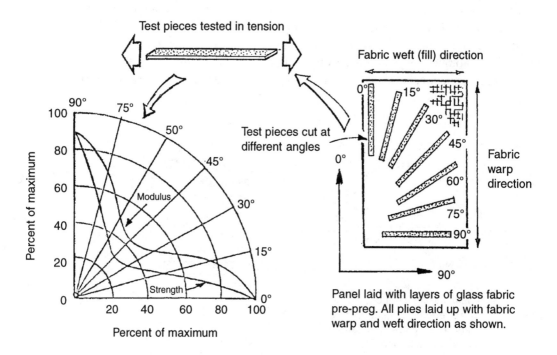

Fig. 5.5 The mechanical strength/stiffness of a composite glass fabric panel varies with fiber
direction. (Courtesy of Aero Consultants (United Kingdom) Ltd.)

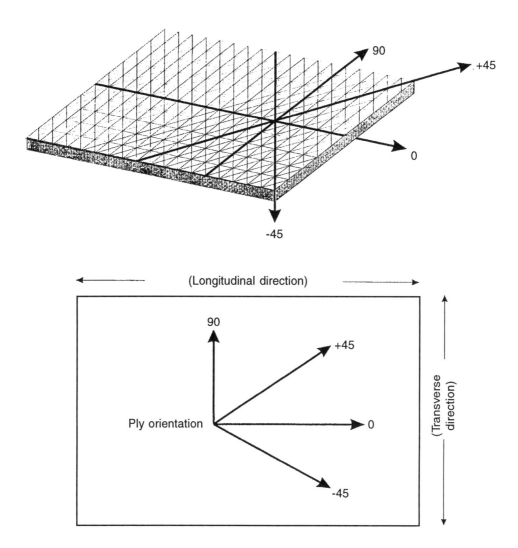

Fig. 5.6 Warp clock. (Courtesy of Abaris Training Resources Inc.)

This warp clock is identical to the one used in Boeing Drafting Standard BDS 1330, pages 8 and 9. Several important notes follow this diagram:

• Normally show the symbol as if looking from the bag side toward the tool side. Coordinate any deviation from this rule with Manufacturing Engineering.

• Normally show the 0° axis of the symbol parallel to the laminate's primary load direction.

• Do not use a mirror image of the symbol.

- On a composite part picture extending beyond one microfilm frame, show a ply orientation symbol in each frame.

- Separate parts bonded together require ply orientation symbols for each part.

- Show ply orientation at the point where it is most applicable. Because it may vary significantly over a contoured part, note the deviation allowed if critical (e.g., ±3°).

Each fabric or tape ply is given a number beginning from the tool side of the part. A ply table on the drawing gives the angle required for each ply in the lay-up. BDS 1330 is quite lengthy and complicated and requires time for study. Always use the latest revision. Other manufacturers may have a different method. Most do not have a direct equivalent of BDS 1330, although similar information is spread throughout their design manuals and is readily available to and used by their own staff. An urgent need exists for an International Organization for Standardization (ISO) standard to cover this subject. The question of ply orientation is covered on all relevant drawings and in all SRMs. Check and use the method specified for each aircraft type and work to any limits given.

5.3.2 Balance and Symmetry

In addition to the orientation of each layer, another significant factor at the design stage is the precise sequence in which each layer is laid up. A lay-up that is not balanced will be bent or twisted or both on leaving the mold or tool face and will not have the intended shape. Balanced laminates and symmetrical laminates have carefully precise definitions, and it is possible for a laminate to be balanced but not symmetrical or to be symmetrical but not balanced. Designers try to produce laminates that are both symmetrical and balanced because warping is undesirable. For a more detailed explanation of this subject, see Refs. 5.1 to 5.5.

AIR 4844 defines a balanced laminate and a symmetrical laminate as follows:

1. **Balanced Laminate:** A composite laminate in which all laminae at angles other than 0 or 90° occur only in pairs (not necessarily adjacent) and are symmetrical around the center line. This type of laminate will have the least tendency to bow after cure. See also symmetrical laminate. A laminate may be balanced but not symmetrical or symmetrical but not balanced. See textbooks on laminate design and Figure 5.7.

2. **Symmetrical Laminate:** A symmetrical laminate can be defined in the following two ways:

   - A composite laminate in which the sequence of plies below the midplane is a mirror image of the stacking sequence above the midplane.

- A laminate is said to be symmetrical when the plies in the upper half of the laminate (for ordinate z = 0) are identical, in terms of ply properties, ply angle, ply thickness, and ply position relative to the midplane, to the plies in the lower half of the laminate (for ordinate z = 0). Symmetrical laminates can be constructed from individual plies whose fibers are singly oriented or multi-oriented. The plies can be all of one composite material, or a hybrid can be constructed in which more than one composite material is used.

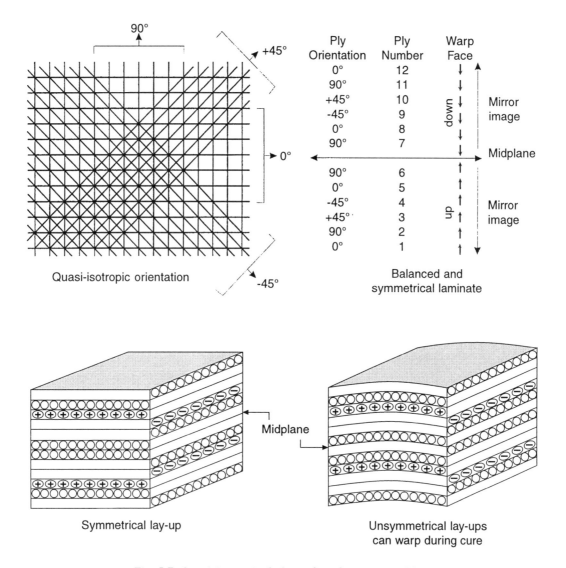

Fig. 5.7 Quasi-isotropic, balanced, and symmetrical lay-up.
(Courtesy of Abaris Training Resources Inc.)

5.3.3 *Nesting and Stacking*

Nesting laminate and stacking are defined as follows:

1. **Nesting Laminate:** In reinforced plastics, the placing of plies of fabric so that the yarns of one ply lie in the valleys between the yarns of the adjacent ply (nested cloth). Nesting applies only to unsymmetrical weaves or to those that have distinguishable warp and fill faces, such as satin weaves. Symmetrically woven plain weave cloth does not have warp and fill faces. Unidirectional tape does not have warp and fill faces. The advantage of nesting is that it provides slightly better compaction between the plies and a lower void/porosity content. To achieve nesting, note the following:

    - If the two plies in question have the same ply orientation, orient them warp face to warp face or fill face to fill face.

    - If the two plies in question have ply orientations 90° apart, orient them warp face to fill face.

    - If the two plies in question are oriented at angles other than 0° or 90° to each other, nesting is not possible.

2. **Stacking:**

    - The lamination sequence in which the warp surface of one ply is laid against the fill surface of the preceding ply. The sequence of lay-up repeats warp-fill, warp-fill, warp-fill, etc., until the lay-up is completed.

    - The lamination sequence in which plies are continuously laid up warp surface to fill surface until the lay-up is completed. From these comes stacking sequence.

3. **Stacking Sequence:** A description of a laminate that details the ply orientations and their sequence in the laminate. See Figure 5.8. This definition of stacking applies to the lay-up of all types of fabric and tape, except that some cases (e.g., symmetrical plain weaves, unidirectional tape, 2 x 2 twills, and basket weaves) do not have warp and fill faces. If a ply with warp and fill faces must be laid a particular way down, then the drawing or SRM should state this clearly and also show how someone who is unfamiliar with this subject can identify the two different faces.

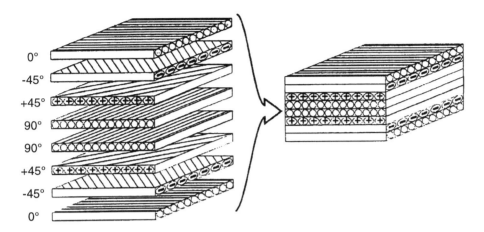

0°
-45°
+45°
90°
90°
+45°
-45°
0°

Fig. 5.8 Stacking sequence. (Courtesy of Abaris Training Resources Inc.)

5.4 Core Orientation

When laying any type of honeycomb core, the ribbon direction of the honeycomb must be oriented correctly to drawing. The ribbon direction is the direction of the node bonds that join the various layers of paper or foil together. This means that the honeycomb has approximately twice as much material in one direction as the other. Designers take advantage of this to orient the honeycomb to provide the greatest strength and stiffness in the desired direction. Consequently, during repairs or modifications, the honeycomb must be oriented in the same way as the original to restore strength and stiffness. The required direction can be found on the component drawing, in the SRM, or in the Component Overhaul Manual.

5.5 Operational Environment

During original design, the materials of construction for each part are chosen carefully to meet the design requirements of loading and environment. Fibers are chosen for strength, stiffness, weight, and temperature resistance; pre-preg resins, wet lay-up resins, and adhesives are chosen for their strength, stiffness, and glass transition temperature (T_g), which decides their maximum temperature of use. The amount of water they absorb will affect the value of T_g when hot and wet. Resins also must be checked to ensure that they do not suffer unduly from softening in the presence of maintenance chemicals such as de-icing fluid, fuel, cleaning detergents, and Skydrol or other hydraulic fluids. Honeycomb materials are selected for strength and stiffness. When making repairs, it is important to ensure that the specified material is used if possible. If it is not available, then it is important to use a material with the same or similar properties. Unless the user has design authority of his or her own, it will be necessary to obtain the manufacturer's approval for the use of alternative materials. The influence of the operational environment may affect the composite either advantageously or adversely. Most effects will be adverse in some form, but heating may occasionally offer the benefits of post-cure and will also serve to dry any moisture or other fluids that may have been absorbed by the resin matrix or paint or other protective coating. Each likely environmental source and its influence on a composite are discussed below.

5.5.1 Temperature

The service temperature experienced by aircraft, ships, trains, and automobiles may vary widely, depending on the location of use and the speed involved. Subsonic aircraft are designed for use at temperatures from -55 to 80°C (-67 to 176°F). Skin temperatures can reach 105°C (221°F) in the sun, but little loading is involved. Trains, automobiles, and ship decks must reach this temperature in some areas if they are not moving. Supersonic aircraft and missiles can operate at much higher skin temperatures, depending on the speed and duration of the flight. Concorde wing skin leading edges operate at approximately 103°C (217°F) at normal speeds and the fuselage skin at approximately 90°C (194°F). Faster aircraft reach even higher temperatures. The forward end of space vehicles, on reentry to the earth's atmosphere, glows red hot and must approach 800°C (1472°F) or more. Ground vehicles can be as cold as approximately -50°C (-58°F) in the coldest regions. Each type of vehicle has its own design code to which reference should be made before designing a new component. When attaching parts of widely different thermal expansion coefficients, establish the range of temperature through which each assembly will be cycled in service. Expansion must either be allowed for in the design, or the choice of materials must ensure that mating parts have similar coefficients. The materials selected also must have adequate mechanical and physical properties over the full range of temperatures expected in service.

5.5.2 Humidity

Under humid conditions, and especially if those humid conditions are combined with high temperature, composite matrix resins will absorb moisture. The amount varies considerably with the resin system and particularly with the curing agent. After many years, a saturation level of moisture absorption is reached. The worst possible case is likely to lead to the amount of water absorbed being approximately half of the figure obtained at the same average temperature by total immersion in water. Tests can be run to find the saturation value, then composites can be immersed until half this value is obtained, and then mechanical tests can be run to find the strength of the material under the specified conditions. Dry fabrics and tapes, nonmetallic honeycombs, balsa wood, and (to a lesser extent) foams can absorb moisture from the atmosphere. It is helpful if these items are dried before use. Aluminum and steel parts (if not made from corrosion-resistant steel) can suffer corrosion in humid environments. Material stocks should either be protected with grease or kept under dry conditions. Acrylic plastics absorb moisture in storage, which increases their susceptibility to solvent crazing.

5.5.3 Contaminants

Contaminants take many forms—from nothing more than unwanted moisture to oils, greases, fuel oil (kerosene), de-icing fluids, hydraulic fluids, exhaust fumes, silicone sprays, and polishes. They include human skin oils if bonding faces are touched prior to bonding, cleaning detergents, and atmospheric pollution from a variety of sources. This list shows why smoking and eating is banned in bonding areas, why clean white cotton gloves must be worn, and why a clean room is specified when parts are to be laid up for repair or manufacture.

5.5.4 Erosion

Erosion occurs on all leading edges of aircraft components. Particular points are nose radomes, wing leading edges, helicopter blade leading edges, tailplane and fin leading edges, and radio and navigation aerials. Causes of erosion are rain, hail, dust, and sand, and the effects of all these are greatly aggravated by the speed of the aircraft. Few problems of this type were reported until the mid-1940s when increasing aircraft speeds resulted in the first serious reports being made in areas where heavy hail is common. All aircraft with operating speeds above 250 miles/hr are subject to rain erosion, and only some of the smallest aircraft now fly at speeds lower than this. Years ago when a Boeing 747 ran into heavy ash in a volcano plume in Indonesia, the center two-thirds of both glass-faced No. 1 cockpit windows were sand blasted, and visibility was reduced to nil. Fortunately, the outer third of each window and the two plastic side windows on each side were unaffected because the ash flowed around them. The most frightening aspect of this case was that all four engines ran down, and it took some time to restart them. The engine cowls showed erosion around the lip, and the fan blades were highly polished. Wing leading edges were eroded slightly, and landing lamp lenses also were sand blasted. This is the most extreme case of erosion to date.

A problem with the leading edges of many engine cowlings is that if the leading edge projects even slightly into the airstream, the edge will become worn and delaminated and pieces eventually will break off. Cowling leading edges will have to be improved in future designs to provide edge protection. On the pointed nose of the Boeing 747, where passenger windows are provided, even the leading edges of the outer window panes show significant erosion if they project slightly into the airstream. Erosion should be given more consideration at the design stage in the future, especially for composite materials.

Factors affecting erosion are:

- Type of material
- Surface finish
- Shape of the component
- Speed
- Raindrop size

Factors contributing to erosion resistance are:

- Surface hardness
- Strength
- Tear resistance, in the case of resilient coatings
- Resilience or high elongation at break

Helicopter blades and some leading edge areas and radomes are coated with polyurethane or neoprene rubber sheets or paint coatings to reduce erosion. In the case of radomes on which rubber boots or overshoes have been used, the thickness has been found to be important. Boots of 0.25 mm (0.01 in.) thickness lasted approximately one year on test, and a thickness of at least 0.5 mm (0.02 in.) is recommended.

Of the composite construction systems, filament winding has been found to be the most susceptible to erosion, whereas woven socks provide the best results. Thus, it would seem reasonable to conclude that unidirectional tape lay-ups would suffer similarly. Improvement in erosion resistance may be achieved by careful preparation. Because of the localized nature of rain drop impact, freedom from voids, bubbles, and dry spots is important, and the preparation of smooth surfaces of high resin content is necessary to maintain erosion resistance. In studies of the effect of bond strength between the plies on rain impact resistance, good interlaminar shear strength was found to be very helpful. This was particularly evident if a coating had been used to protect the laminate. If the coating was penetrated at a local defect, for the better quality substratum, the damage continued to be localized. If the substratum had poor interlaminar shear strength, the mode of failure was by rapid and extensive delamination under the coating. Good interlaminar shear strength requires a tough matrix resin. See Ref. 5.6.

5.5.5 Thermal Stresses

Thermal stresses can occur in composite parts for several reasons:

- As residual stresses resulting from mismatch of thermal expansion coefficients between fibers running in different directions

- In hybrid composites as mismatch between the thermal expansion coefficients of different types of fiber

- As residual stresses resulting from mismatch between the coefficients of the fiber and the resin

- As stresses caused by heating a sheet of composite from only one side (e.g., engine cowling heated from the inside and cooled on the outside, or the outside skin of a supersonic aircraft heated quickly relative to the cooler inside)

- As residual stresses due to cooling a part after cure at high temperature

These stresses can be calculated using modern computer programs. See Ref. 5.1.

It is important to consider these stresses at the design stage after the range of temperature of the part in use has been defined. Tough resin systems with a high strain to failure can help to minimize cracking due to thermal stresses. Hybrids have been found to delaminate if the range of temperature in service is too high.

5.5.6 Hygrothermal Effects

Hygrothermal effects are usually considered together, when perhaps there should be a hygro expansion coefficient measured separately from a thermal expansion coefficient. The reason for this is that the effects of moisture absorption are similar to those of thermal expansion. In the thermal case, there can be both expansion and contraction about room temperature as temperatures rise and fall; in the moisture absorption case, there is only water uptake from zero. Contraction will occur during drying back to zero. This subject is also covered in Ref. 5.1.

5.5.7 Fire Resistance

This factor is important, especially for materials used in passenger cabins. The primary consideration is that passengers can be evacuated safely and remain in a healthy condition after escape. Although fire must be prevented, retarded, or extinguished quickly, other factors cannot be ignored. The most important of these are smoke generation and the toxicity of the fumes produced. Smoke may prevent passengers from finding a door, and all passenger aircraft now have floor-level lighting to help with this. However, if the elimination of smoke results in more toxic fumes, the passengers may be dead or incapacitated before they can reach a door. All three factors must be reduced to acceptable levels if passengers are to escape without suffering from burns, other injuries, or permanent health damage. This limits the choice of materials.

Fire resistance alone is assessed by the following factors:

- Surface spread of the flame
- Fire penetration
- Ease of ignition
- The contribution of fuel
- The oxygen index (i.e., the minimum oxygen content that will support combustion)

The other factors are assessed by:

- The smoke density as measured in a standard test
- The toxicity of the fumes generated

Composites vary from highly flammable to nonburning. Their fire performance depends on the following factors:

- Matrix type
- Quantity and type of any fire-retardant additives incorporated
- Quantity and type of fillers used
- The reinforcement type, volume fraction, and the construction used

The most important of these is the choice of the resin matrix type. The order of flammability of common resins is as follows:

- Polyester—Most flammable
- Vinyl Ester
- Modar
- Epoxy
- Phenolic—Least flammable

Phenolic resins are used almost exclusively in passenger cabins because they have low smoke emission and almost no toxic combustion products. See FAR/JAR 25.603.

5.6 Electrical Requirements

Several types of electrical effects and requirements involve composites, and these are discussed separately below.

5.6.1 Galvanic Corrosion

This form of corrosion occurs when two dissimilar metals (conductive materials) are in electrical contact with each other and are exposed to an electrolyte. Electrolytes are solutes (chemicals that dissolve in water) which, when dissolved in water, produce solutions that conduct an electric current. The behavior of electrolytes is very different from the behavior of non-electrolytes. Some typical electrolytes are sodium chloride, calcium chloride, nickel chloride, sodium sulfate, potassium chloride, and hydrochloric acid. Solutions of these chemicals in water undergo ionization. According to the theory of ionization, first proposed by Arrhenius, electrolytes dissociate into positively and negatively charged ions in solution. These charged ions migrate freely throughout the solution and consequently are responsible for the conductivity of solutions of electrolytes.

Electrolytes can be divided into three types of substances:

1. Acids, which are substances that ionize in solution to produce hydrogen ions (H+)

2. Bases, which are substances that ionize in solution to produce hydroxide ions (OH-)

3. Salts, which are substances that ionize in solution but produce neither hydrogen nor hydroxide ions

We frequently hear of acid rain. Epoxy resins usually are basic (alkaline), and coastal airports suffer from sodium chloride spray. Other salts are present as atmospheric pollutants. Therefore, aircraft must experience all three forms of electrolyte production previously mentioned. Because an electric current can flow through an electrolyte, materials of different electrical potential connected by an electrolyte result in a flow of current from the material of higher potential to the lower. A less noble metal (e.g., zinc) will dissolve and form the anode; a more noble metal (e.g., copper) will act as the cathode. See Table 5.3.

Table 5.3
Galvanic Series of Metals and Alloys (Seawater) (From Ref. 5.7)

	Electrical Potential	Anodic (Corroded End) Least Noble
Magnesium and Magnesium Alloy		
Zinc		
7079 Aluminum		
7075 Aluminum	-0.75	
6061 Aluminum		
5052 Aluminum		
Clad 2024 Aluminum		
3003 Aluminum		
6061-T6 Aluminum		
7178 Aluminum		
Cadmium		
2017-T4 Aluminum		
2024-T4 Aluminum		
2014-T6 Aluminum		
Steel or Iron	Aluminum/Graphite Difference 0.85	
Cast Iron		
Lead		
Tin		
Brasses		
Copper		
Bronzes		
Monel		
Nickel (Passive)		
Inconel (Passive)		
Stainless Steels		
Silver		
Titanium		
Graphite (Carbon Fiber) a Nonmetal	+0.1	
Gold		
Platinum		
		Cathodic (Protected End) Most Noble

The electrochemical series arranges metals in order of their standard electrode potential (Ref. 5.7). This gives a different order from the galvanic series because, in practice, most metals are covered by an oxide film, and this tends to shift the solution potential to more positive values. Different test methods give slightly different orders. According to Ref. 5.7, "So far as corrosion is concerned, the galvanic series is more useful than the electrochemical series, but even small environmental changes can shift the potential in either direction." Depending on the nature of the corrosive environment, the cathodic reaction may involve hydrogen evolution or oxygen absorption. The electronic current flows from the anode Zn through metal to the cathode Cu. It is also clear that the corrosion current flows at the expense of the anode metal, which is corroded, whereas the cathode is protected from attack (Refs. 5.7 and 5.8). It is necessary to minimize the electrical potential differences by material selection or to insulate materials of different electrical potential from each other (e.g., when bonding carbon fiber to aluminum alloy, it is wise to place a layer of glass fiber between the materials to provide electrical separation when moisture may be present). The only other way to prevent galvanic corrosion is to keep the whole assembly totally dry so that no electrolyte exists to carry the current. In the design of some engine cowlings, a film adhesive with a nylon carrier was used to isolate the carbon fiber skins from the aluminum honeycomb core. The use of an epoxy or other resin system with a low water uptake always helps and, if possible, an adhesive which is of a neutral pH or which allows only the leaching of neutral pH extracts when in contact with water or other fluids. Glass fiber and aramid fiber composites do not cause or suffer from galvanic corrosion because they are nonconductors of electricity. However, carbon fiber, although nonmetallic, is a conductor of electricity; relative to various metals, its composites are highly noble. Consequently, if carbon composites are used in contact with metals, galvanic corrosion of the metal will occur if an electrolyte is present. Experience with carbon fiber repair patches on aluminum alloy structures, without a glass fiber insulating layer, has confirmed beyond any doubt that corrosion will occur if moisture is present. If a carbon fiber patch is applied to reduce the stress level in a cracked part, it is also essential to seal the inside face of the crack to prevent moisture penetration through the crack to the bondline. Sealing is necessary even if the patch is nonconducting. This is another good reason for using adhesives or resins with a low water uptake.

Corrosion prevention requires correct material selection and sound engineering design. Important points are as follows:

• Avoid the use of dissimilar materials when an electrolyte is present and galvanic corrosion is probable.

• If two different metals must be used, they should be as close as possible to each other in the galvanic series.

• The anodic material should have as large an area as possible relative to the cathodic material, which should have as small an area as practicable (i.e., nuts, bolts, and rivets).

- Dissimilar metals (conducting materials) should be insulated from one another.

- An insulating coating should be applied to prevent access of the electrolyte to the junction between the different metals. Alternatively, the electrical contact may be broken by insulating fittings.

- The anode metal should not be painted or otherwise coated when forming a galvanic couple because any breaks in the protective coating would lead to rapid penetration of the metal by pitting. If the metal is not painted, uniform corrosion of the anode is much less severe. One example of pitting is stone chips that damage car body paint work. A quick application of corrosion-resistant primer is helpful.

- Avoid the presence of crevices, even in the case of the same metals. Riveted or bolted lap joints favor the formation of crevices.

- Avoid residual stresses in metal parts because these cause a different potential from unstressed areas and assist the development of stress corrosion.

- Avoid the use of insulating layers that absorb moisture because they can absorb electrolyte and become conductive.

- Use corrosion-resistant primers before making adhesive bonds.

For more details of the function of corrosion inhibitors, see Ref. 5.7.

Corrosion is a complex and extensive subject, but engineers must be aware of it and take the precautions specified in SRMs, Component Overhaul Manuals, and other relevant documents. Because it is a subject of concern and interest to every form of transport and engineering activity, further reading is recommended. Corrosion is covered extremely well in Airbus A.320 SRM 51-22-00, and other SRMs should contain similar information.

5.6.2 Electromagnetic Interference

This is another complex and specialized area that must be considered. From the viewpoint of the repair mechanic or repair design engineer, it must be understood sufficiently to ensure that repairs do not reduce the protection provided in the original design either at all, or at least not below the minimum level permitted by the SRM. Aluminum alloy aircraft structures provide adequate shielding in themselves; however, when composite structures are employed (even if they are carbon fiber, which is conductive), they do not provide the same level of conductivity and interference protection as aluminum. However, Ref. 5.9 (pp. 260–272), states that on the Harrier AV8B/GR5, "It was found that carbon/epoxy has the inherent shielding capability required in a fighter aircraft and that, generally, there was little difference in relation to orthodox metal aeroplanes." In this case, carbon fiber/epoxy was found to provide sufficient shielding. Glass fiber and aramid composites do not conduct electricity. If used in areas where electronic components were fitted, these materials would require special treatment (i.e., conductive coatings or surface fabrics of woven metal or metal-coated fiber) to give protection to the electronic systems behind them. Modern aircraft use sensitive electronic control systems, and their protection against interference is essential to safe operation. Where RF/EMI (radio frequency/electromagnetic interference shielding) is included as part of the design, any repair work must ensure that this protection is replaced and tested to ensure that the level of protection provided by the repaired part meets the specification requirements. IATA DOC:GEN:3043 states, "On aircraft made of metallic structure, sensitive avionics equipment is automatically screened from induced current. When designing composites, alternative and adequate means of electronic shielding shall be applied in accordance with ARINC (Aeronautical Radio Incorporated) standards."

5.6.3 Electrostatic Discharge

The friction of high-velocity air passing over a large-surface electrical insulator such as an aircraft radome results in the buildup of static electricity, which periodically discharges by arcing to a metallic portion of the aircraft. In the course of discharging, it can punch a small hole in the radome protective coating, and this can lead to moisture penetration and radar signal strength reduction. This discharging also can result in interference with the aircraft's electronic equipment. The use of an antistatic coating having an electrical resistance low enough to allow the static charge to gradually bleed off solves this problem. Antistatic coatings compatible with particular coating systems are formulated by the addition of a small amount of electrically conductive material, generally carbon, to the base coating solution. A 0.025 mm (0.001 in.) thick antistatic coating applied over the primary coatings will normally fulfill the antistatic requirements. Typical antistatic coating performance requirements are given in MIL-C-7439, which covers elastomeric-type rain erosion coatings. Boeing Material Specification BMS 10-21 covers two types of coatings: Type I is a conductive coating for use on all fiberglass areas other than antennae and radomes, and Type II is considerably less conductive and is an antistatic coating for use only when specifically called up on radomes or antennae. In these cases, it is essential to apply only the specified amount of paint to precisely located specific areas as defined on the part drawing.

Static charge and precipitation static have a common origin, but they differ in degree. Normal air produces a static charge by the friction of air molecules, but precipitation static results from the friction of larger masses such as raindrops, hailstones, dry ice, and snow. This is often known as p-static. It is of much higher voltage than static resulting from air friction because of the larger masses involved, and it is the reason for conductive coatings being applied to nonconductive areas such as radomes and cockpit windows. The static charge buildup when flying through rain, hail, or snow can lead to considerable amounts of interference with radio communications. This often occurs at low altitudes where the various forms of precipitation are most common and where radio communication is most important (i.e., during the approach to an airfield and the landing phase, when clear understanding of instructions from ground controllers is critical to flight safety). Antistatic coatings specified on components must be maintained in a serviceable condition and must not be damaged or degraded by repairs. See Ref. 5.10.

5.6.4 Lightning Strike Energy Dispersion

This is another specialist area in which research is continuing today. In this case, carbon fiber components have been found to require lightning protection. Their inherent conductivity is not sufficient to offer protection from lightning strikes. IATA DOC:GEN:3043 (Ref. 5.11) requires that, "All components located in lightning risk zones, or in areas through which current is likely to pass, shall be designed with sufficient protection to avoid significant damage." See Figure 16.12 in Chapter 16 for the zones on an aircraft which are most at risk to lightning strikes.

FARs and JARs require that lightning protection be provided. Note that lightning strikes can be extremely powerful and damaging. The maximum voltage can be as high as 2 million volts, and the maximum current as high as 300 amps. Instantaneous local temperatures on components as high as 1000°C (1832°F) are possible, and 20,000°C (36,032°F) may be reached in the arc itself. In one case, where the current went through a bolt in a spar of carbon fiber, the hole was greatly enlarged and the composite around the hole had lost all the resin and was as dry as if it had never been impregnated. Epoxy resin begins to char at approximately 400°C (752°F). Thus, it is clear that achievement of instantaneous gasification of cured resin clearly must have required a very high temperature. In another case, on an old type of aircraft with a radome that was not fitted with lightning diverter strips, a lightning strike penetrated the radome and struck the radar scanner. The instantaneous temperature that was produced increased the air pressure in the radome sufficiently to break all the radome attachment latches and to blow the radome away from the aircraft against the smaller pressure of the airstream. Because the aircraft was over land at the time, the radome was recovered and the cause of the problem could be established. A similar incident on another large aircraft resulted in the failure of five out of six latches and a hole approximately 380 mm (15 in.) in diameter being blown in the side of the radome. This was repaired, and a modification to fit lightning diverter strips was applied at the same time. Even protected radomes can suffer significant damage from an occasional severe strike, but experience has shown that radomes with lightning diverter strips suffer less frequently than unprotected ones. When damage occurs, it is much smaller than it would have been if no protection was fitted. When repairing radomes, ensure that the diverter strips are in good condition, that they are connected electrically to the boundary ring and the fuselage, and that the resistance is within limits at each point. For further information, see Ref. 5.12. See also FAR/JAR 25.581, ACJ 25.954, and ACJ 25X.899. These require lightning protection for fuel systems and consideration of electromagnetic shielding for critical electronic equipment and wiring.

5.6.5 Radar Transmissivity

Radomes perform several functions. They are fitted at the nose and sometimes the tail and, for military purposes, either above or below the fuselage, depending on whether they are acting as airborne early-warning systems, battle controllers, or submarine hunters. Radomes affect aircraft appearance and drag, and they keep the weather and airflow away from scanners. However, their most important function is to allow the radar signal to pass through with the minimum amount of loss. If they do not accomplish this goal, then their other functions could be performed less expensively.

The most common designs for radomes are:

- Honeycomb sandwich
- Fluted type, using square fiberglass tubes instead of honeycomb
- Solid laminate type

Radome design requires that the spacing of the sandwich skins in honeycomb and foam-cored radomes should be one-quarter of the radar signal wavelength and for solid radomes one-half of the signal wavelength. Significant departures from these values cause additional reflective losses, which reduce the transmission efficiency. The fitment of new windshear detecting radars requires that transmission efficiency should remain high. This means that new radomes must be made to a higher and more consistent standard than previously considered necessary and that repairs must take more account of transmission requirements. Many radome repair shops already require performance of a moisture check and a wave test before any work begins. This ensures that any previous repairs have not reduced transmission below limits and that any wet areas can be dried or cut out and repaired. Repairs on repairs usually are not permitted, and it is most uneconomical to make a repair and then discover that an area close to the repaired area is defective. This could mean removal of the repair that was done, to make a single repair of larger size, and to avoid needless overlaps that could also result in failure of the wave test after completing the repair. For an x-band radar system, the optimum sandwich thickness is 8 mm (0.31 in.). See Refs. 5.6, 5.10, and 5.13.

5.7 Mechanical Requirements

Parts of this topic have been covered in Section 5.1 of this chapter, and this should be read in conjunction with the following information.

5.7.1 Tensile and Flexural Strength

Aircraft structures must meet the design requirements of FARs and JARs and those of other local regulatory bodies if their requirements exceed them. Each part of the structure must be designed to have sufficient strength when the "1g" loads are multiplied by the factors considered necessary for safety. These are much higher for military aircraft, especially fighters. Each part must be strong enough to resist the maximum tensile, compressive, and bending (flexural) loads that are likely to be applied to it in the worst design case. Torsional loads also may be critical. For example, the fuselage must not twist excessively when maximum rudder deflection is applied, the wings must not twist too much when aileron is applied, and the tailplane must not twist under elevator loads. In extreme cases, especially with ailerons, control reversal can occur (i.e., when aileron is applied severely, the wing may twist enough to cause roll in the opposite direction). These problems were discovered during World War II when combat aircraft were pushed to their limits. Some cases occurred at the end of the war when, before returning to civilian life, a few pilots made what was, in some cases, literally their last flight as they tried to see how fast their aircraft could go. Steep dives were made, and some never recovered from them. Others recovered and reported that above a certain speed the controls operated in reverse. These effects were then evaluated by test pilots, and the stiffness of structures was studied carefully. See also Section 5.1 of this chapter.

Flutter of control surfaces and structures occurs at certain speeds. Vibration frequencies also must be checked to ensure that resonant frequencies are not reached at which structures can vibrate rapidly and fail from fatigue in a short time.

Static Loads: In addition to the various types of loading already discussed, two definitions of loading are basic to aircraft design. These are limit loads and ultimate loads.

- **Limit Loads:** Limit loads are the maximum loads anticipated on an aircraft during its service life. The aircraft structure shall be capable of supporting the limit loads without suffering any detrimental permanent deformation. For all loads up to the limit loads, the deformation shall not interfere with safe operation of the aircraft.

- **Ultimate Loads:** These are the limit load times a factor of safety. Generally, the ultimate load factor of safety is 1.5. The requirements also specify that these ultimate loads shall be carried by the structure without failure.

Both loads must be met when an aircraft structure is tested to destruction before the type is released for service. See FAR/JAR 25.571, 25.603, and ACJ 25.603 for composite materials. See also FAR/JAR 25.613 for material strength properties and design values.

Pressurized fuselages are subjected to a fatigue test in a water tank. Wings and tails are subjected to a program of fatigue loading at a range of stress levels designed to simulate real-life loading in service. Of course, these are different for civil and military aircraft and depend on the intended use. For example, short haul aircraft may fly sectors as short as half an hour and may experience many landings; long haul aircraft may fly sectors of thirteen hours, with an average of approximately seven hours.

5.7.2 Stiffness

For some parts, it may be as important to meet stiffness requirements as it is to meet strength requirements. A part may be strong enough to meet the loads applied to it; however, if it deflects so much under load that it fails to fulfill its own intended function or rubs against another part or prevents its mechanical operation, then the stiffness of that part must be increased so that these problems do not occur. This has happened with leading edge slats in some cases if the rigging is not exactly correct. If stiffness is inadequate, flutter may occur if a resonant frequency is reached due to turbulent airflow, and the part may fracture quickly because of the fatigue cycles applied. Control reversal may occur if the torsional stiffness of a wing is insufficient to meet the loads applied by aileron deflection at high speed. Stiffness is an important design criterion. When a part is repaired, it should be made with as near to the original stiffness as possible. Excessively stiff repairs may attract too much load and fail at the edge of the repair. Inadequate stiffness may cause unwanted flexibility.

5.7.3 Fatigue

Fatigue is the result of repeated loading and occurs most frequently at points of high local stress such as bolt holes and rivet holes or at any sharp changes of section, which should be avoided at the design stage as much as possible. Fatigue failure can occur in smooth test pieces without stress concentrations if the loading is high enough. It occurs more often at the locations mentioned above because design does not always estimate the stress concentration effect of holes and section changes as accurately as is required. As a result, the stresses at these points can be higher than expected. In some steels, fatigue failure of plain specimens generally occurs under cyclic loading at approximately 25% of the ultimate stress after approximately 1 million cycles. (Ref. 5.14.) This varies from metal to metal and alloy to alloy. Thus, in any design work, specific data for the alloy in question under the expected conditions of use should always be used. If a part is more highly stressed than this, it can fail after a smaller number of cycles. Note that the shape of most fatigue curves is such that a fairly small reduction in stress can give a useful increase in fatigue life. Alternatively, if a part is subject to sonic fatigue near an engine and if the loading occurs at several kilocycles per second, the part will be required to operate at even lower stress levels because of the extremely high frequency of loading. Carbon fiber composites can operate at higher stresses than metals before fatigue problems occur. Glass fiber composites suffer from fatigue effects at stresses nearer to those of metals. Phillips in Ref. 1.1 of Chapter 1 states that unidirectional Kevlar 49 composites in tension/tension fatigue at room temperature show a very good performance, superior to "S" glass and "E" glass composites and 2024-T3 aluminum. Only unidirectional carbon composites are superior. The picture is more complex for aramids in flexural fatigue. At a modest number of cycles, they exhibit poor fatigue strength, inferior to "E" glass. This is presumably a result of the poor static flexural performance of Kevlar 49.

Aramids are known to have a poor resistance to compressive loads. However, at a high level of cycles (10^6 to 10^7), the excellent inherent fatigue performance of the aramid results in Kevlar 49 and "E" glass having similar performance. This presumably occurs because at the lower stress levels, the strain will not cause buckling of the fibers in compression. These results clearly show that for each application of composites, the testing done or the test results used for estimating likely performance must be specific to the loading and usage anticipated for the product in question. It is impossible to read across from the performance of one fiber type to another without first checking that it is reasonable to do so. Konur and Matthews have written a valuable review of fatigue in composite materials (Ref. 5.15). See also FAR/JAR 25.571.

Fail-Safe Design: The two basic design concepts are safe fatigue life and fail-safe structure, as defined below.

1. **Safe Fatigue Life**: "The operational period expressed in terms of number of flying hours, number of flights, or number of applications of loads during which the possibility of fatigue failure of the part concerned under the action of the repeated loads of variable magnitude in service is estimated to be extremely remote." Safe fatigue life means that fatigue testing has been done, if considered necessary, usually to two or three times the expected life of the part or more, depending on the design requirement. If the component is replaced by a new one when the fatigue life has expired, no failure is expected. If a part is either totally inaccessible or expensive to replace in terms of both part cost and manhours, it may be necessary to test it to a life well in excess of that expected for the aircraft. Compliance with this approach means that good records must be maintained for any parts with a design life less than that of the whole aircraft. In the case of parts that can be taken from one aircraft to service another (e.g., engines, hydraulic jacks, undercarriages, wheels, or brakes), it is vital that the required data (i.e., flight hours, number of flights, braked landings, touch-and-go landings, cabin pressure cycles, supersonic time, engine cycles, and engine hours) are recorded for each period of time on each aircraft to which the part is fitted and that running totals are maintained to ensure that the part can be called for removal at the correct time.

2. **Fail-Safe Structure**: "A structure which is so designed that after the failure in operation of a part of the primary structure, there is sufficient strength and stiffness in the remaining primary structure to permit continued operation of the aeroplane for a limited period." Fail-safe means that the item can fail, be found cracked or failed on a routine inspection, and be replaced without causing a major catastrophe or severe economic consequences. This is achieved by multiple load paths—if one fails, the other

can carry the load until the problem is found. Examples are separate sections fitted back to back, such as window posts and wing skins in separate panels so that one panel can be cracked while others take the load together with the spars. When wings are used as integral fuel tanks, a crack in the bottom wing skin, which is loaded in tension and most prone to fatigue, will be quickly revealed by a fuel leak.

The fail-safe approach requires good and regular inspection. Fortunately, the Federal Aviation Administration (FAA), Civil Aviation Authority (CAA), and other regulatory authorities require that any faults found in operation must be reported to those authorities and to the aircraft manufacturer via an Incident Report. Depending on the importance of the defect, the manufacturer will issue an Alert Service Bulletin or a Service Bulletin to all operators, requiring inspection or corrective action or both within a specified and usually short period of time. If the matter is very serious, it will be marked as Mandatory. If it is not serious, then a Service Bulletin will be issued which may be optional. Defects found during maintenance, which have not been previously reported and actioned, are handled by Mandatory Occurrence Reports (MORs). These are useful to safety because defects found during routine maintenance, which could have serious consequences in flight if allowed to become worse, also must be reported to the regulatory authorities and manufacturers to allow action to be taken before an incident occurs in flight. See FAR/JAR 25.629.

5.7.4 Impact Resistance (Damage Tolerance)

Damage tolerance also is considered in design decisions. For example, brittle and hard materials may have to be used for some parts, but their resistance to crack growth is much less than that of tougher materials. More careful inspection is required, and the limits on crack length are much lower before a part must be changed or repaired. Tough materials are required for good impact resistance. The impact resistance of a composite material depends on the toughness of the resin matrix, the quality of the interface bond, and the toughness of the fibers. It is a complex problem because in one-shot applications such as body armor, the energy dissipated in disbonding a laminate having poor bonding of the resin to the fiber may actually be beneficial. Such items are replaced after one use. Aramid composites have an outstanding ability to resist impact damage and particularly ballistic impact. They show better ballistic resistance than glass-reinforced plastics (GRP) of equivalent weight and now dominate the body armor market where cost is less important than performance. The properties required for resistance to repeated impact may be different. Impact damage occurs in service from a variety of causes. In Ref. 5.16, Dorey says, "This is the most common form of damage and is likely to occur at some time in the course of the service life of a composite part. However, the location of the impact, the energy of the impact, and the extent of the damage are not so predictable." He adds, "Toughness depends on preferential splitting parallel to the fibers. The extent of this splitting depends on the properties of the fibers and on the matrix, on the interfacial bond strength, and on interactions between the plies in the laminate. Under impact loading, the damage shows similar trends with mechanical properties but also depends on the velocity, energy, and properties of the impacting projectile and in particular on the dynamic response of the composite specimen. Testing at different velocities can produce strain rate effects or changes in failure mode, and this will affect the design of standard impact tests."

Impact damage is often worse toward the back side of the laminate, which makes detection difficult. If impact damage is suspected, it is wise to examine the rear face behind the point of the suspected impact if possible. If that cannot be done, then ultrasonic methods should be used. The stiffness of the local structure affects the amount of damage. Impact work on composite structures containing stiffeners was compared with similar impact in the middle of a bay between stiffeners. In this particular set of tests, it was found that 30 to 50 joules of energy were required to cause barely visible impact damage (BVID) in the middle of a bay where a large volume of material was being deformed. By contrast, only a few joules were required near a stiffener. Residual compression strength after impact is better with toughened resin matrices. The subject of impact damage is complex, and the reference quoted should be studied with other papers if a deeper knowledge is required. See also FAR/JAR 25.571. For bird strike requirements, see FAR/JAR 25.631.

5.7.5 Creep

Creep is the result of steady or repeated application of stress in one direction at a temperature near or above the usable limits of the material. It could occur in Concorde wing and fuselage structure and on other fast aircraft in which the skin temperature is near the limit for the special alloy used. Warnings are provided to pilots to ensure that overspeeding, which could lead to temperatures resulting in creep, does not occur and, if it does occur, that action is taken quickly before damage can be caused. Turbine blades in aero and other engines can suffer from creep resulting from the centrifugal stresses developed at high rotational speeds. Turbine blades also operate at temperatures that are carefully monitored and controlled. Many have air-cooled blades to allow the use of high gas temperatures through the turbine stages. Service temperatures of composites are limited by the T_g of the resin matrix under hot/wet conditions, and composites also would creep if these limits were exceeded for any length of time. Any such effects would be caused by movement at the interface bond or in the resin rather than in the fibers when loaded in a flexural mode. In tension, the effect would be mainly due to creep in the fibers and therefore would depend heavily on the type of fiber used. This shown by Phillips (Ref 1.1 of Chapter 1), who states, "In spite of their high inherent tensile strength, and even in unidirectional configurations, aramid fiber composites have creep rates generally very much higher than for glass or carbon composites."

5.8 Attachments/Joints

The design of efficient joints, whether bonded or mechanically fastened, is a considerable challenge, and research continues to improve the design of composite joints in particular. The aim is always to achieve joints of minimum weight and maximum efficiency. The strength of joints is confirmed by test and static loading, and fatigue testing is performed to confirm that design requirements can be met for the design lifetime. In spite of this effort, problems continue to occur. In Refs. 5.12 and 5.17, Michael Niu makes the important point, "Joints are perhaps the most common source of failure in aircraft structure, and therefore it is most important that all aspects of the design are given consideration when making the structural analysis." He thoroughly covers the design of joints in metals and composites in these two volumes, and a study of them is recommended. In the course of composite repairs, two basic types of joints may have to be made. These are adhesively bonded joints and mechanically fastened bolted or riveted joints. Bonded joints are used for thinner skins; mechanically fastened joints are used for thicker ones. As a basic principle, bonded joints typically are used up to skin thicknesses of 2 to 3 mm (0.080 to 0.012 in.), and mechanically fastened joints are used for thicker skins. The type of joint specified in the SRM should always be used. The thin skins of honeycomb sandwich panels are almost invariably repaired by bonding, whereas solid laminate wing skins usually are bolted or blind riveted. They may be adhesively bonded in addition to bolting if the joint is to be permanent. There are various reasons for each type of joint, but the main one is that the scarf length or overlap for a bonded joint becomes large for thicker skins and, in the case of overlap joints, can actually be heavier than a bolted joint. Note that a scarf joint requires the removal of more material than a lap joint. Use a lap joint if permitted. A large scarf joint would require a considerable amount of accurate work and hence cost. The thicker the skin, the greater the load it carries. A bolted joint does not require the same care in surface preparation. Good bolted joints require careful hole preparation and fastener installation and the use of the correct fasteners torque loaded to the specified values. Bolted joints rely partly on the friction between the faces and mainly on the shear strength of the fasteners for strength. Avoid excessive torque on the bolts, but ensure that the specified value is reached. To avoid crushing the composite, large washers and bolt heads tend to be used with composites, and washers are usually required under the tails of rivets. Nuts and bolts must not be rotated against the surface of any composite part without a washer being placed between the surface and the nuts or bolts.

5.8.1 Bonded Joints

Chapter 9, Section 9.3.9 (scarf joints and stepped lap joints) also mention bonded joints from a practical viewpoint. The use of bonded joints gives lower stress concentrations and weight penalty. However, they cannot be easily disassembled or inspected, good surface preparation is essential, and they are sensitive to environmental conditions. Bonded joints in composite parts are almost always of the scarf type, lap type, or the stepped lap type and in bonded metal joints of the standard overlap type, except for special joints where several sheets of metal are used to allow the load transfer to be more smoothly tapered. Joint design is very important and is discussed below.

- **Joint Design:** This is defined by Kinloch in Ref. 5.18 as, "The geometric features of the joint and the way in which the applied loads are transmitted from one substrate to the other."

- **Joint Efficiency:** This is defined in Ref. 5.18 as, "The fracture load of the joint divided by the load necessary to fracture the weakest of the substrates."

A vast amount of work has been done on this subject, and it is not yet fully understood. The problem has been approached in several ways. In 1938, Von Olaf Volkersen (Ref. 5.19) did some work on riveted joints and realized that the first rivet took most of the load. He performed a type of shear lag analysis (similar to that used for riveted joints) and developed a stress concentration factor

$$Sf = \sqrt{\frac{2c^2 . Ga}{E . t . d}}$$

where   Ga = Shear modulus of the adhesive (N/mm^2)
   c = Half the overlap length of the joint (mm)
   E = Young's modulus of the adherend (N/mm^2)
   t = Adherend thickness (mm)
   d = Adhesive layer thickness (mm)

The apparent shear strength of the joint is failing load/area of the bond, which is proportional to 1/Sf. To increase the shear strength, it is necessary to reduce the shear stress concentration factor Sf because it can be seen from the equation that Sf increases at the same rate as the overlap length. A problem with this equation is that it does not consider the bending effects, due to offset loading, which add tensile stresses to the shear stresses at the ends of the joint. However, if we ignore the bending stresses due to offset loading and consider only the shear stresses using the Volkersen equation, several ways of reducing the stress concentration factor are possible. Not all of these will be possible in every design case.

The four options are as follows:

1. **Increase the Adherend Modulus:** This can be done only by using a stiffer material. The normal choice of metals is aluminum alloys of a number of compositions, titanium alloys of which there are a very small number, or steels. Stainless or corrosion-resistant steels are normally used for aircraft applications. In most industrial applications, the choice is likely to be mild steel. For metals, the order of stiffness is steel, titanium, and aluminum. For composites, the order of stiffness of the more commonly used fibers is carbon, aramid, "S" glass, and "E" glass. For most repair work, a change of modulus is not an option because the repair will have to be made using the original fiber type or the original metal alloy.

2. **Increase the Adherend Thickness:** This also increases the eccentricity of loading, not accounted for by this equation. Likewise, it increases the weight and cost of the part concerned. Excessively thick repair plates or composite layers also will make the repair area too stiff; therefore, this is not an option.

3. **Increase the Adhesive Layer Thickness:** This increases the probability of flaws in the adhesive layer, and only small increases are practical.

4. **Decrease the Adhesive Modulus:** This can be done by using toughened adhesives, usually epoxies or acrylics. Note that, as a general rule, the reduction of Ga, the shear modulus of the adhesive, is beneficial in reducing the stress concentration in any loading system. Reducing the shear modulus will usually reduce the adhesive strength; thus, there are limits to the value of doing this. A chamfer and a fillet of glue also reduce the stress concentration at the ends of a lap joint. See Figure 5.9.

Chamfers can be made on either face and, if on the outside, must be almost razor thin to be effective. The choice of adhesive is important, and the requirements of the SRM should be followed. If an alternative adhesive is used, it must have very closely similar properties to the original and must be authorized by an approved design authority. Changing the adhesive properties is not an option for repairs. Volkersen's work was progressed by Goland and Reissner (Ref. 5.20), who included bending effects. Since then, Adams *et al.* (Refs. 5.21 to 5.24), Crocombe (Ref. 5.25), and many others have used finite element analysis to further refine the calculation of the stress pattern across a lap joint. See Figure 5.10.

Each of these studies has assumed that stress is the most important factor. Some tests show that this may be true for brittle adhesives but that fracture toughness may be more important for tough adhesives. In this case, the speed of failure depends on the propagation rate of a crack. If this is very slow, then failure will occur slowly by a peeling mode if the stress level is maintained and failure will occur when the total bond area is no longer sufficient. Again, the strength of the joint remains constant after the peeling starts; therefore, the stress at the point of slow fracture is the important factor. If the speed of crack propagation is fast, then failure is instantaneous; this has been demonstrated in laboratory experiments.

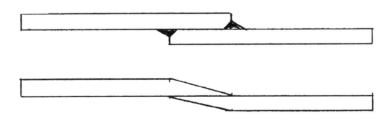

Fig. 5.9 Chamfers and fillets at the ends of lap joints.

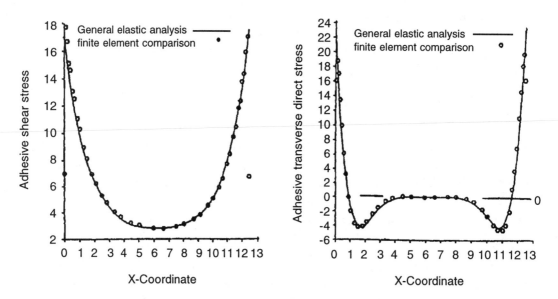

Fig. 5.10 Stress distributions across a lap joint from the general elastic analysis and the finite element comparison. (Courtesy of Dr. A.D. Crocombe of the University of Surrey, United Kingdom.)

These experiments also showed that the failing load in a typical aluminum alloy lap joint was similar, regardless of the overlap length. Also, even with a large overlap, the failure of a joint made with a brittle adhesive was instantaneous, whereas the failure of a similar joint with a tough adhesive was by a peeling mode until the bond area was too small to sustain the load. It is a complex problem, and the behavior of adhesives with different levels of toughness varies considerably. The work of Armstrong (Ref. 5.26) shows that fracture energy, fracture toughness, and elongation to failure can be related to lap joint strength more closely than adhesive tensile strength or modulus. When selecting alternative adhesives and resins, ensure that all properties are closely similar.

What is certain and offered as sound advice by many is as follows:

1. **Adhesives are much weaker than the parts they join.** This is true for metals and most other types of homogeneous materials. It is also true for composites in the fiber direction but not necessarily in the transverse direction or for any properties of the composite which are resin dominated. When composite parts must be joined, tough relatively low modulus adhesives are recommended to spread the load over the surface. Whatever the substrate, adhesives must be used in a mode that makes the best use of their mechanical properties (i.e., they must be used in compression, shear, or tension). Cleavage and peeling forces should be avoided as much as possible. Where some cleavage and peel forces are unavoidable, only toughened adhesives should be used. For most repair work, only shear joints will be used and thus the only option is careful design of this type of joint.

2. **Joint design must make use of the adhesive properties to their best advantage.** If significant loads are involved, it is important to choose strong, tough adhesives and to bond as large an area as is reasonably practical. Good design should make every effort to achieve the following:

    *   Maximize the use of compression, shear, or tension modes of loading
    *   Minimize peel and cleavage forces
    *   Maximize the area over which the load is distributed, whatever the mode of loading

Joint design must avoid high local stresses as much as possible, and low average stresses should be ensured by the design. It is common to load an adhesively bonded lap joint to no more than 15% of the failing load in a lap shear test at the maximum temperature of usage. Laboratory experiments have clearly shown that increasing the overlap length in a lap joint does not significantly increase the actual failing load. Increasing the width of a lap joint increases its strength in direct proportion to the width. However, when joints are used at high temperature (creep loading) or under conditions of fatigue loading, it has been found to be beneficial to use overlaps of 50 to 70 times the thickness of the thinner adherend. For scarf joints, a taper of 1 in 50 with additional layers over each end is recommended. This also is helpful to long-term durability because the longer the overlap, the longer it takes for failure to occur by disbond or corrosion. This means a longer time period before failure, during which problems can be found by inspection and suitable repairs conducted. Typical good and bad joint designs may be found in Refs. 5.22, 5.27, and 5.28. For aircraft and other structural joints, the joint design given in the SRM or equivalent document should always be used. As a general guide for composites, the full skin thickness should have an overlap of 50 times the skin thickness, and the tapered steps should go beyond this. Further information on this subject is available in Refs. 5.29 to 5.32 and in Refs. 5.12 and 5.17.

5.8.2 Mechanically Fastened/Riveted Joints

There are standard rules for the design of these joints in design handbooks and textbooks for metals, and a large amount of work has been done to produce similar information for composites. One basic principle is that rivets should never be used to carry any significant tension load. Such loads should always be carried by bolts.

Mechanically fastened joints in composites should be made in skins with approximately quasi-isotropic lay-ups, and the grouping of similarly oriented plies in the stacking sequence should be avoided. Larger edge distances are required than are normally used with metals, and it is typical to employ larger fasteners at a greater pitch rather than smaller fasteners at a closer pitch as is common with metal joints. Fasteners must be made from suitable materials (e.g., titanium), or corrosion-resistant steel must be used with carbon fibers to avoid corrosion of the bolts. Mechanically fastened joints do not need special surface preparation, are easy to inspect, and can be disassembled without much damage if care is taken. However, they have high stress concentrations at the holes and can be heavy compared to bonded joints. This subject is covered in detail in Refs. 5.12, 5.17, and 5.33 to 5.35. For practical details, see Chapters 10 and 11. For repair work, use the design information provided in the SRM or equivalent document. The joints shown in these manuals have been evaluated by test by the aircraft or component manufacturer and should always be used. These designs also should be used by repair designers, and calculations should be made to ensure that temporary repairs and especially permanent repairs are adequate. The manufacturer should always be consulted if any doubt exists as to the suitability of a mechanically fastened joint. In most cases, a large quantity of useful information is supplied in the SRM. For example, the Boeing 757 SRM 51-40-00 through 51-41-09 is particularly thorough and gives details on all fasteners used for the airframe and engines. Other aspects of bolted and riveted joints covered in the SRM are the following:

- Hole finish
- Hole accuracy
- Oversize holes
- Interference fits
- Cold working of holes in metals
- Countersinking requirements
- Strength of various types of fasteners
- Fastener edge margins
- Details of various types of fasteners
- Torque values to be used when tightening bolts

All of these details indicate that the use of fasteners is another complex subject. It is important that all of these factors are considered and that specific requirements are checked before a repair is done.

5.9 Other Design Requirements

5.9.1 Aerodynamic Smoothness

This is important to minimize drag and to maintain the original profile, especially of wing and tailplane leading edges and engine intakes. The SRM has a section that details the requirements for each area of an aircraft, and it should always be consulted when doing a repair to ensure that these needs are met.

It is also necessary to consider this if a repair is made to an area of the fuselage slightly forward of the pitot static points for the measurement of airspeed. At takeoff and landing, constant reference to airspeed must be maintained to avoid the risk of stalling and to maintain adequate speed for good control response. Measurement of airspeed must be accurate, and great care must be taken to ensure that repairs do not affect the required accuracy.

5.9.2 Weight and Balance

This is particularly important on smaller aircraft where the control surfaces are manually operated. Control surfaces such as ailerons, elevators, and rudders have a total weight and a balance requirement. This is especially true of trim tabs, which are lightweight structures. Sometimes a repair may take the component outside one or both of these limits. If it does, then the repair must either be repeated in a way to keep the part within limits or the part must be scrapped. If too much weight is added near the trailing edge, then the amount of extra balance weight required to be added forward of the hinge line to restore balance may take the weight of the part above the total weight limit. Sometimes there are physical clearance limits to the number of weights that can be added. Paint alone may put a part out of balance, and previous coats of paint may have to be removed to remain within limits. Weight and balance checks must be done after painting. However, it is good practice to check weight and balance before painting so that the old paint can be removed if meeting the weight or balance requirements seems marginal. Always keep as much weight in hand as possible because the same part may require repair again in the future. Also, remember that even on large aircraft having fully powered control systems, it is often necessary to meet weight and balance requirements on control surfaces. The reason for this is that these large ailerons, elevators, or rudders are operated by a single hydraulic jack. Although this is designed as a safe-life item and has been subjected to fatigue testing, the possibility of a rare and unlikely failure must be considered. If such a failure occurs and the part is out of limits for weight and balance, a severe flutter condition is possible. This could cause serious control disturbances or even fin or tailplane failure. This unfortunate result would be a high price to pay for one or more excessive coats of paint. Always check the weight and balance requirements before starting a repair.

5.10 References

5.1   Datoo Mahmood, H., *Mechanics of Fibrous Composites,* Elsevier Applied Science, London, UK, ISBN 1-85166-600-1, 1991.

5.2   Ashbee, K. *Fundamental Principles of Fiber Reinforced Composites,* Technomic Publishing Co. Inc., Lancaster, PA, 1989.

5.3   Hull, Derek, *An Introduction to Composite Materials,* Cambridge University Press, Cambridge, UK, ISBN 0-521-23991-5 (hardbound) and ISBN 0-521-28392-2 (paperback), 1985.

5.4   Matthews, F.L. and Rawlings, R.D., *Composite Materials—Engineering and Science*, Chapman and Hall, London, UK, ISBN 0-412-55960-9 (hardbound) and ISBN 0-412-55970-6 (paperback), 1994.

5.5   Agarwal, B.D. and Broutman, L.J., *Analysis and Performance of Fiber Composites*, John Wiley and Sons, New York, 1980.

5.6   Walton, J.R., Jr. [ed.], *Radome Engineering Handbook—Design and Principles*, Marcel Dekker, New York, 1970.

5.7   Jastrzebski, Zbigniew D., *The Nature and Properties of Engineering Materials*, 2nd edition, John Wiley and Sons, New York, 1976.

5.8   Hess, Fred C., [advisory ed.] and Holden, John B., *Chemistry Made Simple*, W.H. Allen and Co., London, UK, 1967.

5.9   Middleton, D.H. [ed.], *Composite Materials in Aircraft Structures*, Longman Scientific and Technical, ISBN 0-582-01712-2, 1990.

5.10   "Minimum Operational Performance Standards for Nose-Mounted Radomes," RTCA-DO-213, RTCA Inc., 1140 Connecticut Ave. N.W., Suite 1020, Washington, DC 20036.

5.11   *Guidance Material for the Design, Maintenance, Inspection, and Repair of Thermosetting Epoxy Matrix Composite Aircraft Structures,* 1st edition, International Air Transport Association, 2000 Peel St., Montreal, Quebec, Canada H3A-2R4, DOC:GEN:3043, May 1991.

5.12   Niu, Michael C.Y., *Composite Airframe Structures—Practical Design Information and Data*, Conmilit Press Ltd., P.O. Box 38251, Hing Fat St. Post Office, Hong Kong, ISBN 962-7128-06-6, 1992.

5.13   "Maintenance of Weather Radar Radomes," Federal Aviation Administration (FAA) Advisory Circular 43-14.

5.14   Forrest, P.G., *Fatigue of Metals*, Pergamon Press, 1962.

5.15   Konur, O. and Matthews, F.L., "The Effect of the Properties of the Constituents on the Fatigue Performance of Composites—A Review," *Composites,* Vol. 20, No. 4, July 1989, pp. 317–328.

5.16   Dorey, G., "Impact Damage in Composites—Development, Consequences, and Prevention," Proceedings of 6th International Conference on Composite Materials, ECCM 2, London, UK, Matthews, F.L. [ed.], 1987.

5.17   Niu, Michael C.Y., *Composite Airframe Structures, Practical Design Information and Data,* Conmilit Press Ltd., P.O. Box 38251, Hing Fat St. Post Office, Hong Kong; published in the United States by Technical Book Co., 2056 Westwood Blvd., Los Angeles, CA 90025, ISBN 962-7128-06-6, 1991.

5.18   Kinloch, A.J., *Adhesion and Adhesives, Science and Technology,* 2nd edition, Chapman and Hall, London, UK, ISBN 0-412-27440-X, 1990.

5.19   Volkersen, Von Olaf, *Die Nietkraftverteilung in Zugbeanspruchten Nietverbindungen Mit Konstanten Laschenquerschnitten*, Luftfahrtforschung, 1938.

5.20   Goland, M. and Reisner, E., "The Stresses in Cemented Joints," *Journal of Applied Mechanics*, March 1994, pp. A.17–A27.

5.21   Adams, R.D. and Peppiatt, N.A., "Stress Analysis of Adhesive Bonded Lap Joints," *Journal of Strain Analysis*, Vol. 9, No. 3, 1974, p. 185.

5.22   Adams, R.D. and Wake, W.C., *Structural Adhesive Joints in Engineering,* Elsevier Applied Science, London, UK, 1986.

5.23   Adams, R.D., "The Mechanics of Bonded Joints," Paper C180/86, International Conference, Structural Adhesives in Engineering, University of Bristol, June 1986, in *Proceedings of the Institution of Mechanical Engineers*, 1986, pp. 17–24.

5.24   Adams, R.D. and Harris, J.A., "The Influence of Local Geometry on the Strength of Adhesive Joints," *International Journal of Adhesion and Adhesives*, Vol. 7, No. 2, April 1987, pp. 69–80.

5.25   Bigwood, D.A. and Crocombe, A.D., "Bonded Joint Design Analyses," Chapter 11 in *Adhesion 13*, Allen, K.W. [ed.], Elsevier Applied Science, London, UK, ISBN-1-85166-331-2, 1989, pp. 163–187.

5.26   Armstrong, K.B., "The Selection of Adhesives and Composite Matrix Resins for Aircraft Repairs," Ph.D thesis, The City University, London, 1990.

5.27   Packham, D.E. [ed.], *Handbook of Adhesion,* Polymer Science and Technology Series, Longman Scientific and Technical, Harlow, UK, ISBN 0-470-21870-3, 1992.

5.28   Hartshorn, S.R. [ed.], *Structural Adhesives, Chemistry and Technology,* Plenum Press, New York, ISBN 0-306-42121-6, 1986.

5.29   Hart-Smith, L.J., "Adhesive-Bonded Double-Lap Joints," Report CR-112235, NASA, Washington, DC, 1973.

5.30   Hart-Smith, L.J., "Adhesive-Bonded Scarf and Stepped Lap Joints," Report CR-112237, NASA, Washington, DC, 1973.

5.31   Hart-Smith, L.J., "Adhesive-Bonded Single Lap Joints," Report CR-112236, NASA, Washington, DC, 1973.

5.32   Hart-Smith, L.J., "Design of Adhesively Bonded Joints," Chapter 7 in *Joining in Fiber Reinforced Plastics*, Matthews, F.L. [ed.], Elsevier Applied Science, London, UK, 1986, pp. 271-311.

5.33   Hart-Smith, L.J., "Design and Empirical Analysis of Bolted or Riveted Joints," Chapter 6 in *Joining in Fiber Reinforced Plastics*, Matthews, F.L. [ed.], Elsevier Applied Science, London, UK, 1986, pp. 227–270.

5.34   Matthews, F.L., "Theoretical Stress Analysis of Mechanically Fastened Joints," Chapter 3 in *Joining in Fiber Reinforced Plastics*, Matthews, F.L. [ed.], Elsevier Applied Science, London, UK, 1986, pp. 65–103.

5.35   Phillips, L.N. [ed.], *Design with Advanced Composite Materials,* copublished by The Design Council, London, UK, and Springer-Verlag, Berlin, ISBN 0-85072-238-1, 1989.

Chapter 6

Safety and Environment

6.1 Introduction

Safety and environmental issues are becoming matters of increasing concern because mankind has at last become aware of the toxicity of a wide range of materials in common use. (See Ref. 6.1.) Governments worldwide are enacting vast amounts of legislation to try to deal with the situation, and anyone concerned with composites and adhesives should be aware of safety precautions in handling these materials and also of legislation governing the disposal of toxic waste. Local legislation is not the same everywhere; therefore, it is important to know the local regulations and to abide by them or to do better if that is not too difficult. We all must live with the results of what is done; thus, it is sensible to work to a higher standard than required if possible. For example, it does not require much effort to cure adhesives and pre-pregs before disposal, even if such steps are not required by law. In the United States, much of this is covered by the Occupational Safety and Health Administration (OSHA). In the United Kingdom, it is covered by the Health and Safety at Work Act 1974; the Classification, Packaging, and Labeling of Dangerous Substances Regulations 1984; and the Control of Substances Hazardous to Health (COSHH) Regulations 1988. (See Ref. 6.2.) This may be used as a model for European legislation. The COSHH regulations require five basic actions:

1. Assessment of health risks in all work activities

2. Control of exposure to various risks within specified limits

3. Monitoring of exposure levels to ensure that limits are observed (e.g., atmospheric sampling)

4. Health surveillance, which may be required for those working with carcinogens, toxic materials, or radiation

5. Instruction, training, and information

Records must be kept current and may be inspected by staff of the health and safety executive. Stocks of adhesives and similar materials must be stored under suitable conditions. Working containers, to which bulk stocks may have been transferred, must be labeled with the product name, hazard, and warning labels as given on the original package. Waste materials and empty containers must be disposed of in accordance with regulations. Many countries have similar legislation and control authorities, and all users of hazardous materials must be aware of their own local rules. Most waste is either burned or sent to rubbish tips. If toxic material is disposed of incorrectly, then dangerous chemicals can leach out and find their way into groundwater, then into rivers, and finally into drinking water. British Standard BS 7750 Environmental Management Systems provides a means of establishing a documented system for delivering an environmental policy. The European Commission Eco Management and Audit Regulation requires companies to establish an environmental policy and to produce a public statement of their performance compared to their objectives.

Some major concerns about pollution are as follows:

*   Contamination of air, soil, and water by toxic or harmful substances.

*   Damage to the ozone layer in the upper atmosphere. The problem materials here are chlorofluorocarbons (CFCs) and chlorinated solvents. Damage to the ozone layer allows a greater amount of ultraviolet (UV) radiation from the sun to penetrate to ground level. This has already resulted in an increase in skin cancers, especially for those who like to obtain a good tan. Exposure can be reduced by spending less time in the sun, by using more UV-resistant body creams, or both.

*   The greenhouse effect is a result of increased upper atmosphere levels of gases such as carbon dioxide from combustion of fuels and methane from biological sources (mainly human and animal sewage). This greenhouse effect restricts the re-emission of solar heat and leads to global warming, suggesting that a greater effort to control world population is needed to reduce emissions from both sources.

- Interactions of gases in the atmosphere, which result in chemical smogs and acid rain. Solvents, reactive monomers, and curing agents may contribute to smogs and chlorinated solvents to acid rain. Solvents in cleaners, primers, and adhesives or resins contribute to the formation of ozone and smog because they contain volatile organic compounds (VOCs), which can participate in complex chemical reactions in which light energy converts atmospheric oxygen to ozone. It is a case of keeping the right materials in the right places. Ozone is needed in the upper atmosphere to provide protection from the shorter wavelengths of UV; however, near ground level, ozone is harmful to biological processes even at concentrations as low as 0.1 ppm. For this reason, considerable efforts are being made to reduce VOC emissions. Refer to Ref. 6.3 and Figure 6.1.

There are two major approaches to the problem:

1. Find safer materials that can be used with no special precautions. Significant research and effort is going into this approach, but time will be needed to replace all the products in use. Often, a penalty in performance or price is paid if the safer material is used. In the long term, we hope to find that the overall benefit will favor the safer materials.

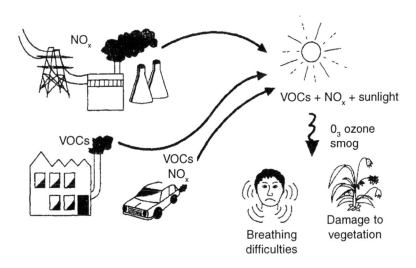

Fig. 6.1 Solvents (VOCs) and smog. (Courtesy of Materials World, *Journal of the Institute of Materials, London, United Kingdom.)*

2. Find safer methods of using the materials we have that are dangerous but useful. Meanwhile, we must ensure that all personnel working with hazardous materials are aware of the dangers and the safety precautions needed to overcome them. Safety equipment and training must be provided in these cases.

A helpful document has been produced by the Suppliers of Advanced Composite Materials Association (SACMA). (See Ref. 6.4.) This booklet contains an extensive bibliography, which may be useful to company safety officers and medical officers. It also produces a booklet on the prevention of dermatitis. (See Ref. 6.5.) Ciba-Geigy has issued a Safety Manual (Ref. 6.6), and many others probably can be found.

Concerns exist about the pollution of air, land, and water. From an individual viewpoint, toxins can enter the body by inhalation (dust, fumes, and vapors), absorption through the skin (skin contact because gloves or protective clothing were damaged or not used), or by ingestion. Ingestion (eating or inadvertently swallowing) toxic materials can usually be avoided, but any contamination on unwashed hands or gloves can easily be transferred to food. Safety equipment and discipline are required to prevent breathing of dust and vapors and skin contact. Overalls should be laundered regularly and frequently to avoid any liquid resin, curing agent, or other contaminant passing through the clothing and contacting the skin. Eye protection often is necessary. Air contamination comes from chemical dusts, chemical gases, and any reaction in which these products are involved while in the air. There is concern about CFCs which contribute to the production of smog by photochemical reactions and are also blamed for depletion of the ozone layer.

Contamination of land and water can occur by careless disposal of toxic waste. Likewise, near power stations and other chemical operations, contamination can occur by particulate fallout from chimney smoke on the land surrounding the factory. It can even be found in neighboring countries when the fallout occurs downwind over large distances.

Several concerns receiving considerable attention are as follows:

- Organic solvents (especially the chlorinated type)

- Hexavalent-chromium-containing compounds

- Asbestos

- Heavy metal compounds

- A number of chemical compounds of which free 4,4-methylene dianiline containing substances appear to be of the greatest concern

Hazards associated with adhesives are shown in Table 6.1 (from Ref. 6.7).

Table 6.1
Hazards Associated with Adhesives*

Adhesive Type	Causes Burns	Flammability	Explosion	Harmful Vapors	Harmful to Skin/Eyes
Solvent Borne		x	x	x	x
Water Borne				(x)	x
Hot Melts	x			(x)	x
Powder Forms			x		
Plastisols, Curing Rubbers	x			x	
Chemically Reactive Types					
Epoxides		x		(x)	x
Polyesters**					x
Formaldehyde Resins (Phenolics)		x		x	x
Polyurethanes		x		x	x
Acrylics				x	x
Cyanoacrylates				x	x

x = hazard; (x) = possible hazard

* Courtesy of *Materials World*, Journal of the Institute of Materials, London, United Kingdom.
** Note that polyesters can also be explosive if the accelerator and curing agent are inadvertently mixed together instead of into the resin as specified in the instructions.

6.2 Workshop Conditions for Good Bonding

For the production of good bonds with either composite or metal components, the required conditions must be provided in the workplace where the critical aspects of the work are to be done. Although much of the preparatory work can be done in less-than-ideal conditions, final surface preparation and lay-up must be done in a clean room or at least under the conditions specified as follows:

- Air temperature should be 18 to 30°C (64 to 86°F).

- Repair surface temperature should be 18 to 30°C (64 to 86°F).

- Relative humidity should be 65% absolute maximum. (See Figure 6.2.) Room temperature and humidity should be maintained within the cross-hatched area. It is not difficult in some countries to achieve 35% relative humidity, which is the ideal. Humidity values lower than this do not significantly improve bonding and are too dry for operator comfort. A good target figure is 35%.

- Dust-free atmosphere is necessary. Provide a positive pressure of clean, filtered air in the clean room or work area to keep out dust and other contaminants. No drilling, grinding, or trimming operations are permitted in clean rooms.

- No smoking should be permitted in clean rooms or bonding areas.

- No exhaust or oily fumes should be permitted. Only electrically powered vehicles should be used.

- No aerosol containers of any type should be permitted.

- No silicone release agents are permitted in any bonding shop, including WD-40 and any other dewatering fluids. Release agents of any type should be applied to tooling outside the bonding area and should be allowed to dry before being brought into a lay-up area.

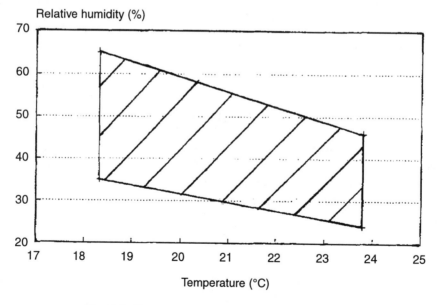

Fig. 6.2 Clean-room temperature and humidity control.
(Courtesy of the Federal Aviation Administration.)

- Clean, white cotton gloves must be worn at all times after surface preparation and prior to bonding.

- Handle parts by their edges. Do not touch at any time the surfaces to be bonded. See Figure 15.5 in Chapter 15.

- Resins should not be used if excessive humidity exists (>85%) because water will condense on the resin and affect the cure and subsequent properties. Regardless of the prevailing relative humidity, containers should always be closed immediately after removal of the required quantity of resin or hardener. This is especially important for the curing agent, which absorbs atmospheric moisture more quickly than the base resin. From an environmental viewpoint, it also reduces the amount of unwanted chemical vapor in the workshop air.

6.3 Respirable Fibers and Dust

6.3.1 Fiberglass

The following is taken from Ref. 6.4 and is slightly abbreviated and modified in parts.

Fiberglass is one member of a family of products known collectively as man-made vitreous fibers because of their synthetic, amorphous glassy nature. Glass fiber will not burn, rot, or absorb moisture or odors. Fiberglass is generally supplied in two forms: wool-type fibers and textile (continuous filament) fibers. Textile glass fibers, the type used in composite reinforcement, differ from the wool type because they are die-drawn rather than spun. Most fiberglass for reinforcement is greater than 6 μ in diameter. This size of fiber diameter does not reach the deep lung areas (nonrespirable fiber). Glass fibers break only into shorter fragments with the same diameter. Textile glass fibers destined for reinforcement applications are coated with a polyvinyl acetate-chrome chloride, polyvinyl acetate-silane, polyester-silane, or epoxy-silane finish to suit the resin system to be employed in the final product. Exposure to glass fibers may cause mechanical irritation of the eyes, nose, and throat. Contact with uncured resins and dry but not cured epoxy compatible sizing could cause skin sensitization in some people.

6.3.2 Carbon and Graphite Fibers

A substance-specific threshold limit value, time-weighted average value (TLV-TWA) for synthetic forms of graphite has been established at 10 mg/m^3 (total dust containing no asbestos and less than 1% free silica); natural graphite is to be controlled to 2.5 mg/m^3

(respirable dust). There are proposals to extend this to all forms of graphite, both natural and synthetic. One manufacturer uses 5 carbon fibers/cm^3 of air as a company limit. As a minimum, the OSHA nuisance dust standard of 10 mg/m^3 total dust, 5 mg/m^3 respirable fraction, should not be exceeded. The principal health hazard associated with the handling of carbon and graphite fibers is mechanical abrasion and irritation. These hazards are a direct result of the physico-mechanical properties of the fibers rather than a toxico-chemical reaction. Skin rash (occupational dermatitis) is common in workers handling these types of materials without proper skin protection, and inflammatory eruptions and drying of the skin have been noted. Allergic or sensitizing reactions are not reported when handling unsized fibers. However, sized fibers or pre-pregs prepared from these carbon or graphite fibers will carry with them the toxicological properties of the resin or sizing material. For full details of tests, see Ref. 6.4.

6.3.3 Aramid Fibers

These are also known as polyparaphenylene diamine terephthalate (PPD-T) and sometimes as polyparaphenylene terephthalamide, or simply as para-aramids. Citing Ref. 6.4 again, the following information is given. An acceptable exposure limit of 2 respirable fibrils/cm^3, 8-hour TWA has been established by Du Pont, the U.S. manufacturer of these fibers. Aramid fiber is not a sensitizer and shows only minimal potential for irritation of human skin. PPD-T dusts show no effects in animal lungs beyond those normally shown with benign dusts. Although too large in diameter to be inhaled, aramid fibers can produce respirable-size fibrils (subfibers) that are peeled from the surface when the fiber is abraded. The complex shape and electrostatic charge that are inherent characteristics of aramid fibrils make it difficult to produce significant airborne concentrations. Nevertheless, machining/sanding/abrading operations that can generate airborne fibrils should be monitored to ensure that the Du Pont acceptable exposure limit of 2 fibrils/cm^3, 8-hour TWA is not exceeded.

6.3.4 Sanding Dust

This can be produced from any of the fibers/resins/fillers/sealants used in composite component manufacture. In general, all dust should be avoided, whether it is physically or chemically harmful or given only the rather general description of nuisance dust. The following TLVs are given for different types of dust.

Fiberglass dust has a TLV-TWA of 10 mg/m^3 at the present time. For carbon and graphite fiber dust, synthetic graphite has a permissible exposure limit (PEL)-TWA of 10 mg/m^3 (total dust), 5 mg/m^3 (respirable fraction); natural graphite is to be controlled to 2.5 mg/m^3 (respirable dust). Aramid fibers have a recommended acceptable exposure limit of 2 fibrils/cm^3, 8-hour TWA.

A general dust description of "particulates not otherwise regulated" carries a TLV-TWA of 15 mg/m^3 (total dust) and 5 mg/m^3 (respirable fraction). The National Institute of Occupational Safety and Health (NIOSH) has urged the adoption of a 3 fibers/cm^3 standard for fiberglass. The U.S. Navy has the same limit for carbon fibers. This is somewhat tighter than the recommended figure for aramids; therefore, a common standard for all three fibers is possible in any workplace. The handling of powder epoxy materials, fillers (i.e., microballoons and silica), glass, carbon, or aramid fibers necessitates taking precautions against the inhalation of dust. Attention must be given both to the establishment of work procedures that minimize the raising of dust from these materials and then to the maintenance of efficient local and general ventilation. This also is needed to remove fumes and vapors. Thus, it can simultaneously serve two useful purposes. (See Health and Safety Executive [HSE] Guidance Note EH 22—Ventilation of the Workplace, available from HMSO, P.O. Box 276, London SW8 5DT, England.) When sanding operations are performed, equipment fitted with dust extraction should be used. Masks also should be worn if doubt exists about the efficiency of extraction. Personnel not wearing masks should be prohibited from a sanding area. It is preferable to perform sanding in a separate room with suitable protective equipment.

6.4 Fumes and Vapors

6.4.1 Resin Fumes

Two-part paste adhesives, matrix resins, potting compounds, and sealants emit toxic vapors, and mixing should occur in a fume cupboard with its own air extraction system. All uncured resins, catalysts, promoters, and hardeners (or curing agents) should be considered and handled as poisons. The actual degree of danger differs for each individual material; however, all are toxic to a significant degree. Most of these materials emit harmful vapors or solvents that can be readily absorbed through the skin. Several are mucous membrane and eye irritants. (See Ref. 6.4.) Note that high temperatures that are generated locally during machining and especially during grinding of composites may also cause toxic fumes resulting from decomposition of the resin or fiber. It is important to have an extraction system to remove vapors and dust. Film adhesives and pre-pregs emit solvent and other vapors during hot curing, and ovens should also be vented to a point outside the workshop by means of powered air extraction systems if recommended by the material supplier. Guidance on setting up local exhaust ventilation (LEV) systems is given in the U.K. HSE Series booklet HS(G) 37 (available from HMSO, P.O. Box 276, London, SW8 5DT, England). Styrene vapor emission is a problem when working with polyester resins, and national and international limits have been set. (See Ref. 6.8 for details or local standards.) Local legislation is changing today as lower emission levels are being specified. It is essential to comply with the latest legislation. Where maximum allowable concentrations are set, some countries also make it an obligation to reduce the levels as far as practical below this maximum. Styrene vapors are heavier than air; therefore, extraction systems must be either very close to the vapor source or near floor level to ensure that vapor levels remain within the required limits. See Ref. 6.9 for a number of designs for vapor extraction systems. Some of these are for special cases, such as boat hulls where vapors can accumulate in the bottom if the system is not correctly arranged.

6.4.2 Solvent Vapors

Cleaning solvents also emit vapors, and breathing of these should be avoided as much as possible. These should be kept in cans with suitable lids, and only the minimum amount necessary for the task should be poured into a beaker. Open containers of solvent should not be permitted to evaporate into the workshop atmosphere. Solvents should be used only in well-ventilated areas or in a fume cupboard. Wear a mask if specified by the appropriate Material Safety Data Sheet (MSDS). In some cases, this may require a full face mask with organic filter. Many solvents are flammable, and precautions against fire should also be taken. No operations causing flames or sparks are permitted near solvents. Use only the solvents specified in the Structural Repair Manual (SRM), and use the safety precautions required by the MSDS. There are plans to phase out chlorine-containing solvents except some not considered reactive in the definition of volatile organic content (VOC) and to reduce the VOC of as many solvents as possible. See Chapter 10 for an example. (Note: The term VOC will be heard many times in the next few years. As previously stated, VOC also denotes volatile organic compounds.) Considerable work is being done to develop water-based primers, adhesives, and coatings. Some success has been achieved, but there is a long way to go. Ketone solvents such as acetone, methyl ethyl ketone (MEK), and methyl isobutyl ketone (MIBK) have a low order of acute toxicity. However, they are mild-to-moderate skin irritants and moderate-to-severe eye irritants. Overexposure to these solvents may cause central nervous system (CNS) depression. Their odor threshold is low; therefore, ample warning is given. Most ketones are highly flammable and have flash points below room temperature. Airborne concentrations must be controlled primarily to avoid fire and explosion hazards. Check local regulations for TLV and PEL.

Chlorinated Solvents: In addition to CNS depression caused by many organic solvents, exposure to high concentrations of chlorinated hydrocarbons has been reported to cause cardiovascular effects (sensitization of cardiac muscle); therefore, the use of epiniphrene (adrenaline) is not permitted. Situations with high stress or excitement should also be avoided by persons exposed to these solvents because of the release of endogenous adrenaline under such conditions. Generically, chronic animal exposures to chlorinated hydrocarbon solvents have also caused liver and kidney changes. It is important to ensure that TLVs are not exceeded when working with solvents.

6.4.3 Sealant Vapors

Many sealants also emit solvent or other vapors and should be treated in the same way as solvents. Likewise, mixing in a fume cupboard is recommended.

In the past, sealants have used and, in many cases continue to use, dangerous curing agents. Care is necessary when mixing and during application. In this respect, sealants should be treated in the same way as adhesives. Efforts have been made to replace those sealants that cure with lead-containing compounds and chromate-containing compounds. Until they are finally phased out, care in using these sealants will be necessary. It is essential that the data sheet and MSDS for each sealant should be carefully read and the instructions followed. In the same way as two-part epoxies, two-part sealants also must be weighed accurately, mixed in the correct ratio, and mixed thoroughly to achieve their intended performance.

6.4.4 Coatings

Protective coatings and paints also contain solvents and should be treated similarly.

6.4.5 Fuel

Special breathing apparatus must be worn when working inside fuel tanks, and all local regulations for this task must be strictly observed. See also Section 6.11 on safety requirements for electrical appliances. Note that safety precautions for kerosene gas turbine fuel (AVTUR) are different from those for petrol for piston engines (AVGAS), which is more flammable. Ensure that appropriate precautions are observed.

6.4.6 Acid Fumes and Splash

These and drops of acid splash may be encountered during various processes involving surface treatment for metal bonding or corrosion protection. Tank anodizing processes are normally conducted in a separate shop such as a plating shop where various forms of metal plating such as nickel, copper, chromium, or cadmium plating are performed. These shops have their own special personal protection requirements. However, it is becoming more common to carry out nontank anodizing and other metal treatment processes such as Alodine (Alochrom) treatment in the repair shop. Therefore, personal protection requirements for these processes must be used in the repair shop. The SRM requirements when using Alodine and similar treatments state,

> Warning: Do not breathe nitric acid gas. It is very dangerous to breathe nitric acid gas. Have a good flow of air or respiratory protection. Keep acid away from sources of heat, fire, and sparks. Get medical aid immediately if gas is breathed or only thought to have been breathed.

When performing nontank anodizing processes, such as phosphoric acid containment system (PACS), phosphoric acid nontank anodize (PANTA), or similar processes, observe all the safety precautions detailed in the SRM. Hydrofluoric acid (HF) is particularly damaging to human skin and may be called up by itself or as a pretreatment before Alodine (Alochrom) conversion coating. Full protection should be worn when using this acid (i.e., goggles giving all-around protection to ensure that contact with the eyes is not possible, masks, and protective clothing). HF antidote, a calcium gluconate gel, must be available when working with this acid. For details of use, see Chapter 15, Section 15.2.3. Take the recommended actions, and seek medical aid immediately. This gel contains 2.5% calcium gluconate and is available from IPS Industrial Pharmaceutical Services Ltd. Healthcare, Manchester, W14 1NA, United Kingdom, and from other sources.

6.4.7 Exposure Limits

The following definitions are provided because they are commonly found in health and safety literature and thus knowledge of their meaning is helpful.

Threshold limit values (TLVs) refer to airborne concentrations of substances and represent conditions under which it is believed that nearly all workers may be repeatedly exposed day after day without adverse effects. The American Conference of Governmental Industrial Hygienists (ACGIH) has established the following three categories for TLVs:

1. **Threshold Limit Value—Time-Weighted Average (TLV-TWA):** The time-weighted average for a normal 8-hour work day and a 40-hour work week, to which most workers may be exposed, day after day, without adverse effects.

2. **Threshold Limit Value—Short-Term Exposure Limit (TLV-STEL):** The concentration to which workers can be exposed continuously for a short period of time (15 min) without suffering from irritation, chronic or irreversible tissue damage, or narcosis of sufficient degree to increase the likelihood of accidental injury, impair self-rescue, or materially reduce work efficiency, and provided that the daily TLV-TWA is not exceeded.

3. **Threshold Limit Value—Ceiling (TLV-C):** The concentration that should not be exceeded during any part of the work day (ACGIH 1988–1989).

TLVs are based on the best available information from industrial experience, from experimental human and animal studies, and, when possible, from a combination of all three. Because of wide variations in individual susceptibility, a small percentage of workers may be affected by some substances at concentrations at or below TLV. TLV assumes that the exposed population consists of normal, healthy adults and thus does not address the aggravation of pre-existing conditions or illnesses. (See Ref. 6.4.)

6.5 Skin Contact

6.5.1 Fibers in Contact with the Skin

Carbon and some other fibers are very stiff and will easily penetrate the skin. These fibers are small in diameter and are difficult to see, even with a magnifier. If a fiber can be felt in a finger or anywhere else, it is important not to break it while it is embedded because this increases the difficulty of locating and removing the fiber. Use a magnifier to find the fiber, and assist in removing it or seek medical help to remove it so that minor surgery does not become necessary. Boron fibers should be treated similarly to carbon fibers as mentioned above, and medical assistance should be obtained if any fibers puncture the skin. Trace amounts of mercury may be found on these fibers at a very low level. The MSDS does not regard boron fibers as hazardous waste. They are much larger in diameter (0.127 to 0.178 mm [0.005 to 0.007 in.]) and are stiffer than most fibers; thus, they can puncture the skin more readily than other fibers. Boron composites are difficult to drill and machine and require diamond-coated drills, cutting tools, saws, and grinding wheels and a good coolant supply. Wear rubber gloves when handling fibers, especially when cutting away damaged areas of composite for repair or handling the damaged component during inspection. Fine carbon and glass fibers can be extremely irritating, and boron fibers can easily puncture the skin. Protective clothing and gloves should be worn. Sticky tape can be useful to lift most of the fibers, but any that have penetrated the skin probably require removal with tweezers. A magnifying glass will be helpful in locating the fibers. Although most composite fibers do not cause the health problems associated with asbestos, it is wise to avoid breathing them. The long-term effects on the lungs depend on the precise fiber length in relation to the structure of the lung as well as on the chemical constitution of the fiber. However, all dust is irritating to some extent and is undesirable, regardless of its chemistry. In some cases, the chemical content of the fiber also may be detrimental. It is good practice to avoid dust as much as possible and to use sanders and cutting tools with built-in dust extractors. Vacuum cleaners should be used for workshop cleaning so that dust and fibers are drawn into the bag. Loose fibers and dust should never be blown from an area with an airline. Disposal should be in accordance with local regulations. If none exist, then strong, sealed bags should be used and the dust should be treated as hazardous waste.

6.5.2 Resins in Contact with the Skin

Skin contact with resins should be avoided at all times for two reasons:

1. With liquid resins and adhesives, any toxic components can enter through the skin; therefore, all contact should be avoided. Many two-part paste adhesives have a viscous base resin and a water-thin curing agent, which make initial mixing slow and difficult

even if the final mix is of a paste viscosity that is easy to apply. Goggles should be worn because the water-thin curing agent is usually toxic and fairly alkaline and splashing into the eyes is a serious possibility because of the stiffness of the resin in the early stages of mixing. Suitable clothing should be worn to prevent splashing the curing agent on the skin. Curing agents often have a pungent odor, and face masks are recommended. Mixing should occur in a fume cupboard in which the fumes can be drawn away. Future development of resins and adhesives could usefully work toward the formulation of curing agents with a nonsplash viscosity.

2. With film adhesives and pre-pregs, contact with human skin oils can contaminate the adhesive and reduce or prevent bonding in the contaminated areas. Small amounts of solvent and toxic chemicals also can be absorbed by the skin, although expected to be much less than with a wet resin. One-part hot-setting paste adhesives should be treated in the same way as film adhesives. Polyethylene or latex rubber gloves are recommended. If the paste viscosity is low, goggles should be worn. See Sections 6.9.2 and 6.9.3.

- If resins enter the eyes, flush copiously with water and contact medical services immediately.

- If resin contacts the skin, wipe off as much as possible with paper tissues and wash with warm soapy water.

6.5.3 Solvents in Contact with the Skin

Apart from being mildly toxic in some cases, solvents remove the natural oils from the skin and leave it very dry. Excessive use of solvents to clean the skin can cause it to crack, as it does with prolonged exposure to frost, and thereby provide entry paths for infection. The skin should be cleaned with plenty of soap and warm water. Solvents should not be used. Solvent-repellent barrier creams are recommended. (See Section 6.5.5 for discussion of barrier creams.)

6.5.4 Selection of Suitable Gloves

Gloves are essential equipment when working with all types of chemicals. The correct type of glove must be chosen for each purpose because the gloves must be made from materials resistant to the particular chemical being used. If liquids are involved, then the gloves must be impervious to the liquid in question. An all-purpose glove does not exist. Refer to the MSDS for each material, or consult your supervisor or company safety officer. Mechanical or thermally resistant gloves may be made from cotton, aramid, or leather; however, they offer no protection against liquid chemicals. Cotton gloves can be used as liners for polyethylene or rubber gloves to absorb perspiration and to keep the hands dry. Leather gloves are recommended when working with boron fibers. Be sure the right glove is chosen for the job. The following list from SACMA is a great help, but check the MSDS for each product used.

Glove Type	Protects Against
Aramid	Cutting and Intense Heat
Cotton	Abrasions
Disposable Plastic (Latex)	Microorganisms, Mild Irritants, Fibers
Natural Rubber (Latex)	Acetone (<1 hour), Epoxies, MEK, Light Work
Lead-Lined	Radiation
Leather	Abrasions, Punctures, Fibers
Metal Mesh	Cuts, Scrapes
Neoprene Rubber	Acetone (<1 hour), Epoxies, N-Methyl Pyrrolidone (NMP)
Nitrile Rubber	Epoxies, Isocyanates
Polyvinyl Alcohol	Methylene Chloride, Toluene, MIBK, Styrene, Tetrahydrofuran (THF)
Polyvinyl Chloride (PVC)	Dimethylsulfoxide (DMSO), Isopropyl Alcohol, Epoxies
Rubber (Butyl)	Dimethylformamide (DMF)
Rubber (Insulated)	Electrical Shocks and Burns
Viton	Methylene Chloride, 1,1,1-Trichloroethane, Toluene

Gloves are the most important form of personal protection against dermatitis.

6.5.5 Skin Creams for Personal Protection

Moisturizing creams and barrier creams protect the skin but must be regarded only as supplements. They cannot replace good personal hygiene and chemical-protective gloves. Moisturizing creams replenish the moisture in the hands. Barrier creams prevent that moisture from escaping and keep mild irritants from penetrating the skin. Two main types of barrier creams are available:

1. **Water-Repellent Creams:** These leave a thin film of beeswax, petroleum, or silicone on the skin. Their primary use is in machine shop operations, where gloves cannot always be worn safely and where water-based cutting fluids are used.

2. **Solvent-Repellent Creams:** These may be supplied in the form of ointments that leave a thin film on the skin or as vanishing creams. Either will typically repel oils, paints, and solvents. These creams commonly contain lanolin, sodium alginates, methyl cellulose, sodium silicate, and tragacanth.

Barrier creams must be used with care because some operations are contaminated by them. They should be used in conjunction with gloves rather than as a replacement for gloves. Barrier creams must not be allowed to contact any surface that is to be bonded. For the most effective protection, the hands should be washed before applying a barrier cream. Barrier creams wear off, and thus the hands should be washed and the cream reapplied several times a day.

6.6 Material Safety Data Sheets

All hazardous materials should have a Material Safety Data Sheet (MSDS) supplied by, or at least obtainable from, the material manufacturer or supplier. The sheet should be identified by the description and part number of the particular product. The MSDS should provide a description of the contents of the adhesive, resin, sealant, solvent, or other material and should provide details of the hazard rating applied to the product. Details of the protective clothing or equipment required to work safely with the product also should be given.

All personnel required to work with hazardous materials have a right of access to the MSDS for each hazardous material involved in their work. Large companies generally have direct access to this information on a computer based in the workshop. They also have one or more health and safety officers who can be contacted for advice.

Material suppliers may provide company medical officers with confidential information on the chemical contents of their products. For commercial reasons, the precise formulation is not given; however, the chemicals involved are named to facilitate medical treatment if this should be needed. The suppliers of hazardous materials have a 24-hour telephone number for access to urgent medical advice.

6.7 Exothermic Reactions

An exothermic chemical reaction is one in which heat is generated during the reaction. In several cases, this heat can be considerable and thus can limit the amount of material that can be safely mixed at one time. See the MSDS for each epoxy system. When epoxy resins cure, heat is given off and this must be contained within acceptable limits. The opposite is an endothermic reaction, which requires the absorption of heat. The SACMA Document cited as Ref. 6.4 states that, "an exotherm is an unintentional runaway or out-of-control chemical reaction of a resin system. Such a reaction can occur when the temperature of the resin increases gradually, but then rises rapidly as the reaction goes out of control. Exotherms can occur not only with neat (hot melt) or solution resins, but also with resin in film or pre-preg forms. However, the greatest potential for an exotherm is when a resin system is present in large, undiluted masses and rapid or excessive heating occurs. Contamination by catalytic agents may increase the exotherm potential."

"Exotherms can be prevented, controlled, or contained. Prevention is accomplished by handling materials properly." (An exotherm can be prevented by ensuring that the mass cured at one time is not large enough for the amount of heat generated to be incapable of being removed by normal air cooling. A problem arises only if the rate of generation of heat by the chemical reaction exceeds the available rate of cooling.) An exotherm is detected in the early stages by an otherwise unexplained increase in temperature, and in more advanced stages by smoke with a characteristic odor. At the early stages of the exotherm, control or containment is very possible. However, at more advanced stages (i.e., if it is emitting smoke), do not approach the material or try to control the reaction unless proper personal protective equipment (e.g., gloves or proper respirator) are worn. Tests have shown that the smoke and fumes emitted in an exothermic reaction are toxic.

The usual effects reported after exposure to smoke and fumes from an exotherm are eye, nose, and throat irritation; coughing; nausea; dizziness; and headache. Any person exposed to an exotherm should see a physician. It is better to read the MSDS and mix only the correct quantity.

Mixing Precautions: Each adhesive or resin data sheet states the maximum quantity that can be safely mixed at one time without causing dangerous exotherm. Some exothermic heat is an unavoidable part of the curing reaction. With large quantities, the heat cannot escape rapidly enough. In extreme cases, the material will catch fire and generate large quantities of smoke and toxic fumes. In some cases, mixing more than 100 g in a small container can do this. One occurrence caused a large amount of smoke, evacuation of the workshop for some time, and charring of the bench on which the plastic container stood. It must have been the hottest curing, "cold-setting" adhesive used in that place.

Read the data sheet first, and if the amount to be mixed is near the recommended maximum, use a shallow aluminum tray to allow maximum cooling.

Remember one of the most important sayings in engineering, "Before all else fails, read the instructions."

6.8 Waste Disposal

Waste disposal is a serious matter and is becoming more serious as more countries enact legislation to protect the environment. Each company must ensure that local regulations are met as a minimum and, where possible, that better precautions than those required are taken. As an example, it would be helpful to ensure that all resins are fully cured before disposal. In this way, less toxic material will be released into the ground. Film adhesives and pre-pregs that are outside their shelf life and have not been tested and re-certified for use should be fully cured before disposal. This can be done during the cure of another job and should not require great expense. Similarly, wet lay-up resins that are out of shelf life or have failed re-certification tests should be mixed in the correct ratio and quantity to avoid exotherm, allowed to cure, and then sent for disposal. All other materials should be disposed of in accordance with local legislation. Company safety officers should be able to provide details for each material.

6.9 Safety Procedures

6.9.1 Emergency Action First-Aid Procedures

These procedures should be followed in cases of various types of emergencies:

* **Skin Contact**: Immediately remove liquids or pastes from the skin by wiping with disposable paper towels (remove powders by brushing). Then cleanse the affected area with resin-removing cream, followed by washing with warm, soapy water.

* **Eye Contamination or Irritation:** If eye contact occurs, flush the affected eye with water immediately with an eyewash bottle or fountain—or with low-pressure running water—for at least 15 minutes. Seek medical attention promptly.

* **Inhalation**: Operators affected by the inhalation of powders or fumes, vapors, mists, or droplets should be taken immediately into fresh air and made to rest while medical attention is called.

* **Clothing:** Remove and isolate contaminated overalls and clothing.

* **Ingestion (Entry by Mouth and Swallowing):** Immediately rinse the mouth with water. If swallowing has occurred, drink plenty of water. Seek medical attention immediately.

* **Fires**: If an epoxy material catches fire, use either a carbon dioxide, dry powder, foam, or vaporizing liquid extinguisher, or apply waterspray. Do not use water jets. Avoid the products of combustion while extinguishing the fire. In a serious fire, self-contained breathing apparatus may have to be worn.

• In all emergency cases, it is helpful to give the doctor the MSDS for the material concerned to enable correct action to be quickly and accurately decided.

6.9.2 General Safety Procedures

Most epoxies are considered harmless to handle, provided that the normal safety precautions are taken. These precautions are as follows:

1. Uncured materials must not be allowed to contact food or food utensils.

2. Uncured materials should not be allowed to contact the skin or eyes.

3. Adequate ventilation and fume extraction should be provided in the working area when mixing or using resins.

4. In addition to a data sheet giving material properties, each material should have its own MSDS. Ciba refers to this as Product Safety Information. Read the MSDS before starting work to ensure that the requirements for each material can be followed using the equipment available. If compliance cannot be achieved, obtain anything that is required before starting work. It is important to check each MSDS because some materials are more hazardous and require more protective equipment than others.

5. Mix the correct amount to avoid exotherm.

6. Take adequate precautions when flammable materials are being used.

7. Equipment should be cleaned immediately after use with the recommended solvent. Remove resins quickly before they have time to cure. This is vital for fast-curing systems. Acetone is suitable for removing epoxy materials.

8. In the United Kingdom, waste uncured materials must be disposed of in accordance with the Deposit of Poisonous Waste Act 1972, the Control of Pollution Act 1974, and subsequent U.K. legislation. Similar legislation in other countries must be followed in those countries. The above information is for epoxy systems, and much of it will apply to most adhesive and resin materials. Other materials may require additional procedures; therefore, check each MSDS.

   Important Note: The required precautions vary considerably from one material to another, and it is essential to read the specific information for each product being used.

9. Other safety matters are as follows:

   • Disconnect tools from their air or electrical power supply before changing drills or cutters. Clamp parts in a vise before drilling, especially if they are small. Serious injury can occur if these parts rotate while in your hand.

- To prevent damage and delamination to the back face of composite parts, always drill against a block of plywood or plastic and use a slow feed rate.

- Carbon dust and chips are able to short electrical equipment and may cause corrosion of aluminum parts. When possible, it is helpful to remove carbon parts to the workshop for repair.

- Do not use compressed air to blow dust from a part after sanding. If the air pressure is high, it can cause delamination of the laminate and consequently more damage. It also throws unhealthy dust into the working environment. Always use a vacuum cleaner to remove dust.

- Do not block access to safety equipment because it may be needed in a hurry.

6.9.3 *Personal Safety Precautions*

- Avoid all skin contact with resins, hardeners, sealants, and solvents. Impervious rubber or plastic gloves should be worn (see Section 6.5.4), with cotton gloves inside if necessary for comfort and to absorb perspiration. Barrier creams such as Kerodex Anti-Solv should be used on the hands and any other body parts that may come into contact with resins. Cleanse the skin at the end of each work day by washing with warm, soapy water. Do not clean the skin with solvents because they remove body oils and dry the skin.

- Wash the hands before and after using the toilet.

- Avoid eye contact. Wear goggles of a suitable type for all-around protection if the viscosity of either the resin or the hardener is low enough to make splashing possible. Prescription glasses alone are insufficient because they do not prevent the penetration of splashes from the side. Eye protection is very important because permanent damage to sight can occur very quickly and can be avoided only by prevention using the correct equipment. An OSHA or similarly approved eyewash station must be provided.

- Avoid inhalation of fumes. Wear a mask if necessary. Ensure that the correct type is used for the hazard concerned.

- Avoid ingestion. Wash hands thoroughly before eating. The consumption of food and drink is not permitted in bonding areas. Food and food containers should be kept far away from all resin and fibrous materials.

- Smoking is not permitted in bonding areas.

- Overalls should be laundered frequently. In the event of significant contamination, overalls should be laundered before use.

6.9.4 Dermatitis

Repeated skin contact with resins can cause dermatitic sensitization. In extreme cases of sensitization, a skin rash can develop from only being near a film adhesive and without actual contact. This must result from a very small amount of solvent vapor emitted by a film adhesive and a very serious level of previous sensitization. I personally know of two cases of such serious sensitization. In both cases, the individuals concerned were two of the keenest and most capable people but had to be given other work. Although rare, it is worth considering that after significant training and experience, two very useful people could no longer be employed in their chosen area of work. In this line of work, it is acceptable to dirty the rubber gloves but not the hands because the hands are too valuable to the worker and to the company. The following are abbreviations of the most important points from Ref. 6.5, but the entire book should be studied.

Dermatitis, or inflammation of the skin, can become a serious condition if preventive care is not taken. The word comes from the Greek words "derm" (skin) and "itis" (inflammation). Skin disorders begin when the outermost layer (epidermis) is damaged.

Dermatitis can be caused by:

* Solvents that dry and remove fat or oil from the skin
* Vigorous, irritating scrubbing
* Cold or dry weather conditions
* Existing cuts and abrasions that invite damage
* Bacteria or viruses
* Corrosives
* Abrasive materials (silica or fibers)

When chemicals come in contact with the skin, one or more of the following effects may occur:

* **Minor Skin Damage:** The epidermis suffers harm, allowing irritants to reach the lower layers.

* **Major Skin Damage:** All layers are harmed, the skin thickens to create "armor," which itself may crack and allow further infection and irritation. Skin ulcers or tumors may also result.

* **Internal Damage:** Some chemicals may pass through damaged or even healthy skin, entering the body to damage the liver, kidneys, bone marrow, or other organs. Aromatic amines, amides, methanol, and phenols are some types of chemicals known to produce internal injury through skin absorption.

Symptoms of dermatitis vary, but the most common forms are:

Early Stages:

- Redness of the skin
- Swelling
- Itching
- A burning or "crawling" sensation
- A "hot" feeling

Advanced Stages (which can result if treatment is not obtained):

- Broken skin or blistering
- Reddening
- Scaling
- Infection
- Eczema or psoriasis

Personal Hygiene: Dermatitis is avoidable with good personal hygiene. The following actions are recommended:

- Bathe regularly
- Wash the hands before and after eating
- Wash the hands before and after rest-room visits
- Wash the hands before putting on and after removing gloves
- Use skin moisturizers
- Wash work clothes separately from other clothes
- Change work clothes regularly

Treating Dermatitis: Do not attempt to treat dermatitis entirely without medical assistance. If unfortunate enough to get dermatitis or to suspect its presence at an early stage, the individual should report it to the supervisor and medical staff immediately. Early treatment can prevent serious consequences, and some people are more sensitive to dermatitis than others. As previously discussed, some people have had to transfer to other work after becoming sensitized; therefore, it pays to avoid reaching that stage. To help the medical staff provide the most effective treatment, tell them the following information:

- What materials have been handled by the individual at home and at work. It is too easy to take fewer precautions when working at home than in a full-time job. The same precautions should be taken at work and at home.

- When the condition started.

- Any change noticed in the condition over weekends, vacations, or other period when the individual was away from the work site.

- After-work activities.

Follow the medical advice and take the treatment offered. Likewise, take all the recommended precautions to avoid future exposure; otherwise, a change of job may be required.

Medical Screening of Personnel: Certain individuals show a greater predisposition to skin irritation than others. This may occur from a wide range of substances, including epoxy resins. The irritation may result from short-term contact or may occur after a long period of time. A few individuals may be affected by certain epoxy materials after initial contact; others may be affected only after long or repeated contact. If skin irritation occurs, seek medical attention as soon as possible. Pre-employment medical inspection may show reasons for excluding persons who have known skin sensitivity, a known skin condition, or a record of similar problems. These persons should be permitted to work with epoxy materials only after formal approval from a doctor and after advice and training in personal hygiene and safety precautions. If significant doubt exists about the advisability of particular individuals working with resins and composite materials because of the probability of skin irritation or sensitization, perhaps these persons should be encouraged to seek a different field of work rather than to undergo lengthy training and then have to make a change.

6.10 Action in the Event of a Chemical Spillage

Ref. 6.6 states, "Materials should be kept in securely closed containers, and particular care should be taken to avoid leakage or spillage. In the event of any such escape of liquid materials, the spillage should at once be taken up with sand, cotton waste, sawdust, paper towels, or other absorbent material, and the contaminated absorbent taken to a safe place pending disposal. The contaminated absorbent should then be disposed of in accordance with Part 1 of the Control of Pollution Act 1974 in the United Kingdom and similar legislation elsewhere. Spillages of powder or granular materials should be covered with damp sand and transferred carefully to containers for removal. The area should then be cleaned by vacuum methods. With the spillage cleared away, the contaminated area should be washed clean with detergent solution, preferably hot. In the event of a serious spillage that enters the public drains or waterways, the local water authority must be informed immediately and the police should be notified."

6.11 Explosion Risk

Ref. 6.6 also mentions a dust explosion risk. "Powder synthetic resin materials, such as coal dust, wood flour, and other finely divided materials, can form an explosive dust cloud. Such clouds are subject to control measures set out in Section 31 of the Factories Act 1961 in the United Kingdom. Techniques for handling powders should be designed to avoid the formation of dust clouds. In areas where dangerous concentrations of dust may occur, all sources of ignition must be eliminated. It is particularly important that storage and handling areas should be kept clean and tidy to prevent the accumulation of dust and that they are adequately ventilated." Precautionary measures for combatting dust explosion risks are given in HSE Health and Safety at Work Booklet No. 22—Dust Explosions in Factories, available from HMSO, P.O. Box 276, London SW8 5DT, United Kingdom.

6.12 Static Discharge and Fire Prevention

6.12.1 Static Discharge Prevention

Ref. 6.6 states, "The handling of powders may raise a static discharge risk. When powders are organic and the quantities handled are sufficient to form a dust cloud, there may also be an explosion risk. See the above paragraph. Handling techniques for powders must be designed so as to minimize these risks." Advice should be sought from the supplier of the powder in question. Processes involving the movement of powder materials are likely to generate electrostatic charges. The electric potential can be sufficient to produce a spark discharge, which may ignite vapor or dust.

Measures to overcome difficulties resulting from static electricity include the following:

- Adequate earthing (grounding) of all conductive parts to prevent the generation of spark discharges

- Temperature control of flammable liquids to ensure that temperatures are below the flash point of the liquid

- Feeding through a screw conveyor into an inerted atmosphere

Guidance on precautionary measures to be taken when emptying powder materials from containers (metal, plastics, or paper), transferring the powders through feed equipment, and their addition to flammable liquids is given in British Standard BS 5958 Part 2, Control of Undesirable Static Electricity—Recommendations for Particular Industrial Situations. This standard is available from British Standards Institution, Sales Department, Linford Wood, Milton Keynes, MK 14 6LE, United Kingdom.

If a process is likely to generate static electricity, advice should be sought from the health and safety executive through the district inspector in the United Kingdom, or similar organizations elsewhere. Ignition is the chief hazard, but static electricity also can give unexpected shocks to personnel. These shocks probably will not cause serious harm, but the involuntary movement they cause may result in an accident.

6.12.2 Fire Prevention

In addition to the precautions to avoid explosions, it is important to avoid fires by observing the following precautions:

- No smoking should be permitted nor any processes involving flames or sparks in any area where solvents are used. This includes sanding operations, which can cause sparks.

- Non-spark-producing tools should be used.

- All aircraft handling, maintenance, and repair vehicles and equipment should be electrically grounded (earthed).

- Bagging films, release films, and peel ply rolls should not be unrolled near solvents. Rapid movement may generate static electricity and produce sparks that can ignite solvent vapors.

- Solvents should be kept in containers with lids to minimize the amount of vapor in the workshop. Workshops should be well ventilated to prevent vapor buildup.

- Resins should be mixed in small quantities and never in amounts larger than recommended on the data sheet. This practice is necessary to avoid fire and toxic fumes caused by exotherm.

6.13 Safety Requirements for Electrical Appliances

These requirements differ according to whether electrical equipment is used in workshops or in hangars either on or near aircraft containing fuel. For example, to be approved for use on aircraft, hot-bonding controllers must be positively pressurized inside to prevent potentially explosive vapors from entering the box where sparks could ignite them. All electrically powered equipment for composite repairs must meet local safety standards. The standard for aircraft hangars is broken down into the following three areas:

1. **High Risk:** This is defined as an area within which any flammable substance (gas, vapor, or volatile liquid) is processed, handled, or stored and where, in normal operation, an explosive or ignitable concentration is likely to occur in sufficient quantities to produce a hazard. Such areas within a hangar are:

- Inside aircraft fuel tanks

- Any pit or depression below the level of the hangar floor such as electrical service pits, open ducts, or trenches

Ensure that electrical equipment in high-risk areas as defined as above and all portable electric lead lamps used throughout the hangar are:

- Certified "Flameproof" Division 1 to British Standard Code of Practice BCSP 1003, Part 1 or BASEEFA Code "Ex.d" or to British Standard BS 5345 for Zone 0 and 1 or other national equivalent

- Certified "Intrinsically Safe" to British Standard Code of Practice BSCP 1003 Part 1 or to British Standard BS 5345 for Zone 0 and 1 or other national equivalent

2. **Remotely High Risk:** This is defined as an area within which any flammable substance (gas, vapor, or volatile liquid), although processed or stored, is so well under control that the production or release of an explosive or ignitable concentration in sufficient quantities to cause a hazard is likely only under abnormal conditions. Such areas within hangars are as follows:

- The entire hangar floor up to a level of 460 mm (18 in.) above the floor, such areas to include any adjacent or communicating areas not suitably separated from the hangar by the fire walls or suitable doors

- The area within 1.5 m (5 ft) horizontally from aircraft engines, wings, or other parts of the aircraft structure containing fuel

Ensure that electrical equipment in remotely high-risk areas is certified Zone 2 equipment to British Standard BS 5345 or other national equivalent. For reasons of flexibility and to avoid complications when aircraft are being sprayed or roller painted in hangars, electrical equipment on all fixed docks, fuselage walkways, tail stagings, and mobile rostrums shall be at least to Zone 2 to British Standard BS 5345.

3. **Other Hangar Areas:**

- Ensure that all electrical equipment in those areas not specified above is of the totally enclosed type or is provided with suitable guards or screens in such a way that arcs, sparks, or hot particles cannot constitute a hazard to any nearby flammable material.

- Ensure that electrical cables are of appropriate construction having regard to the risks involved, and where necessary, provide earth leakage protection.

- Ensure that mobile, portable, or transportable electrical servicing units such as ground power units are designed with regard to safety and preferably to Division 2 of BSCP standards. Plugs, sockets, and connectors for attaching cables to power units should be at least 460 mm (18 in.) above floor level and be of the screened type giving effective protection to the pins while the connector on the cable is being inserted. Properly constructed connectors should be provided and used for connecting ground power unit cables to the aircraft installation.

4. **Labeling Portable Equipment:** To ensure that portable equipment is used only in appropriate areas of a hangar, each portable unit shall be clearly labeled in accordance with local regulations to indicate the standard to which it has been built and hence any restrictions that must be placed on its use. In the United Kingdom, electrical equipment used in aircraft maintenance hangars must comply with Section 31 of The Factories Act 1961. (See Ref. 6.10.) In North America, Article 513 of the National Fire Protection Association, National Fire Codes, Vol. 5, Electrical, normally applies. American "Explosion Proof," Class 1, Division 1, Equipment is generally regarded as equivalent to United Kingdom "Flameproof." These regulations are applicable where aircraft fuel tanks contain kerosene type fuel. If wide-cut fuel has been used, it must be drained before the aircraft is positioned in a hangar meeting these regulations. In the absence of data, it must be assumed that restrictions are even tighter for those aircraft and other vehicles operating on high octane petrol (gasoline) or wide-cut jet fuel. Commuter and general aviation aircraft regulations should be studied.

6.14 ISO Environmental Management Standards

The American Society for Testing and Materials' *ASTM Standards International* newsletter (Vol. 3, No.1, 1995) announced the International Organization for Standardization (ISO) 14000 Series of documents on environmental management systems (EMS). They are similar to the ISO 9000 Series for quality management and will provide management standards for environmental management activities.

The six draft ISO 14000 documents that have been circulated for review are as follows:

1. ISO 14000, Guide to Environmental Management Principles—Systems and Supporting Techniques

2. ISO 14001, Environmental Management Systems—Specification and Guidance for Use

3. ISO 14010, Guidelines for Environmental Auditing—General Principles of Environmental Auditing

4. ISO 14011/1, Guidelines for Environmental Auditing—Audit Procedures—Part 1: Auditing of Environmental Management Systems

5. ISO 14012, Guidelines for Environmental Auditing—Qualification Criteria for Environmental Auditors

6. ISO 14040, Life Cycle Assessment—General Principles and Practices

These documents should be helpful to anyone involved in achieving the required environmental standards in the workplace.

Refs. 6.11 through 6.14 also provide useful information on environmental and safety concerns.

6.15 References

6.1   Teshchendorf, Andrew F., "Environment Considerations for Advanced Materials," *Advanced Materials Technology International*, 1992.

6.2   The Control of Substances Hazardous to Health Regulations (COSHH), Statutory Instrument 1988, No. 1657, HMSO, London.

6.3   "Reducing Emissions of Volatile Organic Compounds (VOCs) and Levels of Ground Level Ozone: A U.K. Strategy," Department of the Environment, London, October 1993.

6.4.   *Safe Handling of Advanced Composite Materials Components: Health Information*, Suppliers of Advanced Composite Materials Association, 1600 Wilson Blvd., Suite 1008, Arlington, VA 22209, 1989.

6.5   *Save Your Skin! A Guide to the Prevention of Dermatitis*, Suppliers of Advanced Composite Materials Association, 1600 Wilson Blvd., Suite 1008, Arlington, VA 22209.

6.6   *Safety Manual No. 37m: May 1989. Handling Precautions for Araldite Epoxy Resin Materials*, Ciba-Geigy Plastics, Duxford, Cambridge CB2 4QA, England.

6.7   Abbott, S.G., "Environmental Issues and Adhesives," *Materials World*, August 1994, pp. 422–424.

6.8   Murphy, John, *The Reinforced Plastics Handbook*, Elsevier Advanced Technology, London, UK, ISBN 1-85617-217-1, 1994.

6.9   "A Guide to Health and Safety in GRP Fabrication," Health and Safety Executive (United Kingdom), ISBN 07176-0294-X, 1987.

6.10   The Factories Act 1961, United Kingdom.

6.11   The Environmental Protection (Prescribed Processes and Substances) Regulations, Statutory Instrument 1991, No. 472, HMSO, London.

6.12   Mellan, Ibert, *Industrial Solvent Handbook*, 2nd edition, Noyes Data Corporation, New Jersey, 1977.

6.13   Scheflan and Jacobs, *The Handbook of Solvents*, Macmillan, London, UK, 1953.

6.14   Whim, B.P. and Johnson, P.G. [eds.], *Directory of Solvents*, 1st edition, Blackie Academic and Professional, an imprint of Chapman and Hall, London, UK, ISBN 0-7514-0245-1, 1996.

Chapter 7

Damage and Repair Assessment

Damage and repair assessment is an important area of repair work. On one hand, there is a desire to minimize the damage size. On the other hand, there is a need to ensure that all the damage or contaminated material is removed to guarantee a sound repair. Several methods of damage assessment will be discussed, followed by a section on damage types and mapping and another section on damage significance. No satisfactory test is available to measure the strength of an adhesive bond; however, it has been found that, if the surface preparation of a composite part is adequate and the surface is dry at the time of bonding, this should not be a problem. Drying should be carried out as specified in the Structural Repair Manual (SRM), as a minimum and preferably longer. When repairs are to be made using hot bonding (i.e., anything above 90°C [194°F]), then a longer period of drying is recommended because water vapor and, above 100°C (212°F), steam can cause porosity in the bond line and a lower bond strength. If the surface is properly prepared (see the discussion about the water break test in Chapter 10, Section 10.1.6) and if the bond to it is good, then other aspects that can affect bond strength are:

- The use of a good adhesive or pre-preg that is within its shelf life and free of any contamination or surface moisture.

- The use of the correct curing temperature and care to ensure, by the use of sufficient thermocouples correctly located, that the temperature is within the required band for the specified curing time over the whole area to be bonded. It is strongly recommended that at least three thermocouples should be used, and the following guideline is suggested: The number of thermocouples should be a minimum of (n + x), where n = 3 and x = the number of square feet (or 0.1 m²) of repair area. Note also that sometimes a thermocouple will break after it has been installed; therefore, it is useful to have one or more spares to ensure that the entire repair does not have to be scrapped because of lack of a thermocouple to control and record the temperature.

- The use of sufficient curing pressure during the entire curing process.

Dye penetrant inspection methods should never be used with composites because these fluids contaminate the surfaces and prevent good bonding. If they are used in special cases, great care should be taken to avoid contact with surrounding areas, and all of the contaminated area should be removed. Dye penetrants must not be allowed to contact any surfaces that may require bonding.

7.1 Visual Inspection

Visual inspection is the first and most obvious method of inspection and damage assessment. Aircraft are designed so that defects too small to be seen are not catastrophic. One major manufacturer states, "Parts and assemblies manufactured from CFRP are so designed that if there is no visible evidence of damage, the structure will not have sustained damage to an extent where aircraft safety is affected. Fatigue life and ultimate loads are sustained, and any minor undetected damage will not propagate to a significant extent during the life of the part. Where surface damage is detected visually, however small, there is the possibility that additional hidden damage may exist. In this case, a more detailed inspection must be performed, using nondestructive inspection (NDI) methods, to determine the full extent of any possible internal damage and to assess the "accept or repair" solution in accordance with the Structural Repair Manual (SRM) requirements. In the case of extensive visible structural damage, this routine philosophy would not apply and such damage would be handled on a case-by-case basis."

For composite materials, the smallest damage size likely to be found has been established experimentally by one company as follows: "Using a number of experienced and inexperienced operators, viewing the surface from a distance of approximately 2 m (6 ft) and using a flashlight to illuminate the area, surface damage with an area of 1.4 mm^2 (0.002 in.^{2}) and a depth of 0.3 mm (0.012 in.) was reliably detected (95% probability). This gives confidence that any significant damage will not remain undetected."

We must achieve this position because, if a defect cannot be seen, no one will know that it exists and thus it follows that no action will be taken. Another problem that has occurred on several occasions is that airline personnel have experienced difficulty in determining the difference between visual indications caused by in-service damage and visual indications that may be caused during the manufacturing process or subsequent finishing actions, filling, or painting." To deal with this, the conditions likely to cause concern have been identified, and a visual inspection procedure has been produced for guidance. A visual inspection procedure will be published in Chapter 51 of the Nondestructive Test Manual.

(The IATA document prefers the term Nondestructive Inspection Manual because the action is an inspection rather than a test.) Written descriptions of different surface conditions, their causes, and their importance will be published together with helpful illustrations. These will assist in classifying damage indications into two types:

- **Type 1:** Indications that may signify associated structural damage and require NDI.

- **Type 2:** Indications that are only surface irregularities (cosmetic) and do not affect the integrity of the structure, with no follow-up action necessary.

For more details, see the NDI manual for each aircraft type.

Airbus states that it is mainly looking for debonding or delamination damage with an area greater than 50 mm^2, 8 mm diameter (0.075 in.^{2}, 0.3 in. diameter). This is significantly larger than the smallest area that is likely to be reliably found, which was mentioned earlier. It is helpful to have a little in reserve, but it should be emphasized that if damage is found, however small, it must be investigated and the SRM must be consulted for appropriate action. For noncritical parts, much larger damages can be accepted before they become of structural significance; however, any damage that is found must be assessed in accordance with the SRM and a decision must be made that the damage can be deferred, temporarily repaired, or removed for a workshop repair. In any case, the damage should be covered to prevent moisture ingress, and action called for at the earliest convenient time to prevent further degradation of the part. It is also true that metallic parts may have cracks that cannot be seen but which must be located by eddy current or x-ray methods. Such cracks may be located in bolt holes or other places that are difficult to inspect. Usually, the OEM issues a Service Bulletin to specify critical inspections of this type; in some cases, the Service Bulletin is classified as Mandatory. These inspections usually result from cracks found during fatigue testing of a complete airframe or of a major component by the airframe manufacturer. In some cases, they are triggered by defects found in service. As with composites, a crack below a certain length may be safe, and the goal of inspection is twofold:

- To detect the presence of a crack
- To measure its length to determine if it is above the critical permitted size

Different metals have different crack growth rates; therefore, repeat inspections will be necessary at a frequency that will guarantee an opportunity to remove the part, oversize the hole, or take other remedial action in a timely manner. Visual inspection, as shown in Figure 7.1, is the first line of defense and allows other methods to be used if any problem is suspected. If paint is damaged rather than only peeled off, a thorough inspection should always be made. For homebuilt aircraft with thin skins, an internal bond adjacent to the skin can be inspected for voidage or delamination by observing the rate of evaporation of dew after allowing such condensation to form on the surfaces of interest on the ground.

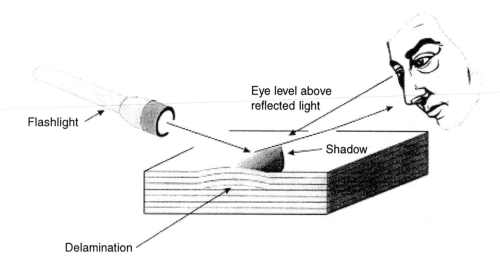

Fig. 7.1 Visual inspection. (Courtesy of Abaris Training Resources Inc.)

In the first instance, a tap test could be used (see Section 7.2). Any sign of impact damage should be carefully investigated and checked to see if subsurface delamination has been caused. This may require the use of ultrasonic or other NDI equipment. Any sign of significantly discolored paint should result in a test for overheating. Visual inspection should be performed mainly at close range (i.e., a few feet or a few inches, or tens of millimeters) and with good light and a magnifier if necessary. One manufacturer calls for visual inspection at a distance of 2 m (6 ft). However, in good sunlight, useful inspections can be made at long range (i.e., 10 to 20 m [30 to 60 ft]) when the sun angle coincides with the part concerned. For example, it is sometimes possible to detect delamination of rudder skins and other parts of sandwich construction from a considerable distance, while standing on the hangar floor, when the sun angle is right. Wavelengths of light have very small dimensions, and it is often remarkable how small a defect can be seen with the right lighting. Small dents in cars often can be found in the same way. If the light is good enough, it is difficult to find a car that does not appear to have some surface defect that is not visible under ordinary conditions. If such an apparent defect is seen, it should always be checked by tap test or ultrasonics to confirm whether a real problem exists. Requirements for routine visual inspections are given in the Maintenance Planning Document (MPD) for each aircraft type.

Advantages of Visual Inspection:

- No expensive equipment is required.

- Airworthiness design philosophy is such that most damage that is of concern is capable of being found visually.

Disadvantages of Visual Inspection:

- Large areas are time consuming to inspect.

- Inspectors have difficulty in maintaining concentration over large areas.

- Composites may have extensive nonvisible damage (NVD). This is particularly important in the case of nonvisible delamination. The critical failure mode of composites is often in compression, and any damage that reduces compression performance is unwelcome.

7.2 Tap Test

A tap test, as shown in Figure 7.2, may be performed using a coin of modest weight or even a thick steel washer of approximately 25 mm (1 in.) diameter and approximately 2.5 mm (0.1 in.) thick. Inspection techniques, other than visual, do not come cheaper than this.

The method has now been automated, at much greater expense, and with somewhat greater accuracy. Two versions are made: one by Rolls-Royce, and another aptly called the "Woodpecker," which is of Japanese design. The method may also be described as audiosonic because it operates in the normal human hearing range. In contrast, ultrasonic methods operate outside the human hearing frequency band.

The tap test method is well loved and widely used because, in the cheaper form, it is simple and available to everyone. All that is required is to tap gently on the suspect surface. Areas of good bond will sound clear and of a higher frequency than disbonded areas, which give a dull sound of lower frequency. By tapping at intervals of approximately 6 mm (0.25 in.), the damaged area can be mapped effectively. If approximately 12 mm (0.5 in.) are cut away beyond this to provide a margin of safety, then this is probably sufficient for fairings and nonstructural items. For these items, the method is often sufficient by itself, especially for initial damage assessment, although it is wise to use more sophisticated methods as a backup to ensure that all damage has been cut away before commencing repairs to more significant parts. The method is especially useful on sandwich structure with honeycomb core. It also works on a solid laminate if the first few layers are delaminated, but it cannot detect defects deeper in the laminate. As with metal-skinned honeycomb panels above a certain skin thickness of approximately 1 mm (0.04 in.), the skin alone gives a good ringing sound and delamination cannot be reliably detected by a tap test.

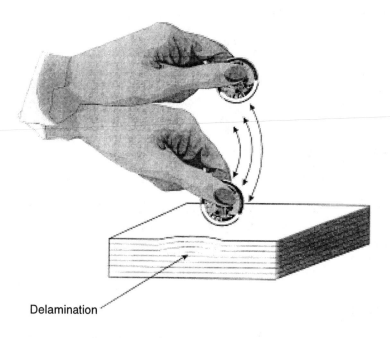

Delamination

Fig. 7.2 Tap test. (Courtesy of Abaris Training Resources Inc.)

The test works only for thin skins of any type, and this limitation must be kept in mind. The larger the area of disbond, the thicker the skin that will give a dull sound. Therefore, a precise limit on the skin thickness for which this method will work cannot be given. The problem with composites is that the fibers are stiff and elastic and spring back after impact if they have not been fractured. They do not form a dent as a metal skin might do to give a clear indication of where inspection is required. A careful visual inspection is necessary to find areas where a more detailed inspection should be made. Damage at the limit of detectability is often called barely visible impact damage (BVID) in the literature. The tap test is more useful for disbond and delamination detection near the surface. It should not be considered sufficient, of itself, to map impact damage. Impact damage, although small on the outside, often indicates much larger damage on the back face in the case of solid laminates. The back face should always be inspected, preferably visually for obvious damage and also ultrasonically to check whether the impact damage that is visible also indicates delamination near the back face.

The automated versions of the tap test do not rely on the efficiency of the human ear, which deteriorates with age and varies from person to person. It may also vary with the state of the individual's health and the background noise level. The advantage of the electronic type is that a change in frequency at which a defect is considered to exist can be set and the area mapped accordingly. With the human operator, this will be a matter of subjective opinion. The same will apply to the electronic version because a human will have to decide the amount of change required to determine that a problem exists, but at least it will be a constant difference after the decision has been made. Nevertheless, this method today is considered as the most useful one for detecting disbond in honeycomb panels. One manufacturer made a careful comparison of all available alternatives and concluded that the tap test is as good as any other test. Many airline personnel would agree with this finding.

Advantages of the Tap Test:

- No expensive equipment is required, at least not for the simple version.

- A tap test provides a quick initial method of investigating the extent of a defect.

- A tap test can be used to detect delamination, disbond, and severe moisture ingress.

- In Ref. 7.1, when commenting on the tap test, Boeing found that, "All heat damage and lightning damage defects were easily detectable and indicated delaminated areas well beyond the center of the defect area."

Disadvantages of the Tap Test:

- It is impractical to cover large areas effectively because it is difficult for an inspector to maintain concentration.

- A tap test is highly subjective.

- A tap test cannot locate small defects (e.g., voids or minor moisture ingress).

- A tap test cannot be used on locations covered by protective coatings and sealants.

- The sonic response is a function of material properties, laminate composition, thickness, shape, and component construction.

- It is not effective on thick metal skins and is able to detect delamination in only the first few surface layers of a thick composite laminate.

Note that the automated versions significantly reduce the subjectivity of the tap test.

A developed form of the automated device provides repeated tapping to allow assessment of material resonant frequencies, which are also a function of material composition and/or defect.

7.3 Ultrasonic Inspection

This type of inspection uses an ultrasonic signal and measures the attenuation of that signal. Ultrasonic inspection is the study of materials or structures using ultrasonic waves, also known as stress waves. These stress waves are mechanical waves or vibrations, and they follow the usual formula:

$$\text{wavelength } \lambda = c/f$$

where c = the velocity of ultrasound
          f = frequency

The velocity varies with the elastic properties of the material under test or inspection. Thus, for a given frequency, the wavelength will vary with the material being investigated. For composites, a frequency in the range 1 to 10 MHz is normally used. These frequencies are outside the audible range because only frequencies up to 20 kHz are audible to humans. A point of interest is that high-frequency (short-wavelength) waves are more sensitive to defects because the theoretical limit of detectable defect size is one wavelength, whereas low-frequency waves can penetrate to greater depths. This means that the choice of frequency to be employed is always a compromise between the sensitivity and penetration required.

Two modes of operation are normally used (see Figure 7.3):

1. Pulse-echo mode, using a single transducer
2. Through-transmission mode, using two transducers

In either case, the transducer(s) must be coupled to the structure via a liquid or solid medium because of the severe impedance mismatch between air and solid materials. At production facilities, this is usually achieved by immersion testing or, for large components, by the use of water jet probes. When inspection is conducted manually, a thin layer of gel may be used as a couplant between the transducer and the structure (pulse-echo mode). This technique is particularly suitable for field work. The large impedance mismatch between air and solid materials means that honeycomb constructions can be difficult to inspect using ultrasonics because the ultrasound will not propagate through the structure except at the cell walls. Transmission down the cell walls is also difficult if the walls are made of an attenuative material such as Nomex (a trademark of Du Pont), which is often the case.

Using a pulse-echo technique, only the top face-to-core bond can be inspected reliably according to one source. By using through-transmission mode, top and bottom bonds can be inspected in a single test. The same basic equipment may be used with either water or gel coupling. Several types of measurement are possible.

The "A" Scan is the display normally seen on the screen of the ultrasonic test set and gives the time history of the echoes received by the receiving transducer.

In the "B" Scan presentation, the vertical axis is the time axis of the "A" Scan, the presence of echoes being indicated by intensity variations. The horizontal axis gives position information, building an image of the component cross section.

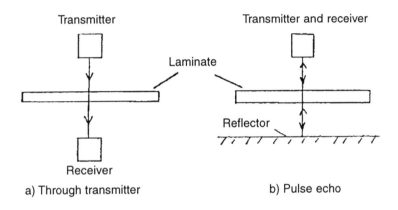

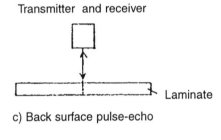

Ultrasonic testing

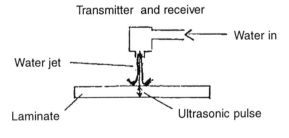

Water jet probe

Fig. 7.3 Diagram of ultrasonic inspection methods.

If the amplitude of a particular echo is monitored at each point on the surface of the work, a "C" Scan can be produced. Measurements at each point are taken using a scanning mechanism, which produces a plan of the defect positions but gives no information about their depth. This method is often used by manufacturers using the through-transmission mode and water jet coupling and has the advantage that a permanent record can be printed. If necessary, this printout could be compared with another scan taken after some period of service or when the possibility of damage was suspected.

Echo Amplitude Measurements at Normal Incidence: The usual method is to monitor the amplitude of the ultrasonic signal in the time domain with the transducer(s) normal to the surface of the structure. In the pulse-echo mode, the amplitude of the echo received from the back face of the component will be reduced by defects in the structure. A delamination whose plan dimensions exceed those of the ultrasonic beam will almost completely remove this echo, whereas porosity will cause attenuation of the signal because of scattering of the ultrasound by the small gas bubbles. Increased attenuation also arises from an increase in vibration damping of the material due to miscuring of the matrix resin or adhesive bond line. In pulse-echo testing, it is also possible to check for the presence of echoes between those from the front and back faces of the component, the presence of significant signals indicating the existence of defects or inclusions. The defect depth can also be found by comparing the transit time of the echo from the defect with that from the back face of the component. This can be a particularly useful form of information if the delamination is located at a depth that cannot be detected by tap testing. This type of test can easily detect delaminations in composites and disbonds in adhesive joints and between the top skin and the core in honeycomb constructions. Foreign inclusions, provided that the material that forms them has a significantly different acoustic impedance from that of the composite, also may be detected in this way. A general problem with ultrasonic testing is that it requires calibration on known standards. See Figure 7.4.

This type of testing is a comparison and does not directly measure any standard unit (i.e., inch, millimeter, amp, or volt). It is a case of comparing the trace of a good standard with the test item and interpreting the meaning of any differences found. This means that the inspector must have a thorough knowledge of the structure being inspected.

Based on the complexity of this type of work and especially the need for accurate interpretation, ultrasonic NDI is a job for only well-trained, experienced personnel.

The general procedure which must be followed is:

1. Calibration of the equipment on known standards
2. Check calibration in the inspection area
3. Inspection procedure
4. Damage identification
5. Reporting

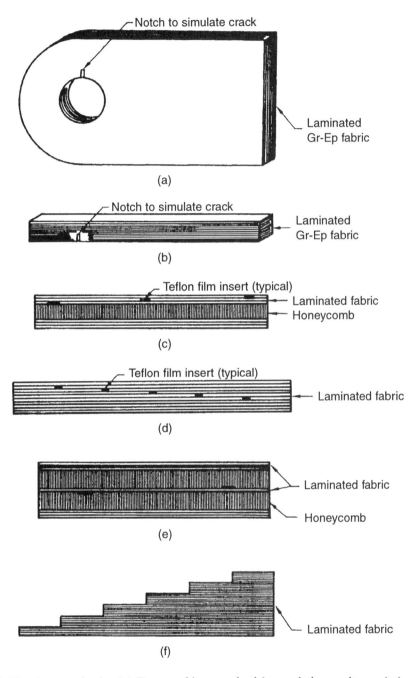

Fig. 7.4 Calibration standards. (a) Fractured lug standard for angle beam ultrasonic inspection; (b) Fractured plate standard for angle beam ultrasonic and eddy current inspection; (c) Delamination/disbond defects in honeycomb panel; (d) Delamination/disbond defects in laminated plate; (e) Disbond defects in septumized honeycomb panel; (f) Thickness standard for ultrasonic pulse echo and thickness gauge. (Courtesy of Boeing Commercial Airplane Group.)

229

Advantages of the Ultrasonic Technique:

- This technique may be used to detect many defects, such as defects in the plane of the sheet, delamination, voids, foreign objects, moisture, disbonding, and cracks perpendicular to the plane of the sheet if the transmitter and receiver are located at an angle with respect to the plane of the composite surface. This technique also may be used to detect delamination between sandwich panel core and skin laminate.

- This method can also indicate the depth of a defect in a laminate.

- The technique is flexible because it may be used in a local portable form or to cover large areas, such as "C" Scan providing a map of defect locations in a part.

- If focused transceivers are used in conjunction with "C" Scan and suitable computing equipment, a three-dimensional image of the composite component and its defects can be built. (Focused transceivers allow some depth resolution.)

Disadvantages of the Ultrasonic Technique:

- A couplant must be used between the transceiver and component. This may be in the form of a gel, or water in the case of "C" Scan, and may contaminate the specimen.

- "C" Scan requires component removal from the aircraft.

- The pulse-echo technique allows detection of damage from only one side of the component.

- The through-transmission technique requires access to both sides of the specimen and is best suited to the production environment.

- Calibration standards are required for each material and thickness. See Figure 7.4 (Figure B-5 from Ref. 7.1).

- Experienced operators are necessary.

- Careful interpretation of data is necessary, particularly with honeycomb sandwiches. For example, disbond between the skin and honeycomb will give a return signal which is greater than a bonded skin and honeycomb because the signal returns from the upper skin lower surface without transmitting down and up the honeycomb wall.

Fokker Bond Tester: This instrument also employs an ultrasonic technique, which uses shifts in the through-thickness resonant frequency of a bonded joint to detect disbonds or poor cohesion due to inadequate curing of the adhesive. The principle is different from the ultrasonic methods mentioned above.

However, the frequencies employed are well into the ultrasonic range, and thus a coupling fluid is required. See Ref. 7.2 and Figure 7.4 (Figure B-5 from Ref. 7.1).

7.4 X-Ray Methods

Conventional x-radiography (x-ray) of carbon fiber-reinforced composites is difficult because the absorption characteristics of the fibers and the resin matrix are similar and the overall absorption is low. Therefore, the volume fraction or the stacking sequence of the fibers cannot be obtained. The properties of boron and glass fibers are more suited to x-radiography, and in these materials, the volume fraction and the fiber alignment have been checked. Inspection of carbon fiber composites is facilitated by incorporating lead glass tracer fibers into the material. Detection of delaminations by radiography is difficult because they tend to be normal to the x-ray beam and thus make little difference to the overall absorption. Some say that it is almost impossible with conventional radiography. Voids (porosity) and translaminar cracks can be detected if they are of appreciable size in relation to the specimen thickness. Foreign inclusions such as swarf in solid composites and in honeycomb constructions the position of core inserts and shims and damage to the core all can be detected. Water in honeycomb structures after long periods of service can also be found by x-ray. Disbonds in adhesive joints made with metal adherends cannot be detected by x-radiography because absorption by the adherends is high and any lack of adhesive makes little difference to the overall absorption. Disbonds in joints made with composite adherends behave similarly to delaminations. Soft (low-voltage) x-rays (i.e., those of lower frequency and longer wavelength than hard x-rays) are generally used with composites in conjunction with high-contrast recording techniques. Low-energy x-ray tubes containing beryllium windows and being capable of operating at voltages between 5 and 50 kV are usually employed. X-ray inspection may be considered as complementary to ultrasonic inspection because it provides indications of defects in planes perpendicular to those detected using ultrasonics.

Advantages of X-Ray Methods:

- X-ray inspection may be used to detect transverse cracks, inclusions, honeycomb core damage, honeycomb moisture ingress (water as shallow as 10% of the cell depth), porosity, and delamination (if a radio opaque penetrant is used).

- When used with a radio opaque penetrant and a radio opaque fiber system, x-ray inspection may provide an accurate picture of damage in a complex structure.

- Stereographic use allows some depth resolution (e.g., 15° source tilt gives ±1 mm [0.04 in.] depth resolution, and 8° source tilt gives ±10 mm [0.4 in.] depth resolution.

- The x-ray technique may be used to measure fiber volume fraction (V_f) and fiber alignment when material characteristics allow (e.g., GRP).

Disadvantages of X-Ray Methods:

- Considerable safety protection is required when this technique is in use. The hangar area requires evacuation, thus imposing a downtime penalty on the production plan.

- The equipment is not easily portable.

- The usefulness of the equipment is limited by access.

- The use of penetrants contaminates the component. Organic penetrants are affected by moisture, which may alter the recorded results. Halogen-based penetrants may result in stress corrosion.

- Some large flaws may not be detected if relative absorptions are similar (e.g., a tight delamination without access for a penetrant). The problem can be partially resolved by using low-intensity sources. DERA, Farnborough, England, has 0.6 kV source equipment, but this is not yet available to airlines and other industrial users. Such low kilovolt sources risk fogging of the image due to the presence of air particles. An alternative and practical solution is to drill two 6-mm (0.25-in.) diameter holes down to the defect and then inject penetrant. This can be worked in under load. However, this technique requires accurate drilling and a knowledge of, or the ability to detect, the depth of the defect. The penetrant also presents a contamination problem and must be removed.

- Experienced operators are required.

7.5 Eddy Current Inspection

Eddy current inspection of composites is limited to those with a conducting phase, and the measurements obtained are sensitive to the volume fraction and integrity of that phase. Therefore, eddy current inspection can be used to check the volume fraction in a carbon fiber composite. This is not always easy, especially with cross-plied laminates. Defects such as fiber breakage and the presence of conducting inclusions also can be revealed. Some work has indicated that the method could be used to determine the stacking sequence in CFRP laminates. Eddy current methods are relatively insensitive to porosity, nonconducting inclusions, and delaminations. Work by Boeing (Ref. 7.1) concluded that, "The eddy current method can detect surface and subsurface fractures and small, sometimes nonvisible, impact damage in fabric or cross-plied tape lay-up. Its optimum application to date is fracture detection in substructure beneath an overlying member such as a laminated skin. As this can be performed with currently available instruments, it would be preferred over other inspection methods." See Figure 7.4 (Figure B-5 from Ref. 7.1) for calibration standards. Fortunately, the method is complementary to ultrasonics in materials such as CFRP because it is sensitive to those defects that are difficult to find using ultrasonics. The resistivity of carbon fibers is much greater than that of metals; thus, lack of penetration depth usually is not a problem.

7.6 Thermography

Two thermographic methods exist today:

1. **Passive:** The response of the test structure to an applied heating or cooling transient is monitored.

2. **Active:** Heating is produced by applying cyclic stress to the structure either in a fatigue machine or in resonant vibration (SPATE system).

In both cases, the surface temperature of the structure is monitored, usually with an infrared camera, and anomalies in the temperature distribution reveal the presence of defects.

The passive technique has been more widely applied than the active technique, and its performance strongly depends on the heat source used, a flash gun generally being most suitable. The conductivity and anisotropy of the composite are also important. For example, in CFRP, the conductivity in the laminate plane is approximately nine times that in the through-thickness direction, which tends to obscure defects that are not close to the surface.

Some work at the Harwell Laboratory in Oxfordshire, England, has shown that the method is made more convenient by using a video recorder to store the rapidly changing temperature pattern after the surface of the structure is heated. In this way, defects in conducting materials whose effect on the temperature distribution is short-lived may be detected. The method can be used with the heat source and the camera on the same side of the structure (pulse-echo technique) or on opposite sides (through-transmission method). The through-transmission method can detect deeper defects than the pulse-echo technique; however, for defects close to the surface, the pulse-echo technique is superior.

Advantages of Thermography:

- Thermography is a quick means of inspecting large areas of structure.

- It can be used to find disbonds in adhesive joints, delamination in composites, and inclusions whose conductivity differs significantly from that of the base material. Any base material will have only one value of conductivity.

- Work at some airlines has shown that thermography is useful in detecting moisture ingress in composites with honeycomb cores. In this case, the camera must be used within approximately 20 minutes of landing from a high-altitude flight, while the moisture continues to be present in the form of ice.

- Thermography is convenient compared with x-ray methods because other staff members do not have to leave the aircraft while the process occurs.

Disadvantages of Thermography:

- Equipment costs are high, and the method is not as sensitive as ultrasonics in detecting disbonds and delaminations.

- Skilled interpretation and knowledge of aircraft structure and systems are required to determine which heat sources and sinks are real defects.

- The aircraft must be accessed soon after landing to maximize resolution of problem areas. More complex and expensive portable equipment is available, which provides its own heat source. This source moves in conjunction with the scanning camera and reduces this problem to some extent.

- Thermography cannot be used with thermally conductive materials such as metals.

- Climatic conditions may be unsuitable for effective inspections.

7.7 Bond Testers

The title "bond tester" usually is reserved for instruments that use the mechanical impedance method. These instruments may be called bond testers, acoustic flaw detectors, or mechanical impedance analyzers. These instruments have been in use in the Soviet Union (now the C.I.S.) for approximately 30 years and are marketed in the United Kingdom by Inspection Instruments. These instruments use measurements of the change in local impedance produced by a defect when the structure is excited in the frequency range of 1 to 10 kHz to detect the presence of defects such as delaminations and adhesive disbonds.

In the United States, Shurtronics Inc. produces an instrument called the Shurtronics Harmonic Bond Tester. It operates on the principle that if a layer of a structure is separated from the base layer(s) by a delamination or disbond, the separated layer can resonate, and, at certain excitation frequencies, it will vibrate at considerably higher than normal amplitudes. This increase in vibration level can be readily detected. If the resonant frequencies of the layers above defects of a size that can be detected lie within the operating range of the instrument, the technique can be a rapid and attractive means of inspection.

These vibration techniques can work at frequencies below 30 kHz; thus, a coupling fluid between the transducer and the structure is unnecessary. The equipment required usually is portable; therefore, the methods are attractive for field use. Also, they are well suited to the inspection of honeycomb constructions and can detect, for example, barely visible impact damage (BVID) in solid laminates. Measurement of the resonant frequencies of a structure can be performed rapidly by analyzing the sound produced when the structure is tapped. (It should be stressed that this measurement is very different from that made in the coin tap test that was previously discussed. The natural frequencies are obtained from a single tap anywhere on the structure; the coin tap test requires a point-by-point investigation of the structure.) Gross defects such as incorrect lay-up in structures fabricated from pre-preg, errors in winding angle in filament-wound components, incorrect fiber volume fraction, and generalized environmental degradation produce readily measurable changes in resonant frequencies. A whole structure can be tested in less than five seconds; therefore, if the detection of small, localized defects is unimportant, the method is very attractive.

7.8 Moisture Meters

Moisture meters should always be used when making repairs to parts made with fiberglass or aramid fabrics or tapes. They also can be used to check aramid honeycombs. The usual type relies on radio frequency dielectric power loss attributed to an increase in the conductivity of the composite due to moisture absorption. Therefore, this type of meter cannot be used with carbon or any other conductive fiber.

Because the conductivity of the fiber is high, any change resulting from moisture cannot be detected. There is a new meter, of Japanese origin, which operates on a different principle that is said to work with carbon composites. However, the price is high, and the meter is not yet in general use. Another meter, called "Hector," is supplied by GMI and measures moisture vapor coming from the specimen.

The most commonly used meter of the existing type is the Model A8-AF (Figure 7.5), produced by the Moisture Register Products Division of the Aqua Measure Instrument Company, 1712 Earhart Court, P.O. Box 369, La Verne, CA 91750-0369. This meter was developed for use with radomes but can be used on any nonconductive composite part. It is battery powered, and the battery condition must be checked before use and the instrument zeroed. The operating instructions are simple but must be followed. The dial has four bands: green, yellow, orange, and red. All composites must be dried until well into the green band before bonding can be attempted.

In the United Kingdom, an alternative instrument known as the M49 Moisture Monitor is supplied by J.R. Technology Ltd., 81 North End, Meldreth, Royston, Hertfordshire SG8 8NU, United Kingdom. This instrument also is simple to use and operates on a similar principle (see Figure 7.6).

Limitations of Moisture Detectors: These moisture detectors emit electrical energy and cannot be used in the presence of metals or other conductive materials, including antistatic coatings containing carbon. Panels having such coatings must be tested from the back side, which typically is not coated with antistatic paint. Caution must be exercised in the case of some panels that contain buried metallic doublers. These doublers will give a false indication of moisture and may cause a panel to be removed needlessly. Conversely, if moisture is genuinely present in the area of the metal insert, this type of meter cannot positively identify it.

Fig. 7.5 Moisture Meter A8-AF. (Courtesy of Moisture Register Products.)

7.9 Interferometry

An excellent paper on interferometry was presented at an SAE TOPTEC meeting entitled "Advances in Composite Repair," held in Seattle in November 1993 (Ref. 7.3). This paper defined the technique as follows:

> Holographic interferometry provides an object image from the properties of reflected light, using their intensity, wavelength, and phase. Phase provides the 3-D effect.
>
> The light source must be coherent, in that it is monochromatic and has simultaneous emission; thus lasers are used. Interference occurs when an object changes its relative position and, after double exposure to laser light, the light's phase has changed. The double exposure occurs before and after object movement. This can be produced by lightly loading the component mechanically or by the application of localized heating. The method works on both metal and composite structures. The method can detect loose rivets, cracks under rivet heads, and weak adhesive bonds in metal structures; in

Fig. 7.6 M49 Moisture Meter. (Courtesy of J.R. Technology Ltd.)

composites, it can detect impact damage, heat damage, weak bonds, and delamination in sandwich structure. (See Figure 7.7, page 8 from Ref. 7.3.) This method appears to be well set for further development and wide application.

7.10 Damage Types

Inspection of composite parts is also conducted at the manufacturing stage, but any defects should have been found before an aircraft structure or component enters service. At this stage, inspections are made for defects such as voids in the matrix, broken or buckled fibers, incorrect fiber type or orientation, or disbond resulting from contamination.

Portable holographic interferometry testing systems

Equipment required:

Portable laser, with optical lens, filters, and tripod
Holographic plates
Darkened area
Photographic processing equipment
Light source
CCD camera (for computer capture)

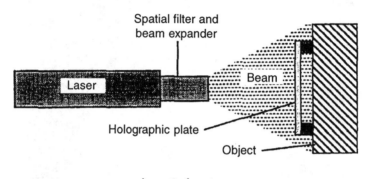

Layout of system

Fig. 7.7 Diagram of holographic technique. (Courtesy of R.B. Heslehurst.)

Incorrect fiber type is usually found at the kitting stage and also by reference to the manufacturer's technique sheet in the case of British Aerospace. Other OEMs probably have similar procedures. After entry into service, several types of damage are possible (e.g., disbond, delamination, cracking, heat damage, chemical damage, moisture ingress, fatigue damage, edge erosion, worn fastener holes, and screwdriver damage). In many cases, the complete area of damage must be mapped to ensure that all damaged material is cut away to enable a sound repair to be made. The most commonly used inspection methods are described above. By using one or more of these methods, the area in need of removal can be defined. The types of damage are given below, and their causes are listed in the order of production and repair, followed by in-service causes in each case.

Disbond: This term describes the separation between one structural element and another (e.g., when skin-to-honeycomb or skin-to-stringer bonds fail) in preference to the term delamination, which is not strictly correct. Possible causes include:

- Backing film left on pre-preg or film adhesive. Fortunately, this is uncommon, but several cases have been recorded.

- Contamination of pre-preg or film.

- Use of time-expired material.

- Incorrect cure cycle (temperature and/or pressure).

- Adhesive omitted at the build stage. This is rare but has occurred.

- Aramid or fabric type honeycomb contaminated or not fully dried before use. If aluminum alloy honeycomb is used for production or repair, it should always be of the corrosion-treated and resin-dipped type or anodized. Many problems of disbond have been found in service when the untreated type has been used. With any type of honeycomb, the use of a film adhesive in addition to surplus resin from the pre-preg can often help to ensure good honeycomb bonding (i.e., adequate fillet size).

- Impact.

- If moisture ingress becomes sufficient to fill the honeycomb, then the freeze/thaw cycles from each flight can cause ice expansion, which can push the honeycomb from the skin in affected areas.

Delamination: This term is reserved for disbonding between individual plies of one structural element. It describes laminate separation into one or more layers. Sometimes delamination may occur simultaneously between several layers of a laminate. On other occasions, a laminate may separate at only one interface. Possible causes include:

- Backing film left on one or more layers of pre-preg.

- Incorrect cure cycle (temperature and/or pressure).

- Use of time-expired material.

- Contamination of pre-preg before or during lay-up.

- Impact.

- Erosion when the leading edge of a part, especially a cowling, is projecting into the airstream without any wraparound layers or metal edge or other form of erosion protection.

- Overtightening of fasteners may cause delamination, especially at panel edges.

- Moisture ingress may have a similar effect.

- Lack of care when drilling a panel.

Cracking: Cracking may be caused by:

- Mismatch of thermal expansion coefficients between composites and adjacent metal parts. This is likely to occur if long composite panels are bolted to long metal parts (e.g., composite leading edge panels to aluminum wing leading edges).

- Local impact.

- Excessive flexure.

- Fatigue.

Heat Damage: Apart from any overshoots in the curing temperature cycle, which should be found from production or repair records and either accepted or rejected at that time, heat damage will be likely to occur only in service. Possible causes are as follow:

- Overheated engine parts causing overheat in nearby areas of cowlings.

- Lightning strikes, which usually affect radomes, tail cones, and the extremities of control surfaces and parts of engine cowlings.

- Overheating in wheel bay areas because of overheated brakes.

- Thermal spikes caused over the entire surface of military aircraft during a supersonic dash. Design will have to cope with the effects of such dashes, which must be regarded as being within normal mission profile, and will have to specify appropriate inspection to cope with exceedences that are unintentional but likely to occur occasionally.

- Fires sometimes occur around engines and auxiliary power units due to fuel pipe fractures or leaking joints. Rarely, but nevertheless on some occasions, composite parts will have to be inspected after fire damage. The extent of damage will have to be defined, and a decision must made to scrap or repair. This may occasionally occur when an aircraft is largely destroyed by fire and some removable parts become available as spares for use on other aircraft if those parts can be repaired and recertified.

Chemical Damage: Possible causes of chemical damage are the following:

- Paint stripper, especially of the methylene-chloride type

- Long-term contact with hydraulic fluids

- Prolonged contact with solvents (e.g., acetone, MEK, MIBK, or chlorinated hydrocarbons)

Moisture Ingress: Chapter 16 covers moisture ingress in detail. As discussed there, the ingress of moisture should be prevented by good design as much as possible. However, parts in service should always be monitored for this problem and action taken if moisture is found.

Fatigue: Fatigue is a rare problem with carbon fiber parts, but it can occur if the parts are heavily loaded as a result of incorrect rigging where repeated loading can take place. It also can result from thermal mismatch on long parts. Although glass fiber composites are more likely to suffer fatigue than carbon fiber parts at equal fractions of ultimate load, they usually are lightly loaded. Thus, fatigue does not normally occur on aircraft parts in service. For more details, see Chapter 5.

Fatigue damage also can occur in repair joints. Work done on scarf-jointed repairs has shown that the fatigue strength and the life of repairs are lower than those of virgin material, but that peeling begins at the ends of the joint long before the joint strength is seriously reduced. This means that normal inspection should reveal any such problems in enough time to perform a further repair. See Ref. 7.4, which states, "The fatigue of an adhesively bonded joint in a composite material is complicated as fatigue damage may be initiated in either the adherends or the adhesive, and the failure path may encompass one or both materials." It further states, "The damage that develops in a composite during fatigue, and its rate of development, depends on the stacking sequence of the laminate and is independent of the load history." The toughness of the matrix also has an effect; therefore, it is impossible to list types of damage that will always occur because many different matrix resins, adhesives, and adherend materials exist.

Erosion and Wear: These topics are covered in Chapter 16.

7.11 Sources of Mechanical Damage

1. Impact damage is common and occurs from dropped tools, careless handling, and impact of ground vehicles, staging, or ladders, in addition to bird and lightning strikes and hail damage in flight. Note that it is necessary to look on the aft face of any impact-damaged structure. The damage tends to spread in a cone shape from the point of impact, and damage on the aft face often is much larger than at the point of visible damage on the outside face.

2. Screwdriver damage can cause severe harm to the edges of parts. Common problems are:

    • Using screwdrivers to open cowling and other latches in winter.

    • Lifting floor panels with screwdrivers or similar objects, especially if this is done without first removing sealant around the edges. Nylon cords usually are provided for this.

- Composite surfaces are softer than metals. Even when screwdrivers are used for their proper purpose, they can easily slip from the recess in a screw and cause damage. Any fastener with a worn recess should be replaced, and likewise any worn screwdriver bit requires replacement. Screwdrivers should be held square to the surface in all directions to reduce the chance of slippage.

3. Lifting the edges of any type of panel with the wrong tool. This may be a screwdriver or similar, but a wide-bladed plastic tool with a shallow tapered edge should be used to spread the lifting load rather than to localize it at a point.

4. Disassembly damage refers to damage that is induced during disassembly for repair. This is particularly a problem when removing metal parts from composite parts (e.g., aluminum alloy hinge brackets from composite trim tabs).

5. Engine disintegration.

6. Tire protector disintegration.

7. Erosion damage occurs at panel leading edges if they are unprotected and are not completely flush. It also occurs on radomes and wing, tailplane, and fin leading edges. Sand and dust storms can aggravate the problem considerably.

8. Stone impact can result from jet efflux from another aircraft that is too close during taxiing or turning on the ground.

7.12 Damage Mapping

Mapping of the area of damage may require several techniques. Because the damage, or disbond, may occur in different locations on a part or arise from different causes, these may require different techniques. In other cases, water ingress may have to be detected. The NDI technique chosen will depend on the type of structure and the type of problem under investigation.

Honeycomb Structures: For example, disbond between the skin and core of a honeycomb or other type of sandwich panel may be easily detected by a tap test if it is located on the outer face of the panel and therefore readily accessible. If the disbond is between the inner skin and the core, more sophisticated equipment will be required to find the problem if the panel must be inspected *in situ* and access to the inner face cannot be gained easily. In this case, the simple tap test is of no value. For parts with a honeycomb core up to 15 mm (0.6 in.) thick, an acoustic impedance method using MIA 3000 equipment provided a solution. For honeycomb thicknesses greater than 15 mm (0.6 in.), the acoustic impedance method was found to be unreliable. In this instance, an old suction-cup method was used on the outer skin, and the deflection of the outer skin and honeycomb was measured. This method can detect only relatively large areas of disbond and is not considered suitable for general use. If possible, access to the inside face should be obtained, and then the inner face can be tested directly or a through-transmission method such as "C" Scan may be used. Moisture in the honeycomb cells can be found by x-ray or thermography.

Solid Laminates (Monolithic Parts): At the present time, pulse-echo ultrasonic methods are considered the best solution for NDI and damage mapping of solid laminates. They can measure the depth of a defect in a laminate. Penetrant-enhanced x-radiography can be used if a surface crack is present to allow the penetrant to reach the delaminated layer. Penetrants are toxic and can cause surface contamination of the composite; therefore, this tends to remain a laboratory method. See Table 7.1 for other methods that can be used.

Table 7.1
Table of Defects and Appropriate NDI Methods

Service Defects			Inspection Method					
	Visual	Tap Test	Acoustic Impedance	Ultrasonic	Thermography	X-Ray	Hologram	Eddy Current
Disbond	x	x	x	x	x		x	
Delamination	x	x	x	x	x		x	
Impact	x			x		x	x	
Cracks	x			x		x		x
Heat Damage	x	x		x			x	
Chemical Damage	x							
Moisture				x	x	x		

This table gives a rough guide only because the methods listed may be effective on one occasion and not on another (e.g. disbond and delamination can only sometimes be observed visually if they are at or close to the surface).

7.13 Assessment of Damage Significance

Because the failure mechanisms of composites are less well understood than those of metals, assessment of the significance of a defect is often difficult after it has been detected by NDI. This leads to conservative testing philosophies, and much research activity is aimed at determining what constitutes an acceptable defect. Always remember that composites are most likely to suffer strength loss in compression, and thus any defect that damages compression strength should be treated seriously. Composites can lose some of their tensile strength, but the loss of compression strength is more serious, especially under hot and wet conditions. Shear forces are reacted by tensile and compressive forces acting together at $\pm 45°$ to the line of the shear loading; thus, compression performance is always important. Any damage or deterioration found at any time, whether during a scheduled inspection or not, should be assessed in accordance with the SRM.

The defects most commonly found in service are listed below, together with their likely effects.

Surface Scratches: The maximum effect of a surface scratch, provided that it does not penetrate more than one layer of fiber, is to reduce the strength in proportion to the loss of that layer. However, note that thin skinned honeycomb structures often have skins of only two or three plies; in some cases, very lightweight panels have only one. In these instances, a scratch through one ply is significant and must be treated accordingly.

Delamination: The effect of delamination will depend on its size as a proportion of the size of the part and where it is located in the laminate. Its primary effect will be to reduce compression strength and hence shear strength. If it is located at the center of the thickness of a laminate, the compression strength will be seriously reduced because it is a function of the second moment of area or moment of inertia of the laminate acting as a column in compression. Delamination of several layers is serious and reduces compression strength to a very low level. The effect on tension strength will be quite small. Delamination is the most serious defect in a composite, apart from cracking, and is also the most commonly found defect. It often occurs at the same time as heat damage.

Translaminar Cracking: Although uncommon, this type of cracking will reduce the tensile strength to zero, if it is a full-depth crack, or in proportion to its depth if it is a partial-depth crack. If the crack is more than one layer deep, it will probably act as a stress concentration and reduce tensile strength by more than the loss of thickness alone. If the laminate is stressed in bending, the loss of strength in the outer layer will mean a loss of strength greater than that due to thickness loss. This occurs because the bending strength is a function of beam depth squared, and the stress concentration effect will have to be added to this.

Bearing Damage (Wear in Bolt Holes): This type of damage is fairly common, especially when access panels are removed frequently for servicing. It may allow countersunk bolt heads to pull through a skin if it becomes too extensive. Shear-out failure also could occur if the edge margin on a panel is significantly reduced.

Matrix Crazing: Matrix crazing should become less common with improved design. It occurs when brittle resins are used with an elongation to failure much less than that of the fiber type in the composite. It has little effect on tension performance but may be expected to reduce compression and shear performance and also to allow fluid penetration.

Aging of the Resin: Some resins degrade in the presence of ultraviolet (UV) radiation. This should affect only the first 0.5 mm (0.02 in.) at most, but it could cause resin crazing and also a reduced resin modulus at the surface. These effects may result in a slightly reduced compression strength of the part concerned. Modern resins should contain UV stabilizers.

Bond Failures: In service, these should result only from impact damage of some sort or as the long-term result of fluid penetration. Otherwise, they should be regarded as manufacturing defects. Disbond is the term used when two pieces of solid laminate that have been adhesively bonded come apart or when the honeycomb core becomes disbonded from the skin in a sandwich panel. Translaminar cracking and bearing damage are almost always associated with a certain amount of delamination. (See Ref. 7.5.)

Heat Damage: Heat damage should be taken seriously because, as previously mentioned, a loss of composite compression strength results from several sources at the same time. The resin is degraded by overheating and may lose some of its components if the level of overheating is severe. Some types of fiber may be degraded. Because of the sudden expansion of moisture, which is always present in resins to some degree (however small), delamination usually occurs at the same time. Severe heat damage (e.g., lightning strike) can cause local temperatures of 700 to 1000°C (1292 to 1832°F) or more and can gasify the resin completely in a small area so that the fabric is completely dry and appears never to have been bonded with any resin. This commonly occurs if the lightning current has passed through a steel bolt. The result is usually an oversized hole with dry fabric around it and a bolt that has received what resembles a spark-eroded undercut. In such cases, the resin should be checked for some distance around and any obviously affected areas removed. Sometimes, the resin can be brittle and crumbly. Those experienced in the removal of heat-damaged material can tell, by the feel of it during sanding, when they have removed the affected material and have arrived at an undamaged area. This feel can be learned only by working on burned material if suitable pieces are available for training purposes. A suitable hardness test or scratch test may help to define heat-damaged areas. Discoloration of paint indicates overheating, and such areas also should be removed. Lightning damage to radomes and other honeycomb panels also should be inspected on the inside. Experience has shown that a small burn on the outside usually indicates delamination/disbond many times larger on the inside. If a lightning strike has exited at a control surface, the operating ball races should be checked because the current passing between the balls and the race attempts to weld them together. Bearings often feel gritty (i.e., they no longer operate smoothly) after a lightning strike, and this should be assessed with a view to replacement. When assessing damage after a lightning strike, the following additional points should be checked:

- The condition of electrical bonding jumper cables and their attachment points.

- Electrical continuity of the entire lightning protection system.

- Condition of any flame-sprayed aluminum coating or woven wire coating.

- The static discharge system. Note that this system is intended to discharge the static electricity built up in flight as a result of air friction. It is not intended to provide lightning protection but can sometimes be damaged in a lightning strike.

The topic of heat damage at lower levels of overheat is also discussed in another paper by Janke in Ref. 7.3 and in a separate paper by Powell. These authors successfully used two specialized techniques to detect heat damage and then performed mechanical tests on specimens to show a correlation of their findings on chemical changes and weight loss with a reduction in mechanical properties. The methods used were diffuse reflectance infrared Fourier transform (DRIFT) and laser-pumped fluorescence (LPF). The effects of short-term, elevated temperature heating on the room-temperature, compressive, interlaminar shear, flexural strengths, and Shore-D hardness properties of dry and wet preconditioned laminates were tested, and the following results found:

- Laminates suffered significant and permanent reduction in room-temperature mechanical strength properties after being subjected to temperatures above the recommended upper service temperature limit of 180°C (350°F).

- Below a certain temperature/time exposure threshold of 288°C (550°F) for 60 minutes, these laminates visually and microscopically appeared to be undamaged but actually lost a significant percentage of their original strength.

- Exposure above 288°C (550°F) resulted in even greater losses in strength. See Figure 7.8.

These results make several important points:

- Exposure to any temperature above the recommended maximum causes some losses.

- Temperatures significantly above the recommended values (i.e., 288°C [550°F]) cause severe strength losses, even when no visible evidence of damage can be found.

- Much higher temperatures (i.e., above 288°C [550°F]) cause even greater losses of strength.

- These higher temperatures are greatly exceeded when lightning damage occurs.

- Hot spots on engine cowlings as a result of overheated engine components could reach temperatures high enough to cause some strength losses. These effects may or may not be visible, depending on the temperature.

- The good news is that two methods of assessing the extent of the damage are now available, and both are portable.

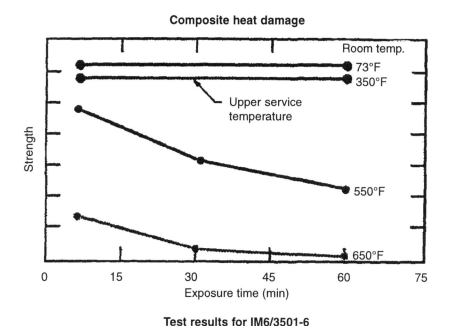

Test results for IM6/3501-6

Fig. 7.8 Strength loss in composite material as a result of exposure at various temperatures above the maximum recommended value.

- The bad news is that control data for each matrix resin in use may be necessary to make a valid check on the strength loss likely to have been caused in any particular case.

These tests were made on overheated areas that were not always visibly damaged, as is the case with severe overheating. They clearly indicate that where overheating is suspected, careful inspection also should be done by appropriate methods even if no visible evidence exists. Such events could occur on cowlings after known overheating of an engine, on an associated component such as a generator or pump, or in wheel bays due to overheated brakes. In these cases, reference should be made to the OEM to establish the likely temperatures involved and the need, if any, for specialized inspection.

Boeing trials (Ref. 7.1) found that heat damage of any consequence results in visible damage, such as scorched, burned, or blistered paint.

Severe damage is indicated by charred resin, loose fibers, and flaked-off skin.

Careful damage assessment is particularly important in the case of overheating, and all available methods should be used, as detailed previously, to ensure that all degraded material is removed before a repair is started.

Chemical Damage: Penetration of the molecular structure of any resin system by the fluids listed under Section 7.10 (causes of chemical damage) may be expected to reduce the modulus of the resin and possibly also its strength. Reduction in matrix resin modulus leads to a reduction in the compression strength of a composite and all other matrix dependent properties. Fluid absorption also leads to a reduction in the matrix resin glass transition temperature (T_g) and hence the maximum temperature at which the structure can be safely used. Methylene-chloride-type chemicals may almost completely remove some resins from a laminate. Cases have occurred where two layers of glass fabric have been completely disbonded from a honeycomb core. Skydrol and other hydraulic fluids soften composites over long periods of time and contaminate their surfaces. Chemically contaminated areas may require complete removal if the part is structurally important, although such parts are usually thick and replacement of the affected surface layers may be sufficient. For sandwich panels with thin skins, it may be necessary to replace complete skins or local areas, depending on the extent of the damage. In other cases, it may be more a matter of specialized solvent cleaning to remove contamination to ensure a sound repair to some mechanical damage.

A scratch test or hardness test may be needed to define the boundary of the damage. If in doubt, seek advice from the manufacturer.

Moisture Ingress: It is particularly important that any damage that could allow moisture penetration into honeycomb areas or any other internal areas should be covered with Speedtape or other suitable material to prevent moisture ingress before a repair can be scheduled. If this is not done, the result may be wet honeycomb and probable further disbond, which will almost certainly mean a larger repair. Moisture ingress in radomes should be corrected as soon as possible because the problem will reduce radar performance. This is usually a result of pin holes in the outer skin, caused by small lightning strikes. These pinholes result in water entering the honeycomb through the skin. In the case of aramid honeycombs, the water can slowly penetrate the cell walls and spread across a wide area. When the area is large enough, signal transmission is adversely affected. Another known problem is water penetration into the honeycomb core of the trailing edge wedges of aircraft control surfaces such as elevators, ailerons, and rudders. These often are made of thin fiberglass skins and honeycomb to reduce weight. Again, pinholes in the skin, sometimes due to the use of only one or two layers of skin and a minimum amount of resin, result in the honeycomb becoming partially filled with water. The problem in this case is the weight and balance effect of the added water, which may take these parts outside the manufacturer's specified limits. These balance limits are necessary because, in the unlikely event of a failure of the operating jack, the control surface could flutter and cause control problems and structural overloading. Moisture absorption reduces matrix resin modulus and hence the compression strength of a composite part. Resin tensile strength also may be reduced. Design allows for this because research has determined that the amount of water absorbed over the service lifetime of a composite aircraft is likely to reach a maximum of approximately half the saturation water uptake of the resin system used that would occur by total immersion for a long enough time. See Ref. 7.6.

Moisture penetration into honeycomb parts is the major problem, and any findings of this should be addressed according to the SRM. The maintenance of the specified paint scheme also is important because a good coat of paint helps to reduce the rate of moisture absorption into a composite and also provides some protection against UV radiation, which is intense on high-altitude flights.

Erosion and Wear: Erosion and wear should be rectified as soon as possible after discovery in order to minimize the problems of repair and to prevent further damage. The aircraft or component manufacturer should be notified, and a design solution should be requested. (See Chapter 16, Design Guide for Composite Parts.) Experience has shown that erosion is one of the most common problems with composite parts. It occurs because cowlings, doors, and other parts are not always rigged perfectly flush with the part in front of them and also because designers have not yet considered this problem sufficiently. A composite panel is similar to a pack of cards turned edgewise to the wind. The panel will delaminate in the same way as a weathered rock but faster, because the windspeed in flight is much higher and the temperature extremes are greater. Leading edges of composite laminates require a metal edging strip, or at least some wrap-around layers of fabric, to prevent edge delamination which quickly spreads as water enters and freezes, thereby wedging the layers apart. As the delamination becomes longer, each layer will act as the vibrating reed in a musical instrument, and delamination will grow until parts crack off. Repair and minor local modification becomes essential as soon as erosion begins to cause delamination.

Impact Damage: Impact damage often is described as a form of damage, but it is probably more accurate to describe it as a source of damage. Unless an impact is severe enough to cause disbond, delamination, or cracking, it is inconsequential. If an impact causes damage, then the damage will take the form of disbond, delamination, or cracking and can be treated as such. In extreme cases, holes may be punched through the skin; this has been seen even on laminates of 3 mm (0.12 in.) in thickness.

7.14 References

7.1 Phelps, M.L., In-Service Inspection Methods for Graphite-Epoxy Structures on Commercial Transport Aircraft, Boeing Company, prepared for Langley Research Center under contract NAS1-15304, Reference NASA CR-165746, November 1981.

7.2 Matthews, F.L. and Rawlings, R.D., *Composite Materials-Engineering and Science,* Chapman and Hall, London, UK, ISBN 0-412-55960-9 (hardbound) and ISBN 0-412-55970-6 (paperback), 1994.

7.3   Heslehurst, Rikard B. and Baird, John P., "Field Holography for Repair Patch
        Structural Verification, Australian Defence Force Academy," in "Advances in
        Composite Repair TOPTEC," 1–2 November 1993, Seattle, WA, Society of
        Automotive Engineers Continuing Professional Development Program,
        400 Commonwealth Dr., Warrendale, PA 15096-0001.

7.4   Robson, J.E., Matthews, F.L., and Kinloch, A.J., "The Fatigue Behavior of Bonded
        Repairs to CFRP," 2nd International Conference on Deformation and Fracture of
        Composites, The Institute of Materials, London, UK, March 1993.

7.5   George, D.L., "An On-Site View of the Inspection and Repair of Carbon-Fibre
        Composite (CFC) Aircraft Structures, *British Journal of NDT,* January 1985,
        pp. 22–26.

7.6   Wright, W.W., "A Review of the Influence of Absorbed Moisture on the Properties
        of Composite Materials Based on Epoxy Resins," Royal Aircraft Establishment,
        Tech. Memo MAT 324, December 1979; now Defence Evaluation and Research
        Agency (DERA), Farnborough, UK.

Chapter 8

Source Documents

This chapter introduces the wide range of documents provided by the original equipment manufacturers (OEMs) and regulatory authorities to enable airlines to meet the safety requirements of the regulatory authorities established by various countries or groups of countries. Examples include the Federal Aviation Administration (FAA) in the United States, which works under the Department of Transportation, the Joint Aviation Authorities (JAA) in Europe, and similar bodies in other parts of the world. These bodies produce Federal Aviation Regulations (FARs) and Joint Airworthiness Requirements/Joint Aviation Requirements (JARs), respectively. They also produce Advisory Circulars (ACs) and Airworthiness Directives (ADs). FARs and JARs establish design requirements for structural strength, performance such as the ability of a twin-engined aircraft to continue a takeoff at full load on one engine, fire safety, and crashworthiness. These organizations cover all aspects of flying, but those listed above are the most relevant to composite repairs. End users are responsible for being aware of these documents and their purpose and limitations. This is essential to ensure that maintenance is performed within the framework of legislation that has been enacted to guarantee as much as possible the safety of public and private transportation systems. In shipping, the equivalent organizations are The American Bureau of Shipping in the United States, and Lloyds Register of Shipping in the United Kingdom. These organizations issue design requirements for ships and smaller craft.

Although this chapter and the entire book have been written around aircraft requirements, the repairers of composite structures and parts in ships, trains, and road vehicles may find the information helpful and should work within any similar legislation applicable to their forms of transportation. These documents are subject to change to meet newly discovered situations, and their names may also change. Therefore, it is important to remain up to date and to use only the latest issues of all documentation.

8.1 Revision Systems

Most of the documents described below are subject to revision systems (i.e., they are updated periodically as required). The documents for a new aircraft type are initially updated more frequently (i.e., every few months); as the type becomes older, it may reduce to every few years. Each document contains a list of all the revisions made and the dates of those revisions to ensure that it is always possible to know how many revisions have been made and when those revisions were made. All revisions are maintained by the OEM, but each airline should have only the latest version available to its staff. All revisions must be incorporated as soon as possible after receipt to ensure that up-to-date information is always available to those directly involved in aircraft maintenance. Between major revisions, airlines are permitted to make temporary revisions (TRs) to documents specific to the types and models of aircraft they operate, provided that approval from the local regulatory authority is obtained. TRs usually are issued by OEMs and are fully incorporated into the relevant document at the next revision. An index of TRs also must be kept with the manuals, microfilms, or compact discs (CDs) to which they refer in order to enable maintenance staff to easily find new TRs.

8.2 Effectivity

Most aircraft are produced in a number of different models, and most airlines request some changes from the standard design. Therefore, it is important to check the effectivity of each part in the Illustrated Parts Catalog (IPC) and of each modification to ensure that the particular part or instruction applies to the specific aircraft on which work is being done. This is especially important if an airline, often a smaller one, has purchased a number of aircraft secondhand. These aircraft may be all of the same nominal type and model; however, if they were originally bought by several different airlines, they may have significant differences and these must be considered. Each aircraft that is built has a line number and a customer number allocated by the manufacturer and a registration number given by the country in which it is registered. This registration number can change each time an aircraft is sold. The line number never changes; it identifies the aircraft for life. IPCs may be customized, especially for large operators, and an effectivity code will be given to each aircraft. Service Bulletins detailing modification and/or inspection procedures should supply the line number and registration number to ensure that effectivity codes can be checked and that the work can be applied to the correct airplane.

8.3 Drawing Numbering Systems

Manufacturers tend to change their drawing systems occasionally, certainly as far as their part numbering is concerned. This means that the part numbering system may differ from one aircraft type to another, even from the same manufacturer. Also, differences exist among manufacturers. The part numbering system for an aircraft type should clearly identify the type concerned and each individual part. It is important to understand the drawing numbering system for the type on which work is being done. This will ensure that the minimum time is taken to find the correct information. Some parts may be used on several aircraft types; therefore, a few parts may have drawing numbers that relate to an earlier aircraft type from the same company. In the front of the catalog, the IPC lists all part numbers featured in it. The IPC also gives guidance to help understanding of the drawing numbering system for that type. Likewise, it clearly identifies parts supplied by outside vendors, such as electrical components, hydraulic components, switches, and valves. These will have part numbers allocated by the vendor from his own drawing numbering system.

Production Drawings: These drawings are produced primarily by Computer Aided Design (CAD) methods and usually include the parts list. In the case of Airbus Industrie aircraft, the parts list is now on a separate microfilm. In many cases, only one drawing will exist for a part, although that drawing may consist of several sheets. All variations to the design of that part will carry a dash number (e.g., -50). This method has economic advantages for the OEM because it avoids the need for having a separate drawing for each design variation. However, maintenance organizations and other end users often have problems with this procedure. Because all the information is supplied on one drawing, the end user may have difficulty in deciphering which pieces of information are relevant to the work required on a specified dash number of a particular part. This information may also be needed to confirm the identification of a part. Although it is appreciated that the OEM's staff are the major end users of production drawings, the drawings often are needed by airlines for the incorporation of modifications and sometimes for the local manufacture of a relatively simple part, such as a floor panel, if spares are unavailable or if local manufacture is the normal method for such parts and the airline has regulatory authority approval for this type of work. The user may even be required to produce a drawing with only that user's type specific information on it. Otherwise, the part produced or modified may have unnecessary holes or may even be made to the wrong shape. Although the method is economical for a large OEM because it avoids the necessity of having a new drawing for each customer, it is not user friendly.

Most OEMs use Drawing Change Notices (DCNs) for rapid changes in production. These should be incorporated on the main drawing within a specified period (i.e., three months). If this is not done, then several DCNs can be outstanding at any time. Revised drawings are supplied to airlines at regular intervals, where a contract exists to do so, and the DCNs should also be supplied if they have not been incorporated at the time the new microfilms are produced. Problems occur if this is not done, and two actions are essential:

1. OEMs should make considerable effort to ensure that the number of DCNs requiring incorporation on the main drawing is held to a minimum.

2. Airlines should verify that they have all outstanding DCNs before making or modifying a part, if they have the approval to do so, and that the drawing they have is the latest issue.

Ply Charts: All drawings for composite parts, including those drawings in Structural Repair Manuals (SRMs), have a warp clock or diagram showing the 0° direction for each composite ply or layer. These drawings also have a cross section of the panel and ply tables, which specify the type of material for each ply and its direction of orientation. See Figure 8.1.

Repair Drawings/Repair Schemes: Repair drawings and repair schemes may be made by the airline or repair station if it has design authority from its regulatory authority and if the repair exceeds the limits of the SRM or is not covered by the SRM. If the repair is large enough to require OEM assistance, the repair drawing may either be made by the OEM for the airline or suggested by the airline and sent to the OEM for approval.

8.4 Internal Documents

Each airline has its own internal documents for maintenance and repair. These are usually as follows, although their names may vary among airlines.

*   **Design Deviation Authorizations, Engineering Instructions**, or similar titles permit minor variations from stated limits or the use of an alternative material or process. These documents are usually issued for a single application and may have a time limit, after which they become invalid.

*   **Repair schemes** may or may not include one or more drawings and may cover general repairs of a minor nature or large repairs. They may be produced under delegated design authority or with the assistance of the OEM, and they may be for "one-off" repairs to damage or for repetitive repairs to cracks, wear, or deterioration on all aircraft in a fleet. They are usually classified as permanent repairs. However, if they have time limits or are to be classified as temporary, then this must be clearly stated on the repair scheme to ensure that appropriate production planning action can be taken.

*   **In-house modifications** may be called company modifications and may be desired to ease maintenance, to change some aspects of the aircraft interior, or to strengthen a part to increase its service life. In some cases, such modifications may be adopted later by the OEM and may become a Service Bulletin.

*   **Alternative parts procedures** are used to authorize the fitment of an alternative part and hence the purchase of improved or equivalent parts from another source. It is imperative that it be established that such parts are fully acceptable in every way before

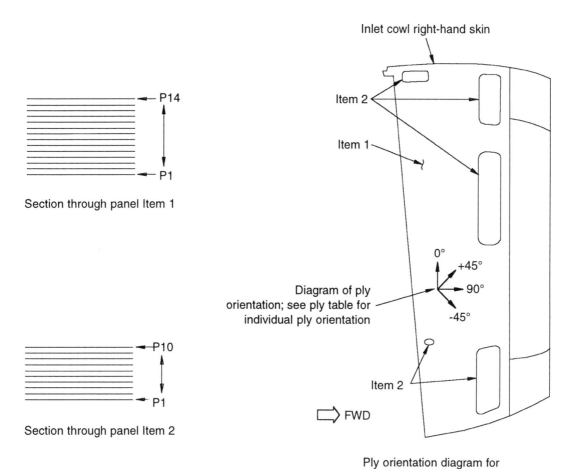

Item No.	Ply No.	Material	Ply Orientation	Note Refer Sheet 1
1	P1, P5, and P14	B	±45°	A E
	P2 to P4 and P6 to P8	C	0°	A E
	P9, P10, and P13	D	±45°	A E
	P11 and P12	D	0° or 90°	A E
2	P1, P2, P5, and P8 to P10	D	±45°	A F
	P3, P4, P6, and P7	D	0° or 90°	A F

Fig. 8.1 Typical cross section showing ply lay-up, warp clock, and ply table. (From Boeing 757 SRM 54-10-01, page 7, dated 20 October 1992. Courtesy of Boeing Commercial Airplane Group.)

approval is given, and this must include obtaining the part from an approved and reliable supplier. Several cases of the sale of bogus parts have occurred in recent years. These bogus parts may be manufactured to an unacceptable standard or made from substandard materials. In other cases, parts with an expired fatigue life have been sold as new parts, and these have resulted in several fatal accidents. Constant vigilance is necessary to ensure that all parts fitted to aircraft meet the required standards.

- **Alternative materials procedures** are utilized to authorize fully equivalent materials to be used if necessary. As with alternative parts, it is vital to establish that any alternative material that is approved meets all the requirements of its intended use and that it is purchased from an approved supplier.

8.5 Material and Process Specifications

To ensure that the required standards are met, all materials used and all processes employed on aircraft and other critical work must meet accepted specifications. In either case, this means that sufficient research has been done to establish the formulation and processing of the material and/or the processing required by the end user to guarantee that the required properties are achieved. This work must then be converted by OEMs or standards-developing organizations into specifications that must be met by suppliers of material and those using processes of various types. These standards-developing organizations include the International Organization for Standardization (ISO), Society of Automotive Engineers (SAE), Centre Européene de Normalisation (CEN), American Society for Testing and Materials (ASTM), Deutsche Institut fur Normung (DIN), U.K. Ministry of Defence (MOD), British Standards Institution (BSI), and many others. The SRM and component drawings call out materials and processes by the OEM company specification number or another specification considered acceptable by the OEM. Sometimes, the material supplier designation may be used (e.g., Redux 312 or AF 163). Repair work must use materials and processes to these specifications or approved equivalents.

U.S. Military Specifications: MIL specifications are U.S. Military Specifications. Each carries a number such as MIL-Y-83379A, which is the specification for aramid yarn, roving, and cloth. The "A" indicates the first raise in issue after the first and original issue. Issue letters continue alphabetically each time the specification is amended. Military specifications are commonly used in civil aircraft manufacture because the military often lead in the application of new technology. If the new material is then brought into civil use and a new specification is not needed, it is common practice to use the appropriate MIL specification. Similar practice exists in the United Kingdom, especially for metal alloys; however, the reverse is beginning to occur, and the BSI is taking much of this role.

Company Specifications: One of the major problems of aircraft repair today is that most airframe and engine manufacturers tend to write their own specifications, especially for adhesives, pre-pregs, potting compounds, and sealants. These are found in the various manuals as Airbus Industrie Material Specifications (AIMS) and Airbus Industrie Test Methods (AITMs), Boeing Material Specifications (BMS), Douglas Material Specifications (DMS), Fokker TH Series, Rohr Material Specifications (RMS), and others—almost *ad infinitum*. This leads to a vast amount of duplicated testing by the suppliers, who may have to test the same batch of material to three or more slightly different requirements using several slightly different test methods. These material specifications usually conclude with a Qualified Products List, which gives details of approved suppliers. Fortunately, a large amount of work is being done by the Repair Materials Task Group of the Commercial Aircraft Composite Repair Committee (CACRC). Starting with repair materials and (it is hoped) progressing to all composite materials over the next few years, we have Airbus, Boeing, Douglas, Fokker, Rohr, General Electric, Pratt and Whitney, and Rolls-Royce working together on one committee. This group has already agreed on a common set of test methods, a vital part of this process. The first common specification for a wet lay-up repair resin for carbon-fiber components has been issued.

Federal Specifications: An old U.S. system, Federal Specifications, remains in use and covers a range of materials including hydrocarbon solvents. A typical number would be TT-N-95B Naptha, Aliphatic. Another old U.S. system is Federal Test Method Standards (FTMS). These standards continue to appear on old drawings.

National Aerospace Standards: Another system that has been in use for many years and remains in operation is the system of National Aerospace Standards. These standards also cover a wide range of parts and are most easily remembered because they cover nuts, bolts, bushes, screws, special fasteners, and rivets. A typical example is NAS 1739 for mechanically locked stem, blind rivets (e.g., "Cherrylock"), which are purchased to NAS 1740, the procurement specification.

ISO Specifications and Test Methods: We hope that this organization will provide an increasing number of specifications for worldwide use, and all companies are to be encouraged to use an ISO specification when one is available to suit a particular need. In those fields spanning all forms of transportation, an urgent need exists to use ISO specifications when possible. Virtually all forms of transportation, especially aircraft, are sold around the world; therefore, the use of ISO standards would considerably simplify repair and maintenance. Fortunately, 15 years of standardization work on composite materials is approaching fruition today, and we can expect many ISO Standard Test Methods to be finalized in the near future. A typical specification is ISO 4585R Textile Glass Reinforced Plastics—Determination of Apparent Interlaminar Shear Properties by Short Beam Test. Considerable efforts are being made to align the requirements of many test methods produced by ASTM with similar ISO methods.

SAE Specifications: The Society of Automotive Engineers (SAE) is a major producer of specifications for numerous industries. In the aerospace field, SAE develops and produces the following:

- Aerospace Material Specifications (AMS)
- Aerospace Standards (AS)
- Aerospace Recommended Practices (ARPs)
- Aerospace Information Reports (AIRs)

Other Specifications: Many other specifications exist, and most countries have their own national standards which are becoming increasingly aligned with ISO, even to the point of being given similar or identical numbers, with some prefix. The United Kingdom has British Standards; Germany has DIN standards. France has the Bureau Veritas, and other countries have similar systems.

In the United States, the Suppliers of Advanced Composite Materials Association (SACMA) has its own composite test methods. These are good, but it would be helpful if the suppliers and end users followed the same methods. Much testing also could be saved if these methods were ISO methods. Boeing and other companies have their own specifications for bolts and other fasteners. This is understandable for special items, but ISO standards for commonly used parts would be helpful.

8.6 Original Equipment Manufacturer Documents

Three documents are common to all OEMs. These documents are Service Bulletins, Service Newsletters, and Telex Authority, as discussed below.

Service Bulletins: Service Bulletins are the means by which modifications or rework procedures are recommended or required by OEMs. Several levels of priority are given to Service Bulletins, according to the urgency of the work required. Some of those below are issued by the FAA, and some are issued by the CAA. Considerable work is being done to harmonize FARs and JARs. In the future, they may agree on common terminology for the priorities given below.

- **Mandatory:** These apply to CAA and French Direction Generale Aviation Civil (DGAC) design certificated aircraft. The task required must be accomplished and within the time scale specified for safety reasons. The task could be performed before next flight or a longer time scale, but it is usually done within a specified number of flight hours or flight cycles.

- **Alert:** These Service Bulletins apply to OEMs responsible to the FAA and other certifying authorities and indicate that a high level of urgency exists, at least for some aircraft in a fleet that may be high-time aircraft. Again, this level of urgency indicates that safety is involved. Most Alert Service Bulletins are made mandatory, either in full or in part, by means of an FAA Airworthiness Directive or a CAA Additional Directive.

- **Non-Alert (Routine):** This level means that the operator may choose to incorporate the modification or not, according to the operator's own needs. This category can be phased in more easily with routine maintenance. Non-Alert Service Bulletins are usually intended to increase the life or functional reliability of a troublesome part or to simplify maintenance or access. The airline must decide whether the benefit of the modification, either in cost savings or in convenience, outweighs the cost of incorporation. Some modifications incur a weight penalty, which should also be considered.

Service Newsletters: Service Newsletters are issued by OEMs to address troublesome components or systems. The objective is to make airline staff aware of a problem and a solution, if one has been found. Service Newsletters often concern ways of making adjustments or give warnings and advice on troubleshooting.

Telex Authority: Telex Authority is probably the fastest means of obtaining documented approval from an OEM for a particular repair design or other proposed action. After discussion of a problem, Telex Authority is a good way to obtain written authority for an agreed action. It can be quoted in the aircraft Technical Log as authority for the action.

8.7 Regulatory Documents

All regulatory authorities issue their own documents specifying requirements or giving advice. The following are the most important regulatory documents.

FARs and JARs: These abbreviations stand for Federal Aviation Regulations (FARs) and Joint Airworthiness Requirements/Joint Aviation Requirements (JARS), respectively. Although they are similar and use the same paragraph and chapter numbers, they have some differences, and aircraft registered in the United States and Europe must comply according to the country of registration. For example, a European-built aircraft registered in the United States must meet both standards. Likewise, the same would apply to an American-built aircraft registered in Europe.

Advisory Circulars: Advisory circulars (ACs) are issued as required by the FAA or JAA when it is considered that some issue should be discussed or when standards must be set or improved in a particular area. An example is AC 145-XX concerning composite repairs, which has been discussed with the airlines for several years and has recently been issued. Although the title is Advisory, compliance is expected.

Airworthiness Directives: Airworthiness Directives (ADs), issued by the FAA, are instructions that must be complied with in the time scale given. They are issued when urgent action is required concerning structural alterations, operating procedures, or maintenance procedures. In extreme cases, they could require grounding of an aircraft type pending investigation.

Additional directives may be issued by the FAA for foreign-built aircraft if a higher design standard is required in a particular area than that of the manufacturing country. Most ADs are preceded by a Notice of Proposed Rule Making (NPRM), which outlines the proposed AD and the reasons for it. All affected parties may question the need for and/or the validity of the proposed requirement.

Airworthiness Notices: These notices are issued by the CAA, as required, to provide information on specific topics.

8.8 Air Transport Association of America ATA 100 System

This system should be studied and remembered because it provides the indexing basis for all aircraft manuals. Although similar systems were used, this system has become the international standard and all of the following manuals are indexed on this basis. A number is allocated to each major part of the aircraft (e.g., Chapter 53 for fuselage), and this is later subdivided. The advantage of this system is that anyone working with aircraft can be sure that the same topic will be found in the same place in the same manual for each aircraft type. In spite of best efforts, this is not always the case; however, it does limit the variations to a very small number. One example is undercarriage doors, which are found sometimes in Chapter 52 for doors and sometimes in Chapter 32 for undercarriages. Apart from this and probably a few more exceptions, it should always be possible to find information about the fuselage in Chapter 53, wings in Chapter 57, and so on. A few minutes with the AMM or IPC index will show how the system works.

8.9 Aircraft Maintenance Manual

The Aircraft Maintenance Manual (AMM) is an important manual and sets the standard to which maintenance must be done. As is the case with all OEM manuals, the AMM is arranged according to the standard of ATA 100 (prepared by the Air Transport Association of America), as mentioned above. This manual provides information on the removal and assembly of parts and also rigging adjustments, electrical and pressure testing, system troubleshooting, etc. It covers maintenance, as its title indicates. Repairs are found in the SRM, Overhaul Manual (OHM), or Component Maintenance Manual (CMM). Guidance is given on the use of this and the other OEM manuals in an introduction section to each one.

8.10 Component Maintenance Manual

The Component Maintenance Manual (CMM) covers specific large parts such as radomes and engine or cowling parts. It is arranged in ATA order and contains its own Parts List and suppliers' addresses and a List of Materials and their suppliers. The Airbus A-300 radome is found in CMM 53-51-11. Similarly, the Fan Reverser Support Assembly-Repair for the GE CF6-80C2 Series Engines is found in the CMM 78-32-02.

8.11 Overhaul Manual

The Overhaul Manual (OHM) is produced only for those components such as engines, gearboxes, pumps, doors, and other major components for which an in-depth overhaul is considered to be necessary on a regular basis or when indicated by general wear. The routine maintenance specified in the AMM and CMM are also required, but these components need a deeper overhaul to change bearings, bushes, seals, and other parts that require removal of the component from the aircraft and also considerable time and tooling to enable the work to be done. Test rigs may also be necessary. After a complete overhaul, a part is recertified for a further period of service, and a Certificate of Return to Service (CRS) is raised.

8.12 Illustrated Parts Catalog

The Illustrated Parts Catalog (IPC) is exactly as its name describes: a catalog with illustrations to assist the identification of each part. It is often customized so that only the aircraft operated by a particular airline are covered. This minimizes confusion because some types have a large number of variants and may have several variants in one airline. The effectivity code is important in this manual. The code for each individual aircraft must be known to ensure that the correct part number can be ordered. This manual will identify the part numbers before and after each modification and should also give the Service Bulletin number. The IPC also states when a new part is interchangeable with an earlier one and when only the new part should be used. In some cases, after a modification has been performed, only the new part will fit. In other cases, both the pre-modification and post-modification items will fit an aircraft that has been modified. Therefore, it is essential to establish whether the aircraft is to pre- or post-modification standard in order to prevent fitment of an incorrect part that could inadvertently modify or demodify the aircraft from its recorded standard. If the modification is intended to improve the life of a part, the use of a pre-modification part could cause maintenance schedule problems if the schedule has been revised to suit the longer life of the post-modification part. To assist in finding the correct part, the illustrations carry small identification numbers to correlate with the parts list. The IPC is an essential and useful document, and its introduction should be read to understand all the information it provides. Remember that the IPC gives authority to fit a part.

Therefore, the IPC must be correct in itself and be maintained to an up-to-date standard to ensure that only the latest issue can be used and that it should be correctly interpreted by the staff using it. This last requirement may mean that staff must be properly trained to use the IPC.

8.13 Structural Repair Manual

The Structural Repair Manual (SRM) is an important manual to those involved in repairs. It is also arranged in ATA 100 order; however, the SRM includes only Chapters 51 to 57 inclusive (i.e., those chapters covering structures). Repair procedures of wide or general applicability are given in Chapter 51; more specific repairs are provided in the ATA chapter for the part in question. All repairs and damage limits given are approved by the relevant regulatory authority and therefore are approved data.

The SRM is written at various levels similar to the order of importance of Service Bulletins. For example, damage limits are given for several levels:

1. **Allowable Damage:** Allowable damage covers two types of damage:

   - Damage that can be removed without adding any parts or material (i.e., repair that can be accomplished by blending out scratches, gouges, or small edge cracks), or by blending out corrosion and applying suitable protective treatments and finishes. The SRM must be consulted for allowable damage limits for each part.

   - Minor damage that with simple protection, such as Speedtape or in the case of radomes a nonconductive plastic tape to keep out water and other fluids, can be carried back to base or for a few flights. This damage must be repaired at the first convenient opportunity or within a specified time.

2. **Temporary Repairs:** These repairs must be removed either at a convenient time prior to the expiration of the permitted time limit, or at the latest by the end of the permitted time, and replaced with repairs conducted to more stringent requirements. Temporary repairs are sometimes unavoidable if a quick repair must be made to enable an aircraft to take off on schedule. Temporary repairs should be avoided if possible because their removal may result in larger repairs later.

3. **Interim Repairs:** These repairs can be allowed to continue in service indefinitely after they have been inspected on several occasions over a long period of time. They should be used in preference to temporary repairs because a second repair does not have to be made unless problems exist with the first repair. These repairs also have size limitations, as explained in the next paragraph.

4. **Repairable Damage:** This damage also has several levels in the case of repairs to composite parts.

- A small area of damage may be repaired by wet lay-up techniques using room-temperature curing resin systems.

- A large repair is permitted if a hot-curing (intermediate temperature, approximately 95°C [203°F]) wet lay-up resin is used.

- A much larger repair, often without limit to its size, is permitted if the original material is used and cured at the original cure temperature.

The SRM or OHM gives precise step-by-step details for each repair and specifies materials and techniques to be used. These are both generally helpful manuals, with the only complaints being that they cannot cover every case precisely and that the size of permitted repairs is often smaller than needed by the airline. Often they do not give a range of alternative materials, which can be a problem if the specified material is neither in stock nor available at short notice. Repairs beyond the limits given must be covered by local design authority or referred to the OEM for approval.

8.14 Engine Manual

Sections of this manual contain composite repairs to engine cowlings and other parts and may have to be consulted.

Chapter 9

Structural Repair Manual (SRM) Repair Method Selection

9.1 Component Identification

The first task in any repair is to correctly identify the part by number and its modification state. Any alteration should result in a change of dash number to the main part number or to a raise in issue of the original part number. If a higher modification state is achieved by a major alteration or redesign, then the new part or assembly usually has a totally new number. Each part of an assembly will have its own part number. Sometimes it is not easy to be sure which of the numbers marked on a large assembly is actually the part number for the complete assembly (e.g., a flap or elevator) and which is the part number of a large skin or surrounding frame. In such cases, refer to the Illustrated Parts Catalog (IPC), make good use of the illustrations, check the effectivity code for the particular aircraft, and establish beyond doubt which is the complete assembly number and which is a number of one of the parts. These large parts may also have their own assembly and detail part drawings. A small aluminum data plate is usually bonded or pop riveted to a large assembly. This should carry the complete assembly part number, a serial number, date of manufacture, inspector's stamp number and issue letter or number or other indication of modification state, and the name of the manufacturing company. This may not be the same company as the aircraft manufacturer because a large amount of work is subcontracted around the world at the time of this writing. For example, it is common for the fuselage to be made by one company, the wings by another, the rudder by another, and the elevators by another. Each company is usually free to choose its own pre-preg and adhesives. This can create problems for the repairer if the SRM does not allow several alternative materials to be used, as mentioned in previous chapters. The following example may help to emphasize the importance of the effectivity code and modification state. Early models of the Boeing 737 had aluminum elevators and rudders, a second version had fiberglass skinned honeycomb panels, and a third version had carbon fiber skinned honeycomb panels. A correct repair cannot be made until the part concerned has been fully identified. Sometimes this may require considerable effort, but it is necessary.

9.2 Damage Classification

The next step is to classify the damage into one of several types. Details of allowable damage and various levels of repairable damage are given in the SRM. These four damage types are:

1. Allowable/negligible damage

2. Damage for which a temporary repair is acceptable but a time limitation is given, after which a permanent repair must be made

3. Repairable damage

4. Damage that is not repairable or is beyond economic repair

9.2.1 Damage Terminology

Damage terminology is covered in the Society of Automotive Engineers (SAE) Aerospace Information Report AIR 4844 Composites and Metal Bonding Glossary. The more important definitions relating to damage are given below.

- **Abrasion:** The wearing away of a portion of the surface by either natural (rain, wind, etc.), mechanical (misfit, etc.), or man-made (oversanding, etc.) means; penetrates only the surface finish. In a composite, does not damage the first ply.

- **Burned:** Showing evidence of thermal decomposition or charring through some discoloration, distortion, destruction, or conversion of the surface of the plastic, sometimes to a carbonaceous char.

- **Charring:** The heating of a composite in air to reduce the polymer matrix to ash, allowing the fiber content to be determined by weight. A similar effect caused in service by lightning strike, fire, or other source of overheating.

- **Chemical Attack:** Damage to the resin matrix by accidental contact with or unauthorized use of chemicals.

- **Crack:** Fractures in either matrix or both matrix and fibers. An actual separation of material. Does not necessarily extend through the thickness of the composite, but can be stopped by differently oriented plies.

- **Delamination:** Separation of the layers of material in a laminate, either local or covering a wide area. Can occur in the cure or subsequent life of a part.

- **Disbond:** An area within a bonded interface between two adherents in which an adhesion failure or separation has occurred. It may occur at any time in the life of the structure and may arise from a variety of causes. Also, colloquially, an area of separation between two laminae in the finished laminate. (In this case, the term "delamination" usually is preferred). See also *Debond*.

- **Erosion:** Destruction of metal or other material by the abrasive action of liquid or gas. Usually accelerated by the presence of solid particles of matter in suspension and sometimes by corrosion. In the case of aircraft, it can be accelerated by hail, heavy rain, dust, and especially the occasional sandstorm or volcanic ash blown to high altitude.

- **Fracture:** The separation of a body. Defined both as rupture of the surface without complete separation of laminate and as complete separation of a body because of external or internal forces (i.e., some layers are completely broken, or the whole part is completely broken).

- **Impact:** Damage from a foreign object (other than ballistic). Usually a dropped tool or collision with a vehicle or item of ground equipment. In flight, may be caused by hail impact or bird strike.

- **Scoring:** A type of wear in which the working face acquires grooves, axial or circumferential, according to whether the motion is reciprocating or rotary. Also applied to a similar effect on the rigid or nonmoving member. A groove which is smooth and has significant width compared to depth. A blunt scratch.

- **Scratch:** An elongated surface discontinuity, which is extremely small in width compared to length. Shallow mark, groove, furrow, or channel normally caused by improper handling or storage.

9.2.2 Critical Areas

As mentioned in Chapter 5, some areas of a part may be more critical than others (e.g., hinge fittings, spars and ribs in a spoiler or flap), and the SRM takes account of this by restricting the amount of damage permitted in clearly marked areas. The SRM must be consulted to ensure that the limits for each area are correctly observed.

9.2.3 Allowable/Negligible Damage

Rather than meaning that nothing should be done, this category means that the repair can be delayed to a convenient time. The damaged area may still require covering with Speedtape, filling with potting or sealant, or another treatment defined by the SRM or good sense. It is important to prevent moisture ingress into honeycomb areas and the spread of damage in order to minimize the size of the repair that must be made. General composite and bonded metal repairs are covered in the SRM Chapter 51, and allowable damage limits for particular parts are given in the SRM chapter appropriate to that part (e.g., SRM Chapter 53 for radomes). Section SRM 53-10-72 lists the materials used in this part and their specifications. A diagram and table give the limits of allowable damage. Similar information for engine nose cowls, fan cowls, and translating cowls is provided in SRM Chapter 54.

9.2.4 Repairable Damage

This damage may have several levels of repair detailed in Chapter 8, each with its own size limits, depending on the temperature and pressure available for the repair. Some lightly loaded parts may have no limits to the size of repair when room-temperature curing wet lay-up systems are used. Others may have some limits and, in areas where it is considered necessary because of the service temperature or other reason, repair with room-temperature resins may not be permitted. Again, the SRM details the limits for wet lay-up repairs and hot-curing pre-preg repairs. In some cases, four levels may be given, such as:

1. Room-temperature wet lay-up
2. Warm-curing 95 to 120°C (200 to 250°F) wet lay-up
3. 120°C (250°F) pre-preg
4. 180°C (350°F) pre-preg

Detailed procedures and size limits are given for each type of repair. For example, SRM Chapter 54 Nacelles and Pylons covers repairs to engine nose cowls, fan cowls, and translating cowls. For the Boeing 757 RB211-535 engine cowl, SRM 54-20-01 page 205, Table 1, illustrates the different repair limits, depending on the type of repair selected. This page is included as Figure 9.1. In this case, warm-curing wet lay-up repairs made at 95 to 120°C (200 to 250°F) have a size limit, but 180°C (350°F) repairs do not.

Damage	Permanent Repairs	
	Wet lay-up 200°-230°F cure (51-71-02)	350°F cure (51-71-01)
Cracks	Clean up damage and repair as hole	Clean up damage and repair as hole
Holes	10.0 inch maximum diameter not to exceed 50% of smallest dimension across honeycomb panel at damage location. Use two extra plies per face-sheet repaired. [A]	No size limit
Delami-nation	Cut out and repair as hole	
Nicks and Gouges	If fiber damage or delamination exists, repair as hole	
Dents	Dents result in fiber damage and delamination. Use repair data to determine dent repair.	

Notes:
- Refinish reworked areas per 51-20 of the maintenance manual.
- Refer to 51-10-01 for aerodynamic smoothness requirements. Where the repair exceeds the limit shown in 51-10-01, consideration should be given to the loss of performance.
- Refer to 51-70-16 for hole drilling and machining of composite structures.

[A] One repair per square foot of area and a minimum of 6.0 inches (edge to edge) from any other damage, fastener hole, or edge of panel.

*Fig. 9.1 Repair data for 180°C (350°F) cure honeycomb panels.
(Courtesy of Boeing Commercial Airplane Group.)*

9.2.5 Overhaul or Replace

If damage is large, a decision may have to be made regarding whether the part should be scrapped or repaired. This can be difficult because composite parts, as with all aircraft parts, are expensive, and the first thought is always, "Can this part be repaired?" As a rough guide, it is often considered that if the damage area encompasses more than 50% of the part area, then it may be more economical to scrap the part. However, there are times when even AOG (Aircraft on Ground) priority supply may take too much time, and, if the part is small, it may be possible to rebuild that part more quickly. If a spare cannot be obtained for any reason, then repair, even at greater cost than buying a spare, may be less costly than having the aircraft out of service.

When the part is repairable, another problem is the time scale that is available. Many aircraft parts, such as large composite wing panels, are drilled to a jig along each edge but not to one complete jig. This is done to cope with assembly tolerances on the substructure. Often, the result is that the part will fit only the airplane from which it was drilled and no others. Thus, repair must be possible in the available downtime for refit of the part to the same airplane. If not, then the part could be repaired, but it would be of little use to any other airplane.

9.2.6 Other Considerations

When making repairs to composite or metal parts, other factors must be taken into account and these may dictate the type of repair. For example, a flush repair may have to be made in some cases for aerodynamic reasons and in other cases for mechanical clearance reasons. Either way, the type of repair, and hence its cost and downtime, may be determined by factors other than the production of a mechanically sound structural repair. In the case of radomes, the repair must be made while maintaining the correct sandwich thickness and using nonconducting materials of acceptable dielectric constant to ensure an acceptable level of radar signal transmission. Control surfaces must be repaired within total weight and balance limits, including the weight of the paint.

Mechanical Clearance: When a repair is done, say, to the leading edge of a control surface or an end rib, it is important to ensure that if the repair is external, it has adequate mechanical clearance from other structure to allow free movement when it is refitted.

Similarly, when wraparound layers are added at the leading edges of cowlings and other composite panels to repair or prevent leading edge erosion, a little material must be removed before adding these layers to ensure that the repaired part will fit. Any part that has an external repair and is close to another part, especially a moving part, should be checked for mechanical clearance. When making these checks, allowance should be made for flexure that may occur as a result of flight loads. Cases have been reported in which clearance that was adequate when on the ground was insufficient in flight because of structural deflections. See, for example, Boeing 757 SRM 51-70-05, page 23.

Weight and Balance: Some aircraft components must be within certain weight and balance limits. These components are control surfaces such as ailerons, elevators, and rudders and their associated trim tabs. Rotating engine and other parts are a separate case requiring specialized equipment. Balance limits will be specified in the SRM and must be maintained. In the case of small aircraft with manual controls, the weight and balance requirement is necessary to ensure that the control forces that must be provided by the pilot are within acceptable limits. Another reason, which also applies to large aircraft with powered controls, is that if the operating rod or jack breaks, the control surface must be accurately balanced sufficiently to ensure that flutter does not occur. Weight and balance requirements are specified in the Aircraft Maintenance Manual (AMM). Special rigs are required to support these parts while balancing is done. See also Boeing 757 SRM 51-60-01 Aileron Rebalance Procedure and also 757 SRM 51-70-05, page 23.

Aerodynamic Smoothness: Aerodynamic smoothness is another important factor that must be considered. Where this is important, it will be mentioned in the SRM chapter for the part concerned and the appropriate reference will be given. This reference will provide permitted profile deviations. If in doubt and no reference is given, check with the SRM Chapter 51-10-01. Some locations of importance are wing leading edges, engine intakes and cowlings, and the fuselage slightly forward of the pitot static ports. Badly shaped repairs in front of these can give false airspeed readings.

Drainage Hole Replacement: Many control surfaces and other parts have water drainage holes, and it is possible to cover a drainage hole during a repair and then forget to redrill it before releasing a part for return to service. This item should be included on a checklist of jobs to be done to ensure that any drain holes that have been covered by a repair are replaced. The holes should be drilled in the original location if possible or close enough to it to perform their function. See Boeing 757 SRM 51-70-05, page 23. This also refers to the AMM 51-41. Drainage holes are usually 10 mm (3/8 in.) in diameter.

9.3 Repair Methods

Several possible repair methods are available, and each has its place in the maintenance of aircraft in good condition for revenue service. Damage varies from negligible at one end of the scale to beyond economic repair at the other, and the facilities available vary from almost nothing at remote outstations to the ability to completely remanufacture some parts in large and well-equipped airline or repair station workshops. The choice of repair method must consider all these factors and also the time available before the aircraft is required for service. Even in the largest and best-equipped airlines, all the options below will be needed at some time to address problems. Sometimes a new part will have to be fitted simply because sufficient time is not available to repair the old part.

9.3.1 Speedtape

Speedtape, as it is popularly called, is a thin aluminum foil that can be supplied in a range of thicknesses, usually from approximately 0.025 to 0.075 mm (0.001 to 0.003 in.). This foil has a backing sheet that must be removed to reveal a layer of pressure-sensitive adhesive. The foil then can be applied to the surface in question (after suitable cleaning) to prevent water ingress or to prevent loose fibers from lifting according to the need. This form of repair is temporary, usually for one flight or a series of flights returning to base. It provides no strength but prevents minor damage from worsening and thus minimizes the size of the eventual permanent repair. Speedtape must not be used on a radome in the scanning area because of its electrical conductivity, but it may be used on a radome aft of the scanning area. In the scanning area, a plastic self-adhesive tape must be used. Suitable plastic tapes are supplied by 3M, such as nonmetallized 3M Scotch Brand 850 or 853, and Permacel, which supplies its P95 polyester tape. Other companies supply similar plastic materials and their own versions of Speedtape.

9.3.2 Resin Sealing

Resin sealing is another form of temporary repair in which a repair resin is painted over the damage to retain loose fibers and to prevent the damage from increasing in size.

9.3.3 Potted Repairs

Potted repairs are made when a small hole found in a honeycomb panel is within the limits of acceptable damage. The hole can be filled with a suitable potting compound of the room-temperature curing variety; this will prevent water ingress until a permanent repair can be made. It may also be covered with Speedtape to prevent water absorption into the potting compound.

9.3.4 Bolted and Bonded Doublers

This type of repair may be made in two forms:

1. As a temporary repair for a return to base, in which it may consist of a metal skin on the outside and large-diameter washers on the inside, with bolts going through the skins and honeycomb core of a large wing or fairing panel that has been damaged or has suffered severe disbonding for another reason. This type of repair makes a mess of the whole panel and sometimes results in the panel being scrapped. If not scrapped, then extensive and expensive permanent repairs are required. However, this type of repair has one great advantage: it enables the aircraft to return to base with minimal delay. The cash value of doing that may outweigh the cost of a new panel or extensive repairs (i.e., 400 hotel bills can be expensive). For details of typical "Time-Limited Metal Overlay Repairs to Composite Panels," see Boeing 757 SRM 51-70-18.

2. This type of repair may be permanent if it is made to a solid laminate wing skin or similar panel. In this case, anodizing or other acceptable treatment of the metal plate will be necessary after final shaping and drilling of all holes. Permanent bolted repairs are usually made to thick skins, 2 or 3 mm (0.08 or 0.12 in.) or more in thickness, in which case they are of lower weight and cost than a bonded repair because of the large

tapered scarf jointing required for a thick panel repair. Repairs of this type are precisely detailed in the SRM because they are used on important load-bearing structure. See Airbus, Boeing, Douglas, and other SRMs for specific repairs of this type.

9.3.5 Pre-Cured Doublers Versus Co-Cured Doublers

Pre-cured doublers are sometimes preferred because the pre-preg can be cured in an autoclave to the original standard and then bonded with a hot-curing film adhesive to the repair area. A problem with this approach is that great care is required and usually tooling to ensure that the profile of the patch matches the profile of the surface to be repaired. It is a simple process but only for very flat surfaces. The normal glue-line thickness is between 0.125 and 0.25 mm (0.005 and 0.010 in.). This does not allow a margin for error in the shape of a surface. The idea of this type of patch is that a stock of pre-cured patches can be held and time can be saved in performing the repair. This ignores three problems:

1. It is impossible to predict the size of a repair in advance.

2. Every curved surface is likely to be different, and aircraft have few external surfaces that are completely flat.

3. The glue line cannot accommodate the likely errors in profile.

Use of a co-cured patch also has advantages. For example, a patch can be laid down on the repair area either as a series of pre-preg layers on a film adhesive layer to make a tough joint or as a series of wet resin impregnated layers for a wet lay-up. The three primary advantages are:

1. In either case, the fabric can follow the surface contours of the repair area.
2. A thin glue line can be achieved.
3. The patch can be cut to suit the size of the repair.

Patch repairs are usually made to thin skinned honeycomb panels when the skin consists of only two or three plies. When making patch repairs, it is common to make the overlap 50 to 70 times the thickness of the skin. For composite patches, typical practice is to taper the patch by making each ply 25 mm (1.0 in.) larger than the previous ply for cold-setting repairs and 12.5 mm (0.5 in.) larger for hot-setting repairs (Boeing). Airbus uses 12.5 mm (0.5 in.) in both cases. Because many fabrics are approximately 0.25 mm (0.010 in.) thick, a 12.5 mm (0.5 in.) lap gives 50 times the ply thickness. Another method is to make the patch overlap 50 times the total composite skin thickness and then to run the tapered steps beyond this. See Figure 9.2.

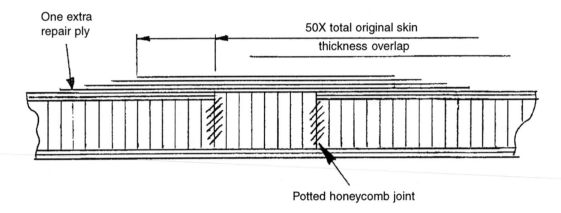

Fig. 9.2 Composite patch overlap. Note that ply drop-offs are carried beyond the 50X skin thickness overlap.

9.3.6 Room-Temperature Wet Lay-Up

This is the preferred method, when permitted, because it has several advantages:

1. It allows the fabric or tape to adopt the shape of the surface being repaired.

2. It requires no heating except to speed the cure.

3. It will always reach full cure after the required period of time.

4. Vacuum pressure is sufficient for a good bond.

5. Additional fabric layers can be used to compensate for any loss of performance, in many cases.

6. It seldom requires much tooling because the structure will not deform at room temperature.

However, this method has two disadvantages:

1. The high-temperature performance is not as good as the original, but this may not always be needed.

2. The resin must be accurately weighed and mixed.

9.3.7 Elevated-Temperature Wet Lay-Up

This method has all the advantages of room-temperature wet lay-up with the additional advantage that OEMs will allow larger repairs. Its only disadvantage is that heat is required, and a hot bonding controller must be used and thermocouples must be located correctly to ensure that the minimum temperature for cure is maintained over the entire repair area for the full time of cure. These systems usually cure at 95°C (200°F), and some support tooling may be required.

9.3.8 Pre-Preg Repairs

Pre-preg repairs can take several forms:

- Low-temperature pre-pregs curing at approximately 95°C (200°F) as for the hot-curing wet lay-up repairs. Development work continues in this area.

- 120°C (250°F) curing pre-pregs may be used to repair parts made at either 120°C (250°F) or 180°C (350°F).

- 180°C (350°F) pre-pregs may be used to repair parts made at 180°C (350°F). These pre-pregs cannot be used and must not be used to repair parts made at 120°C (250°F) or lower.

Advantages of Pre-Preg Repairs:

- The OEMs prefer pre-preg repairs and sometimes require them, and they allow larger repairs if pre-pregs are used.

- No resin mixing is required.

- It is clean and easy to work with pre-preg repairs.

Disadvantages of Pre-Preg Repairs:

- If a repair is made at the original cure temperature, problems occur with honeycomb structures. When the repair area is heated to cure the pre-preg, the air pressure in the honeycomb cells surrounding the repair rises to a higher pressure than atmospheric and considerably above atmospheric if water vapor exists in the cells and the temperature of cure is 180°C (350°F). The external pressure is usually only vacuum, normally approximately 10 psi (0.7 bar), and the result (found many times by painful experience) is that the skin is blown from the honeycomb around the repair. This means that a larger repair then must be made at 120°C (250°F), or tooling must be provided to apply additional pressure above vacuum. It would be beneficial if all composite parts were designed and made so that repair at 25 to 50°C (45 to 90°F) below the manufacturing temperature was regarded as the standard method. This serious problem then could be avoided completely.

- Hot curing is required, and achieving a uniform temperature over the whole repair can be difficult if there are any heavy members such as ribs, spars, or hinge fittings acting as heat sinks. The difficulty sometimes experienced in achieving a uniform temperature across the whole repair area is not always appreciated. Almost every repair is a unique "one-off" job, and all problems cannot be anticipated. Several thermocouples should be used and frequently checked to ensure that no area falls below the specified temperature and also that no area rises significantly above it. If part of the repair area is at the bottom end of the permitted temperature range, then extra time should be allowed for cure. No area can be allowed to run below the minimum temperature for cure, and over-temperature will make the resin more brittle. Insulating blankets may be added to cool areas, or lamps or hot air blowers can be used to warm them. Regardless of the method employed, the pre-preg and repair adhesive layers must remain within the specified temperature limits for the required length of time. A slight over-temperature of 5 to 10°C (9 to 18°F) is acceptable; running below the specified minimum temperature is unacceptable.

9.3.9 Scarf and Stepped Lap Repairs

Scarf Joints: Scarf joints are normally used when good joint strength is required, and they must be made carefully. See each specific type SRM for details. Generally, a taper of 1 in 50 is used for scarf joints, except around panel edges where the skin is much thicker to accept countersunk fasteners. At panel edges, a scarf angle of 1 in 20 is sometimes acceptable. See Figure 9.3.

Additional plies are laid over the ends of the laps to minimize peel effects. Scarf joints tend to fail at the ends after long periods of time at high loads in fatigue tests. This is an advantage because the beginnings of failure can be seen in adequate time to perform further repair. In normal service, loading should not be high enough to produce this effect. See Figure 9.4. Note that Boeing practice is to lay the smallest patch first, whereas Airbus practice is to lay the largest patch first. Both methods have been thoroughly tested, and both produce good joints. Work to the relevant SRM. In glider scarf repairs using composite materials, the large patch is invariably put in first to aid the finishing process. With wet lay-up techniques, variation in consolidation and therefore laminate thickness would make it easy to cut through outer laminates during the finishing process if the laminates were laid up by the classical "small layer first" method.

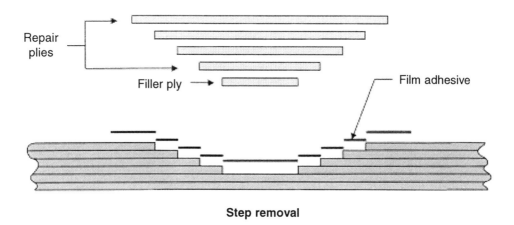

Step removal

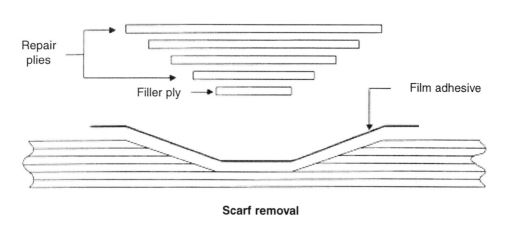

Scarf removal

Fig. 9.3 Scarf joints and stepped lap joints. (Courtesy of Abaris Training Resources Inc.)

Stepped Lap Joints: Stepped lap joints often are used and must be made carefully and accurately. (See Figures 9.3 and 9.4.) They are not too difficult on fiberglass and aramid skins; however, the individual layers are more difficult to see with carbon composites, and extra care is required to avoid cutting through one ply into the next ply. Overlaps usually are 12.5 or 25 mm (0.5 or 1 in.) per ply, and each repair ply should be oriented to lap onto a ply of similar orientation. Stepped lap joints often are used on radomes where it is important to minimize the number of extra plies to ensure good radar transmission.

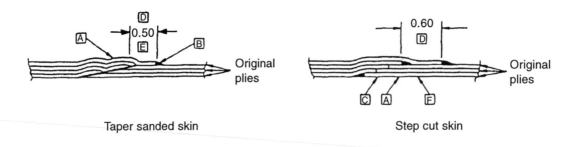

Taper sanded skin   Step cut skin

Boeing methods

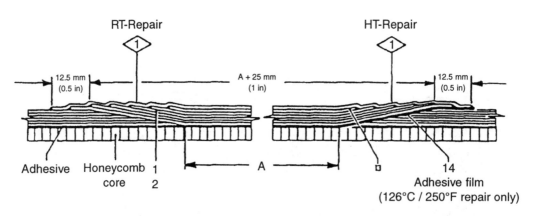

Airbus methods

Fig. 9.4   Boeing versus Airbus scarf joint lay-up methods for repair of damaged panels.
(Courtesy of Aero Consultants (United Kingdom) Ltd.)

9.3.10 Composite Repairs to Metals

Glass Repair to Aluminum: This has been an accepted method of repair for thin aluminum skins in Boeing SRMs for many years. A table usually states the number of layers of glass fiber required to repair a crack or dent in a trim tab or control surface.

Boron Repair to Aluminum: This method has been used in the United States and Australia for several years. (Ref. 9.1.) Examples are repairs to cracks in wing skin stringers on the Lockheed Hercules, cracks in lower wing skins around a fuel drain on the Mirage fighter, repairs to cracks in the magnesium alloy landing wheels of the Macchi trainer, and fuselage patches on the Lockheed P3 Orion. In early 1993, Federal Express, after consulting with Boeing, installed 25 "decal" boron doublers 150 x 200 mm (6 x 8 in.) on undamaged structures in areas of future repair interest on two Boeing 747 aircraft. These doublers have had more than 1,700 flights, have been inspected frequently, and have performed excellently. (See Ref. 9.2.) Textron has funded research at Boeing to assist in the design of these repairs. As a result of much successful military and civil experience and research, it is expected that they may eventually become a standard repair technique and, when they do, they will be found in the relevant SRMs.

Carbon Repair to Aluminum: This method was tried on the wing leading edge panels of the Concorde when cracks developed many years ago. Trials began in 1981 and continued until around 1990. These repairs were successful, with some lasting seven years or longer. Eventually, cracks developed around the patches, and larger metal repairs had to be made. However, none of the repaired cracks suffered any extension.

This method is cheaper than boron and can even use offcuts of dry fabric or pre-preg material. It is important to use a layer of glass fiber as an insulator between the carbon and the aluminum to prevent corrosion, and a tough adhesive should be used together with a good metal surface preparation. The method can be used with dry fabric and a good, tough, high-temperature wet lay-up two-part adhesive, with a suitable high-modulus film adhesive interleaved between layers of dry fabric (homemade pre-preg) or with pre-preg material and a tough film adhesive layer to bond the patch to the metal. The first two methods have been successfully flight proven on the Concorde. The last method has been flight proven with boron.

9.4   References

9.1   Baker, A.A. and Jones, R., *Bonded Repair of Aircraft Structures*, Martinus Nijhoff Publishers, ISBN 90-247-3606-4, 1988.

9.2   Belason, E.B. and Shahood, T.W., "Bonded Boron/Epoxy Doublers for Reinforcement of Metallic Aircraft Structures," *Canadian Aeronautics and Space Journal*, Vol. 42, No. 2, June 1996, pp. 88–92.

Repair Techniques

10.1 Preparation

Preparation for all repairs must begin with thorough cleaning of the part to be repaired. The entire part must be carefully cleaned. This is necessary to ensure that the work area, the clean room and lay-up area, and the repair area of the part itself are not contaminated by oil, dirt, or grease from other parts of the item under repair. Cleaning is covered in detail by SAE ARP 4916 Masking and Cleaning of Epoxy and Polyester Matrix Thermosetting Composite Materials. This SAE document, produced by the Repair Techniques Task Group of the Commercial Aircraft Composite Repair Committee (CACRC), should be studied in detail. The following limited excerpts on cleaning materials and equipment have been taken from it.

Cleaning Agents: When performing any cleaning operation with solvents or other solutions, take a clean cloth or swab, dip it into clean solvent or solution, remove as much dirt as possible, and then discard the cloth or swab. The objective is to clean the part rather than to spread the dirt. Only clean cloths or swabs may be dipped into clean solvent. Cleaning should continue until clean swabs remain clean. This is essential when making the final solvent or distilled water wipe before bonding. Although it may seem pedantic, the cleaning operation cannot be taken too seriously where structural bonding is concerned.

1. **Solvents:** The following solvents are the most commonly used for cleaning composites and other parts:

    *   **Methyl Ethyl Ketone (MEK):** This common wipe solvent is also used as a diluent in some sprayable epoxy adhesives and primers and in some Thiokol-type sealants. Efforts are being made to replace MEK as a cleaner because of its rating of serious flammability, with a flash point of -7°C (20°F) and some evidence of toxicity to animals. MEK may be procured to ASTM D-740.

- **Methyl Isobutyl Ketone (MIBK):** Although a less commonly used solvent that is similar to MEK, MIBK has a higher flash point of 18°C (64°F) and a lower evaporation rate. MIBK may be procured to ASTM D-1153.

- **Acetone:** Acetone is a commonly used wipe solvent. Efforts are being made to replace it as a cleaner because of its rating of serious flammability, with a flash point of -20°C (-4°F) and its high evaporation rate. Acetone may be procured to ASTM D-329.

- **1,1,1 Trichloroethane:** Also known as methyl chloroform. This was a commonly used wipe and vapor degreasing solvent. Because of its ozone-depleting properties, this chemical was prohibited in most countries by 1995. It may appear in old books and, if used, should be treated with care and in accordance with local regulations because overexposure can cause liver and kidney damage. Use an alternative if one is available.

Tulstar Products Inc. (2100 South Utica Avenue, Suite 200, Tulsa, Oklahoma) is now offering in 1997 two new solvents which are nonflammable and less toxic replacements for 1,1,1 trichloroethane and CFCs. The first of these is Leksol, which is supplied as a non-CFC precision cleaning solvent that will work in any application for which 1,1,1 trichloroethane has been used in the past. Leksol is a nonflammable, high-solvency material that does not affect aluminum, magnesium, ferrous metals, or many plastics and elastomers. The data sheet details those plastics and elastomers that are affected. Leksol is not a carcinogen as are many other vapor degreasing solvents.

Tulstar also supplies Borothene solvent, which is chemically chlorobromomethane. This is completely nonflammable and replaces methylene chloride, 1,1,1 trichloroethane, trichloroethylene, and perchloroethylene. It is a CFC replacement solvent. Both are fast-drying solvents. Chlorobromomethane is expected to receive VOC exemption in the fall of 1997, according to information released in the U.S. *Federal Register.* Borothene would be expected to damage plastics, adhesives, and resins severely as does methylene chloride. It is said to be very stable with most common metals such as aluminum, copper, zinc, carbon steel, and stainless steel.

Full details should be obtained from the supplier before use of Leksol or Borothene, and approval should be obtained from the aircraft OEM for each application (i.e., paint stripping, metal cleaning, composite cleaning). Note that a number of chemical companies are developing less toxic and more environmentally friendly solvents for industrial and other purposes. These may be expected to come into use in the future as they achieve approval. Check each SRM for the latest list of approved materials.

- **Isopropyl Alcohol (IPA) or Isopropanol:** IPA is a wipe solvent used as a less hazardous replacement for MEK, acetone, or 1,1,1 trichloroethane. Rated as flammable with a flash point of 11.7°C (53°F), it is toxic by inhalation and ingestion. IPA may be procured to Federal Specification TT-I-735A.

2. **Citrus-Based Cleaners:** This group of cleaners is based on the chemical d-limonene. Typically, these cleaners do not contain any hazardous ingredients as defined by the Occupational Safety and Health Administration (OSHA) and are considered nontoxic. Initial evaluation has shown this group to be effective cleaners, with the solvent dissolving organic contaminants and the water dissolving salts. Some cleaners may leave a residue, particularly if used in the concentrated form, and they are not given a subsequent water rinse. Although some evaluation has been done, these cleaners currently are not recommended for cleaning immediately prior to bonding.

3. **Emulsion-Type Cleaners:** This group of cleaners contains a combination of emulsifiers and surfactants. The surfactants lift the oil from the surface to form a weak emulsion. An emulsion is defined as, "A suspension of fine particles or globules within a liquid." Many contain some chemicals—usually glycol ethers or propylene glycol—that may have special waste disposal requirements. Initial evaluation has shown these cleaners to be effective on light oil types of contamination. Some may leave surfactant residues. Although some evaluation work has been done, these cleaners currently are not recommended for cleaning immediately prior to adhesive bonding. Emulsion cleaners used with pressure spray equipment should conform to AMS 1528. Cleaners of the emulsion type used for hot- or cold-water or hand wipe cleaning should conform to AMS 1532.

4. **Alkaline Detergent Cleaners:** Alkaline detergent cleaners contain an aqueous solution of alkaline salts that form a stable emulsion with oil-type contaminants. Care should be taken in the use of these cleaners on parts with aluminum core or fittings because residues of alkaline cleaners will cause corrosion of the aluminum. Although some evaluation work has been done, these cleaners are not recommended for cleaning immediately prior to adhesive bonding. Cleaners to AMS 1526 should be specified when using water-miscible detergent cleaners with pressure spray equipment.

5. **Wipe Cleaning Cloths:** The choice of cloths used in wipe cleaning, especially that prior to bonding, is important to the strength of the resulting repair. Cloths should be chosen with care and should be free of silicone. Cloths soaked in solvent should be treated as hazardous waste.

6. **AMS 3819 Cleaning Cloths:** These cloths are designed for wipe cleaning prior to bonding. They are lint free and resistant to solvents. Cloth to this specification is highly recommended. Grade A is good for all uses, including final wipe cleaning

operations before bonding or drying cleaned parts after the water break test. (The water break test is explained in Section 10.1.6 of this chapter.) Grade B may be used for most cleaning operations, but Grade A is recommended for final wiping prior to bonding.

7. **Cotton Swabs:** For cleaning small inaccessible places or small repair areas, cotton swabs are useful. The swabs must not be impregnated with oils or creams such as lanolin because they will contaminate the part. The stick that holds the cotton shall be made of a material that is not dissolved or softened by the solvent being used. Wooden sticks are resistant to the solvents specified. Other types should be checked by immersing the whole swab in the solvent for at least 24 hours and then checking for signs of softening when compared with a swab that has not been immersed. Cotton wool pads of a similar type may also be used for slightly larger areas.

8. **Cleaning Brushes:** Brushes used for removing dirt from parts that are being steam or water wash cleaned may be natural bristle or have bristles made from nylon or polypropylene fibers. Brushes with any kind of metal bristles (wire brushes) should not be used.

9. **Steam Cleaning Equipment:** All types of steam cleaning equipment shall be set up and used such that the temperature of the part being washed does not exceed 50°C (122°F).

10. **Hot Water Wash Equipment:** Temperature limits are the same for hot water wash equipment as they are for steam cleaning. Pressures of 30 psi (2 bars) are typical and acceptable. However, note the following caution from SAE ARP 4916.

    "Some types of hot water wash equipment are pressure washers with pressures from 250 to 1000 psi (17 to 70 bars). These high-pressure washers shall not be used on composite components because of the risk of damage to the components."

11. **Cold Water Wash Equipment:** For cleaning solutions that do not require hot water, ambient-temperature water can be used with the same equipment as used for hot water washing, with the same stipulation that high-pressure equipment must not be used.

12. **Vapor Degreasing Equipment:** This was done with 1,1,1 trichloroethane, now phased out. See replacement in Section 10.1. Other alternative methods are being studied and will be added to SAE ARP 4916. The only application for vapor degreasing in composite repairs is the cleaning of repair plugs of aluminum honeycomb. To vapor degrease aluminum honeycomb satisfactorily, the cell walls should be as near vertical as possible to allow the condensate to run off. A maximum angle of 30° from the vertical is recommended. A basket should be capable of supporting the honeycomb to prevent distortion during the cleaning cycle. A better solution is to keep the honeycomb in its original wrapper until it is needed to ensure that cleaning is unnecessary. If the honeycomb must be machined before use, then cleaning will be required.

Cleaning, correct surface preparation, and drying are the three most important factors involved in producing a strong durable bond to a composite part. Therefore, it was necessary to cover these factors at length here, although prior to the start of work, they should be studied in even greater detail in the documents quoted.

10.1.1 Paint Removal

Numerous methods have been tried for paint removal, and an International Air Transport Association (IATA) Task Force has worked on the subject for some time. However, for repair work, hand sanding using a suitable grade of silicon carbide paper gives the best results. Grade 320 is often recommended, but slightly coarser grades may sometimes be needed. One Structural Repair Manual (SRM) suggests Grade 150. Sanding should be performed carefully to avoid damaging the first fiber layer. Paint removal and sanding to give a good bonding surface should not penetrate the outer resin layer. Paint stripping a complete aircraft is another matter that will not be discussed here. See SAE MA 4872 Draft—Paint Stripping of Commercial Aircraft—Evaluation of Materials and Processes, IATA Guidelines for evaluation of aircraft paint stripping materials and processes.

10.1.2 Disbonding Methods

10.1.2.1 Disbonding Adhesively Bonded Metal Parts

Several basic techniques are used for separating bonded metal parts, and these merit careful application. When intentional separation of bonds is needed in repair, it is usually found that these areas are the best bonded of all areas. At this point, the "appliance of science" becomes necessary, and the following three characteristics of adhesives and resins can be used to assist separation:

1. Adhesives become softer and weaker when they are heated.

2. Adhesives become harder and more brittle when they are cooled.

3. Adhesives behave in a more brittle manner at high rates of loading, and in a more ductile manner under low rates of loading.

For disbonding metal parts, two methods may be used:

1. Surround the part with dry ice (carbon dioxide at a temperature of -78°C [-108°F]), and split the bond with a shallow wedge. This method takes advantage of two adhesive/plastic characteristics:

- Adhesives behave in a more brittle manner, the faster the rate of loading. Sharp taps with a suitable size of hammer can give a high rate of loading.

- Adhesives behave in a more brittle manner as temperature is reduced.

In this case, the dry ice makes the adhesive more brittle, reduces its fracture energy, and makes splitting with the wedge easier. This method requires no temperature control but needs plenty of dry ice because aluminum alloys are good conductors of heat. These adhesives are required to operate down to -55°C (-67°F); therefore, the amount of embrittlement is uncertain but has been found to be effective in some cases. A well-tried procedure is to cover the metal skin with chunks of dry ice and then to cover this with an insulating blanket for a few minutes.

Warning: Dry ice should be handled only with insulated gloves and while wearing safety goggles. Frostbite may result if dry ice comes into contact with skin or eyes. In case of contact, obtain medical assistance immediately. Provide ventilation for carbon dioxide, in accordance with local regulations.

2. Apply heat to the bond line to reduce the mechanical strength of the adhesive. The temperature should not generally exceed 150 or 180°C (300 or 350°F) because this may affect the heat treatment condition of aluminum alloys. Various types of hot air guns are available:

- In one type, the air temperature is not controlled. Temperature indicators must be used on the metal surface to avoid overheating.

- In another type, the air temperature can be set to the required figure. This type is preferred because indicators are not needed.

It is suggested that 120°C (250°F) curing adhesives having a glass transition temperature (T_g) of approximately 120°C (250°F) (or lower if they contain moisture) might best be removed by heating to, say, 140°C (284°F) and wedging or peeling away the skin. (The T_g is the point or region at which the adhesive softens and weakens as the temperature rises.) This is recommended because this level of T_g can be exceeded without overheating aluminum alloys. For 180°C (350°F) curing adhesives, exceedence of this temperature would damage the heat treatment of the aluminum alloy; thus, in this case, the dry ice embrittlement process is recommended. Another factor is that these 180°C (350°F) curing adhesives are normally more brittle than the 120°C (250°F) types; therefore, embrittlement to a degree that will assist fracture is more likely.

Although the use of adhesives curing below 120°C (250°F) is uncommon in manufacture, these are used frequently for repairs and would be expected to respond better to the heating process than to the dry ice embrittlement method.

In addition to these two techniques, a third and common technique is used when the skin being removed is to be scrapped. In this technique, the skin is cut with a knife (usually ground from an old heavy hacksaw blade) or with a carefully set oscillating saw into strips approximately 25 mm (1 in.) wide. The strips of skin are then wound off with a roller in a similar way to opening a can of sardines. See Figure 10.1.

This is an effective method for skins of 1 mm (0.04 in.) or thinner and can be used to peel skin from skin or skin from honeycomb core. Note that great care is required when using either of these two methods to cut through a skin. It is only necessary to cut partially through the skin, leaving approximately 0.125 mm (0.005 in.) of metal. This can be torn easily. It is quite possible to cut into the doubler below, or even worse, into a machined fitting to which the skin is bonded. Cutting into underlying honeycomb should also be avoided but is not as serious because it can be replaced or locally potted. Ideally, the

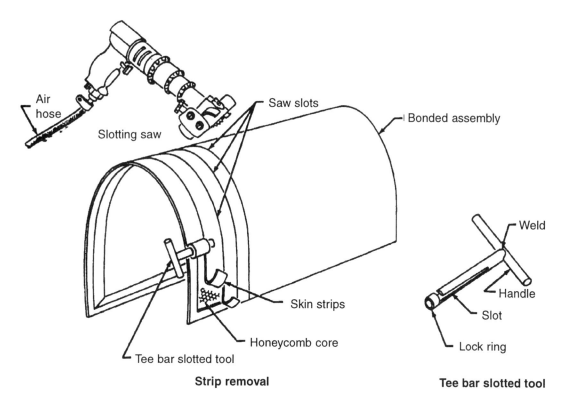

Fig. 10.1 Sardine-can method for removing strips of metal skin. (Courtesy of the U.S. Air Force.)

cutting tool used should be fitted with some sort of stop, which should be adjusted to the required depth after checking the drawing to establish the thickness of the skin being cut. This may be more complex than it appears if the skin is chemically milled and has steps of different thicknesses from the spar to the trailing edge. A drawing then will be necessary to provide the precise locations of the different thicknesses on the same skin.

Removing a bonded metal repair failure can be difficult because the newly bonded area surrounding the failed area can be stronger than the existing component or original bond area. It is highly probable that the action of peeling the repaired area from the existing skin will further disbond the surrounding skin area. This is especially true when removing repairs bonded with new and tough adhesives such as AF 163-2K from components originally bonded with AF 130 or Redux 322. The same would apply in any situation in which the repair adhesive is tougher than the original adhesive. Freezing has little effect in this case because it embrittles both glue lines. Heating may affect the original skin temper condition.

The safest and fastest course of action is to cut around the periphery of the repair and then to remove all of the repair plate plus the repair overlap area as a whole. This may be done in strips or in one piece. If this method is used, then heating or freezing is acceptable to assist the removal of skins from the core area within the cut boundary. (See Figure 10.2.) The size of the next repair may then exceed SRM limits. If so, approval from the OEM or other design authority must be obtained for the proposed repair.

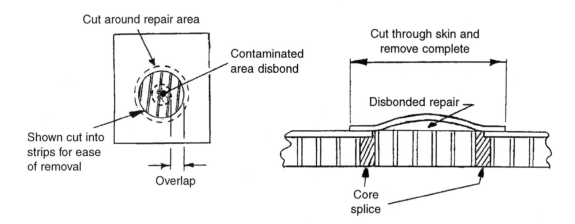

Fig. 10.2 Removal of an existing repair.

10.1.2.2 Disbonding Composite Parts

The methods described above for disbonding metal parts present problems when applied to composites because of the following:

- Most matrix resins are fairly brittle, and failure could occur away from the bond line in another ply. This is especially likely if two layers of composite have been bonded with a tough adhesive but the matrix resin in the composite is brittle.

- Composites are not as thermally conductive as metals, and time may be needed to cool or heat the bond line. In the case of cooling, the outer layers could become more brittle than the bond line.

- Cooling or heating by a large amount could create significant internal stresses resulting from the different thermal expansion coefficients of the fiber and resin.

Composites should not be heated above the T_g of the matrix resin. They should not be cooled below -55°C (-67°F), which is their minimum design operating temperature.

Tough adhesives will be difficult to cool sufficiently to make them adequately brittle. Experience has shown that two layers of pre-cured composite, bonded with a separate adhesive, can be parted using a shallow wedge. Any wedges used should be hard plastic and shallow angled. When tapping the wedge with a hammer, many small taps are preferred to a few heavy blows. Some trial-and-error experimentation on a scrap component is recommended. This technique has been used successfully to disbond composite spars from composite skins; however, failure sometimes occurs in one of the composite plies instead. In this case, the damaged surface must be repaired with an extra layer before rebonding. Experience has also shown that a solid composite cannot be split with wedges to give a clean surface along any one ply, and this should not be attempted. When making stepped lap joints, the common technique is to cut approximately three-quarters of the way through a fabric or tape layer and then to lift it with a sharp chisel, peel back the layer, and break it off at the cut line. This can be done easily with fiberglass and somewhat less easily with carbon fiber and aramid fabrics. This is a useful advantage of brittle matrix resins. If they were significantly tougher, this technique could be more difficult. Stepped lap joints would then have to be machined using special equipment such as that supplied by GMI, or scarf joints would have to be made. See Figure 10.3.

10.1.3 Damage Removal

10.1.3.1 Damage Removal from Metal Parts

This section describes methods that have been effectively used to remove damaged areas from bonded metal parts. If the damaged material is removed carefully, the subsequent repair can be made to the minimum possible size. This may be helpful if the SRM sets

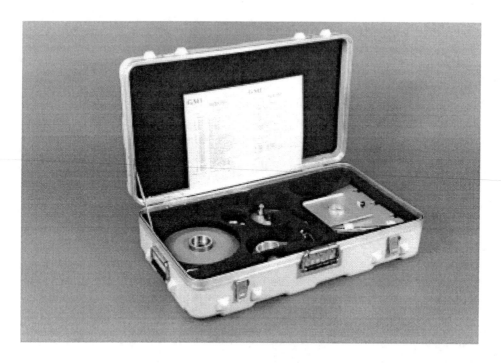

Fig. 10.3 GMI step cutting and scarfing machine with a trade name of "Leslie."
(Courtesy of GMI.)

limits on the size of repairable damage. Care should be taken to avoid extending the damage during preparation for repair. Proper tools must be used to cut the core and skin neatly with smooth edges and well-rounded corners. Some adhesives, especially those used for high-temperature applications, may peel easily. This is an asset to removing damaged skin or doubler material, but care must be taken to prevent the peeling from causing delamination beyond the intended repair boundary. Paint strippers should not be used unless all adjacent areas that might be affected have been carefully masked with metal foil tapes to prevent the chemical strippers from damaging areas such as bond lines and open core areas. Adhesives usually should be allowed to remain on surfaces that will be rebonded if the original adhesive appears to be in good condition and is firmly bonded. This will provide an excellent surface for the repair bond. It is now recommended that for 180°C (350°F) curing adhesives, the surface should be reactivated with cyclohexanone followed by drying for one hour at 80°C (176°F) and application of a new primer coating. This primer coating should be of the correct thickness. The thicknesses of primer coatings must be carefully controlled to avoid reducing bond strength. It is better to err on the side of using too little rather than too much. The simplest way to ensure the correct primer thickness on any surface is to obtain a color chart from the manufacturer. The manufacturer's color chart will show three areas: too thin, correct thickness, and too thick.

10.1.3.2 Removal of Skin and Doubler Material

Numerous types of mechanical equipment are available to cut, slit, rout, or saw sheet aluminum, titanium, or stainless steel. Equipment commonly used includes the high-speed router, jitter, or saber saw and the slotting saw. These are described in Chapter 13. The high-speed router is often found to be convenient because it can be used against a template to define the outline, and the depth of cut can be adjusted easily. Typical router application to remove damaged material is shown in Figure 10.4. The router template can be made from either 3 mm (0.125 in.) aluminum or hard plastic. It can be attached to the part surface with strong double-backed tape. In making the template, appropriate allowance must be made for the diameter of the cutter and the thickness of the guide bushing, as shown in Figure 10.4. The router template can be made from either 3 mm (0.125 in.) aluminum or hard plastic. It can be attached to the part surface with strong double-backed tape. In making the template, appropriate allowance must be made for the diameter of the cutter and the thickness of the guide bushing, as shown in Figure 10.4.

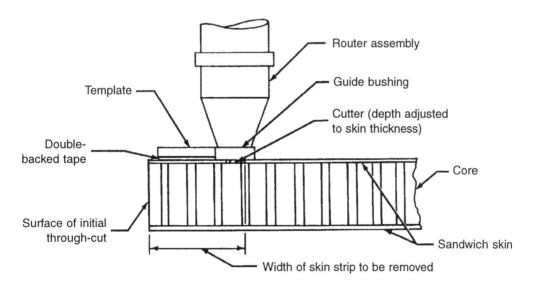

Fig. 10.4 Use of high-speed router to remove a strip of sandwich skin.
(Courtesy of the U.S. Air Force; see also Ref. 10.1.)

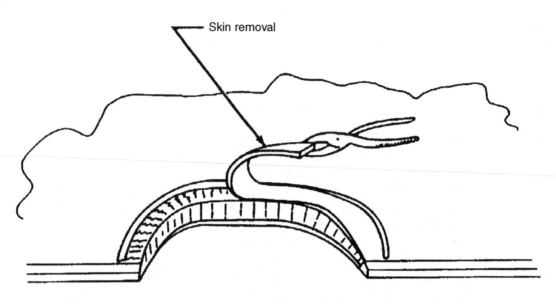

Skin removal

Fig. 10.5 Peeling a skin strip from honeycomb core.
(Courtesy of the U.S. Air Force; see also Ref. 10.1.)

After routing, the template and double-backed tape are removed, and a beveled-edge putty knife may be used to lift the edge of the skin. The skin may then be gripped with pliers or vise grips and peeled away. However, pull strips of skin slowly, and wear protective gloves. When the skin is finally removed, it springs back suddenly and may cause injury if suitable protection is not worn. See Figure 10.5.

After mapping the damage as described in Chapter 7, cut out the skin damage to a regular shape with rounded corners using 25 mm (1 in.) radii minimum if possible. Ensure that the damaged area plus an extra 12.5 mm (0.5 in.) is removed. Smooth the sawn edges with a fine sanding wheel to produce smooth edges and corners with smooth radii. Use a diamond-coated oscillating saw for the straight or slightly curved edges and a diamond-coated tank cutter to give good corner radii. Remove the honeycomb core if it is crushed or contaminated.

10.1.3.3 Damage Removal from Composite Parts

This may be done in ways similar to those used for metals, except that diamond saws give a cleaner cut and are less likely to cause splintering. Corners of cutouts should be radiused in the same way as metal sheets, and the skin may be removed in strips. Although the skin cannot be bent and peeled in the same way as metal, it can be lifted away from the honeycomb core in strips.

10.1.3.4 Removal of Core Material

Removal of light-density core of any type may be accomplished using a sharp pocket knife or a sharpened putty knife or chisel. See Figures 10.6 and 10.7. For the condition shown in Figure 10.6, vertical cuts are made through the core proceeding along the edge of the routed skin. The knife is then slid along the inner skin surface to make the horizontal cut. The core is removed down to the adhesive bond line.

In the case of denser core, machining techniques may have to be used. A core removal operation is shown in Figure 10.7. An abrasive disk is being used with a portable high-speed air motor. A valve stem cutter may also be used for this operation.

Following removal of the core, the adhesive surface may be smoothed using a portable high-speed motor and a sanding disk. See Figure 10.8.

Care should be taken not to remove the adhesive layer, thereby exposing the bare metal. If the adhesive layer is allowed to remain in place and is sanded lightly to ensure that it is clean, no metal treatment is required before bonding new honeycomb or other core material in place. (See also Section 10.1.3.1 of this chapter.) After this smoothing procedure, use a vacuum cleaner to remove the sanding dust and debris. Dry wipe with clean cheesecloth.

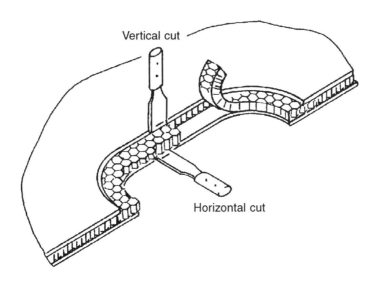

Fig. 10.6 Removal of core with a sharpened putty knife.
(Courtesy of the U.S. Air Force; see also Ref. 10.1.)

Fig. 10.7 Damaged core being removed with abrasive disk mounted in a high-speed air motor. (Courtesy of the U.S. Air Force; see also Ref. 10.1.)

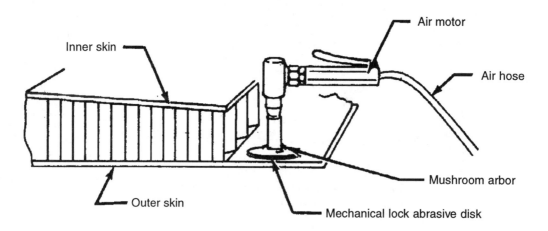

Fig. 10.8 Smoothing an adhesive surface with an abrasive disk. (Courtesy of the U.S. Air Force; see also Ref. 10.1.)

10.1.3.5 Removal of Surface Corrosion

The existence of corrosion on any part or surface should be considered the same as physical damage and must be removed, or the part must be replaced. In general, light corrosion on skins, doublers, or stiffeners can be removed, and the surface can be refinished. If corrosion has reached an advanced stage of pitting and scale, this is more serious and parts may have to be replaced. Check the SRM for the limits for material removal on each part and for the treatment required. When aluminum honeycomb core is affected by corrosion, it loses its luster and appears white or gray in color. In more advanced stages, it becomes brittle and tends to flake. Aluminum honeycomb with any visible corrosion should be replaced because it will not bond well to any adhesive that must be applied to it to bond a new piece of skin or core. Furthermore, its strength will be reduced if the corrosion has any depth.

10.1.4 Moisture and Contamination Removal

Contaminated honeycomb should always be removed because resins and adhesives will not bond to contaminated surfaces. If the honeycomb is aramid or other nonmetallic and is wet only with clean water, it may be dried and allowed to remain in place. Even if the honeycomb appears slightly contaminated, it should be removed and replaced. If metallic honeycomb is wet, it is probably also corroded to some degree and should be replaced. Water can penetrate deep into metallic honeycomb only through fractured honeycomb node bonds; this is also a good reason for replacement. A moisture meter may be used (see Chapter 7) to assess the dryness of fiberglass, aramid skins, and nonmetallic honeycomb. Continue drying until the meter reads in the green band. Carbon fiber skins present a problem because no moisture meter available at moderate cost can indicate the moisture content of this conductive material. Drying to a recommended schedule is all that can be done; this schedule is given in most SRMs as follows:

Dry for one hour minimum at a temperature of 66 to 78°C (150 to 170°F) maximum. The rate of temperature rise should not exceed 3°C (5°F) per minute. To minimize the risk of delamination, avoid both too rapid a rate of temperature rise and the use of high drying temperatures. As a guide, the drying should never occur above the wet T_g of the resin matrix. Slower drying using lower temperatures for longer times is preferred if time permits, and the above rates should never be exceeded. Good, durable adhesive bonds depend on this drying procedure being conducted as a minimum. See Refs. 10.2 and 10.3 and SAE ARP 4977 Drying of Thermosetting Composite Materials.

10.1.5 Surface Preparation of Composites–Repair Sanding and Ply Determination

Sanding is a skilled process requiring care and experience. When step sanding, it is especially important to cut only partially through each ply and to avoid damaging the underlying ply to which the repair ply must be bonded. Individual plies can be seen easily if fairly thick glass or aramid cloths have been used; however, it is more difficult with carbon, especially if a dark-colored or black resin system is also employed. The task is much easier with clear resins and glass plies. For these reasons, the SRM and/or component drawings should be studied to establish the number of plies and the fabric type and thickness of each layer before commencing work. With carbon composites, it may be better to use special tools (e.g., those supplied by GMI or equivalent) to enable precise control of cutting depth. The strength of repair joints depends heavily on retaining the structural integrity of the overlap area to which repair patches are bonded. Aramid fibers can be a problem when cutting or sanding. A good video and booklet that should be studied is supplied by Du Pont External Affairs (1111 Tatnall St., Wilmington, DE 19898), and only the special tools and techniques given therein should be used. A useful safety video is also available from Du Pont. Special shears should be used for cutting Kevlar and other aramid fabrics and pre-pregs, and these should be reserved only for aramids because glass and carbon fibers blunt them rapidly. Special drills and routing tools also are needed. Sanding is a delicate operation, and it is essential to sand only the resin and not the fibers, which "fuzz" easily. Aramid fibers can easily fibrillate into smaller fibers, which appear as fuzzing. Paint should be removed carefully down to the primer with a hand-held pneumatic orbital sander. Final sanding should be done by hand using 240 to 600 grit wet and dry paper in a wet sanding operation to avoid damage to the fibers. It is better to cut aramid plies using a sharp utility knife rather than a router, but routing can be done using a split helix router at 20,000 RPM. In the video, the message constantly repeated concerning drilling, routing, and bandsawing is that sharp tools, high cutting speeds, and low feed rates are essential for a good finish. Overheating of the cutting area must be avoided. Many SRMs detail the types of drills and cutters required for each type of fabric, and these should always be used to ensure best results. A good resin content is desirable in the surface layers for all types of fabric to ensure a good surface finish and an excess of resin so that the surface can be sanded without damaging the fibers. Edge sanding of panels is best done on a belt sander. Sand from the outer face to the middle at 45°, then from the inner face to the middle at 45°, then at 90°. Always work the fibers toward the middle of the laminate. For rough sanding, 80 to 180 grit is suitable. Finish by hand sanding, using 240 to 600 grit.

10.1.5.1 Abrading

Abrading may be used to remove paint or to prepare a composite surface for bonding. In either case, care is required to ensure that the first layer of fabric is not damaged. Abrasion should not go below the surface resin coating. Silicon carbide paper of 150 grit may be used to cut through paint, but 320 grit is often sufficient for abrasion prior to bonding. Use clean abrasive paper, and remember that a new sheet cuts faster than a worn one.

If working on old or home-built airplanes, gliders, or other wooden constructions, a problem can occur with plywood which is worth mentioning here. This problem is called case hardening and is not fully understood. It means that the glue does not adhere well to the surface of the plywood, and the remedy is to sand only along the grain of the wood. Because of bond failures, sanding along the grain was made mandatory by the British Air Ministry in 1942. See Ref. 10.4 by Dr. Norman De Bruyne, inventor of Aerolite glue, Redux 775, and honeycomb. See also any technical notes supplied by the plywood, glue, and airplane manufacturers and The Forest Products Research Laboratory (United States). Ciba data sheets for Aerolite glue call for thorough sanding. The following notes from the *PFA Handbook—The Handbook of the Popular Flying Association* (Crown Printers, Morriston, Swansea, 1980) may be helpful:

> "The hot-press method of plywood manufacture results in glazed, hard outer surfaces due to local crushing of the surface fibers by the press-plates, release agents from the press-plates, or from glue penetration. Light hand sanding of ply surfaces to be glued is a satisfactory method of correcting these unfavorable conditions which would otherwise interfere with the adhesion of glues used in home-built aircraft construction. All aircraft plywood must be sanded before use. Oversanding must be avoided."

The *PFA Handbook* states that the reduction in the thickness of the outer veneer should never exceed 10%.

All sanding of wood should be done along the direction of the grain. The *PFA Handbook* also states that when working with timber for home-built aircraft, gliders, or other purposes, freshly cut (i.e., planed) wood surfaces that are smooth, uniform, and free from crushed surface fibers are generally the most satisfactory for gluing purposes. Sanding and other means of roughening the surfaces are unnecessary and inadvisable on freshly planed surfaces. Sanding wood tends to block the pores in the grain pattern with dust, inhibiting the absorption of adhesive into the grain and causing adhesion to loose dust particles rather than the wood. The critical element seems to be to promote glue absorption into the wood to form a strong mechanical bond, with a large effective gluing area. Note the need to sand glazed plywood and to avoid sanding planed pieces of timber. Dust should be removed with a vacuum cleaner before applying glue.

10.1.5.2 Taper Sanding/Scarfing

This is a popular method of jointing skins, especially with carbon fiber, for which high joint strengths are required and step sanding without special tooling can be more difficult than it is with glass and aramids. The SRM will state something similar to the following: "Taper abrade the edge of the cut-out allowing 12.5 mm (0.5 in.) for each damaged ply around the repair area." For commonly used fabrics of approximately 0.25 mm (0.01 in.) thickness, an overlap of 12.5 mm (0.5 in.) corresponds to a taper of 1 in 50, which has been shown by research to give a strong joint and also a reasonably good fatigue life to such a joint. Accurate cutting of this taper angle is required and is a highly skilled job. To avoid excessive glue-line thickness in places, resulting in strength loss, the taper must be cut accurately on both the structure and the patch to ensure that a good fit and a thin glue line is obtained all around the joint. It is better to cut the taper around the repair and then to lay the repair plies in position. This avoids the need to match the angle of the patch. Pre-cured patches are not used frequently because of the difficulty of matching both the taper angle and any curvature of the repair area. GMI also supplies special tooling to simplify this operation. (See Figure 10.3.) Similar equipment is supplied by ATACS (which is called a scarfer) and probably by other companies. Tapers as steep as 1 in 20 may be accepted at panel edges where the thickness is greatly increased to allow for the heads of countersunk fasteners. Tapers of 1 in 30 may be accepted for lightly loaded parts; however, the goal should always be 1 in 50 unless the SRM permits a steeper angle. Note that for scarf joints, Boeing puts in the smallest patch first, and Airbus puts in the largest patch first. See each type SRM for precise details. Both methods have been proven acceptable by research; however, for strict compliance with the SRM, the correct procedure should be used according to the manufacturer's SRM. See Figure 10.9.

10.1.5.3 Step Sanding and Cutting

This process requires great care and is usually limited to fiberglass and aramid materials where the layers are more easily identified visually than they are with carbon fiber. Step sanding can be done on carbon fiber, but it is easier with special tooling such as that supplied by GMI of Paris, France (see Figure 10.3), which has accurate depth control. The object of this process is to cut a series of steps so that each repair layer can lay in a step to give a suitable length of overlap. (See Figure 10.9.) The SRM will specify the overlap length required. In this process, it is essential to the strength of the final repair that each layer of fiber be cut slightly more than halfway through its thickness and then be broken away and peeled off or machined away accurately with special equipment. If a cut passes through a layer and touches the next layer, then the next layer's strength may be seriously reduced or completely destroyed and the repair layer will have no strength. Sanding aramids is difficult and leaves a fluffy surface. Work carefully to the SRM instructions if working with this material. SRMs usually specify repair of aramids with fiberglass because fiberglass is easier to handle and is more likely to be held in stock.

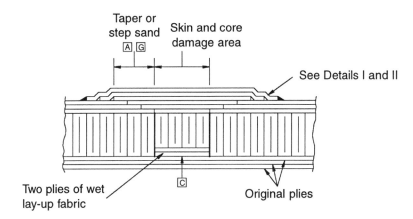

Taper or step sand A G · Skin and core damage area

See Details I and II

Two plies of wet lay-up fabric

C

Original plies

Section through typical repair (wet lay-up only)

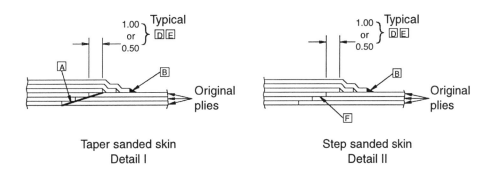

Taper sanded skin
Detail I

Step sanded skin
Detail II

Notes:

A	Taper sand around repair area over distance of 0.50 or 1.00 inch for each existing ply G.
B	Do not expose or damage filaments in untapered area when sanding.
C	Sanding must not expose or damage the filaments in bond ply (ply bonded to core).
D	Each ply must overlap at least 0.50 or 1.00 inch past edge of proceeding ply E.

E	The minimum overlap requirement of 0.50 is typical for most repairs. Some heat-affected zones made of 350°F (177°C) original cure temperature require 1.00 minimum overlap and are identified in component repair charts.
F	Remove damaged plies in steps of 0.50 or 1.00 inch for each existing ply E.
G	Taper sand surfaces in areas of critical aerodynamic smoothness. Refer to SRM 51-10-01.

Fig. 10.9 Stepped lap joints and taper sanded (scarf) joints.
(From Boeing B.757 SRM 51-70-17. Courtesy of Boeing Commercial Airplane Group.)

10.1.6 Water Break Test

The water break test is used on composite and metal surfaces to check that no contamination remains and that the surface is in an acceptable condition for bonding. It must be performed, and the specified surface preparation treatment must be repeated if the test fails. This test is automatically performed on aluminum panels as part of the rinse process after chromic or phosphoric acid anodizing. For composite panels, the test must be specified after final abrasion of the surface to be bonded. The water break test is defined in SAE AIR 4844 Composites and Metal Bonding Glossary as follows, "In this test, water is applied to the prepared surface and should remain in a continuous film over the whole area for at least 30 seconds. [See Figure 10.10.] If the water separates into droplets or the film

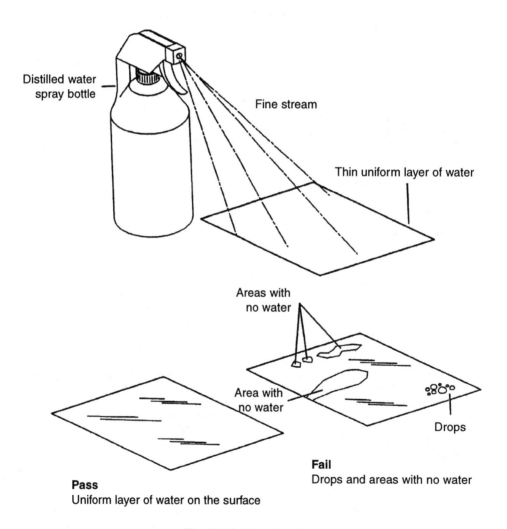

Fig. 10.10 Water break test.

is not continuous, then the cleaning operation has to be repeated until this requirement is met." SAE ARP 4916 states that distilled or deionized water should be used, and details a sequential procedure for this test. It should be used with caution on bonding surfaces adjacent to honeycomb because water entering the honeycomb can be difficult to remove.

Note: The water break test works well on normal composite abraded surfaces. However, if the surface is polished to a smooth finish, passing the test can be difficult, even with a clean surface. Conversely, if the surface is rough, the surface tension of the water film may be interrupted, thereby making assessment of the water breaks more difficult. Good sensitivity can be obtained on composite laminates abraded with grits in the range of 80 to 400 grit. Fortunately, this is the normal range used for abrading composites. It is necessary to dry the surface after this test using the SRM procedure, but care must be taken that the surface is not touched or otherwise contaminated during the drying procedure. Drying with heater lamps is recommended and should occur in a clean room. As with metals, no surface that is to be bonded should be touched between cleaning and bonding, not even with clean, white cotton gloves.

10.1.7 Metallic Surface Preparation

Metals joined either to themselves or to composites by adhesive bonding require good chemical treatments to achieve sound durable bonds, especially if they are to be used in a wet environment. The importance of these processes cannot be overstated. Aluminum alloys are usually anodized using phosphoric acid, Boeing Process BAC 5555, or chromic acid, followed by the unsealed rinsing process. This means that they must be rinsed in clean water below 60°C (140°F), preferably using a spray rinse. A hot rinse at 96°C (205°F) is used when the pores in the surface oxide must be sealed to give corrosion resistance to a part that does not require bonding. This hot rinse must never be used when the surface is to be bonded because the adhesive must enter the pores to achieve a good bond.

To illustrate the importance of anodizing, some personal research may be worth mentioning here. Wedge tests to ASTM D-3762 were conducted on 2024-T3 clad aluminum alloy after periods of immersion in distilled water using several methods of surface preparation. The best preparation used was an unsealed chromic acid anodize. Two well-known, high-performance, two-part, room-temperature curing paste adhesives were used and remained well bonded after seven years of total immersion. The worst preparation was glass paper abrading. On this type of surface, one of these adhesives lasted a week, and the other lasted a day. The only difference among these lengths of performance was the surface preparation. Stainless steels and titanium alloys require special chemical treatments, and these are detailed in OEM process specifications. When bonding is to be done, these treatments must be employed. See the SRM for the treatments required and also Chapter 15.

10.2 Typical Repairs

10.2.1 Edge Band Repairs

Repairs to composite panel edge bands may be necessary for several reasons:

- Worn or damaged bolt holes
- Edge damage
- Lightning damage
- Edge delamination

They are needed frequently because edges are regularly subject to damage.

Typical repairs for panel edge bands may be found in several sections of the Boeing 757 SRM 51-70-03, 51-70-05, 51-7-06, 51-70-07, and 51-70-17. Figure 10.11 shows the taper sanding method of repair using wet lay-up at temperatures from room temperature to 66°C (150°F).

A separate repair for the repair of damaged attachment holes is shown in Figure 10.12. Similar repairs are shown in the other SRM chapters listed above for higher repair temperatures.

10.2.2 Repair of Damage to Core and One Skin

This is the simplest and most common type of repair to a honeycomb panel. It is identical to the case for both skins, except that only one side must be repaired. The repair is relatively easy, especially when the damage size is small, because no tooling or profile support is needed unless repairs are done at or near the original cure temperature. (See Figure 10.13 in Section 10.2.3.) For more details of core replacement, see SAE ARP 4991 Draft—Core Restoration of Thermosetting Composite Materials. The original type and density of core must be used if honeycomb replacement is necessary. Slightly higher density core of the same type usually is acceptable.

10.2.3 Repair of Damage to Core and Both Skins

The normal method used when access is available to both sides is shown below. An alternative method is available when access is possible from only one side. This type of repair to 120 and 180°C (250 and 350°F) cured graphite/aramid/fiberglass reinforced epoxy laminates and nonmetallic honeycomb structure using the 93 to 110°C (200 to 230°F) wet lay-up repair method is covered in the Boeing 757 SRM 51-70-17 Figure 12, reproduced here as Figure 10.13. Similar repairs using higher-temperature curing systems are covered in other sections of this SRM. See also SAE ARP 4991 Draft.

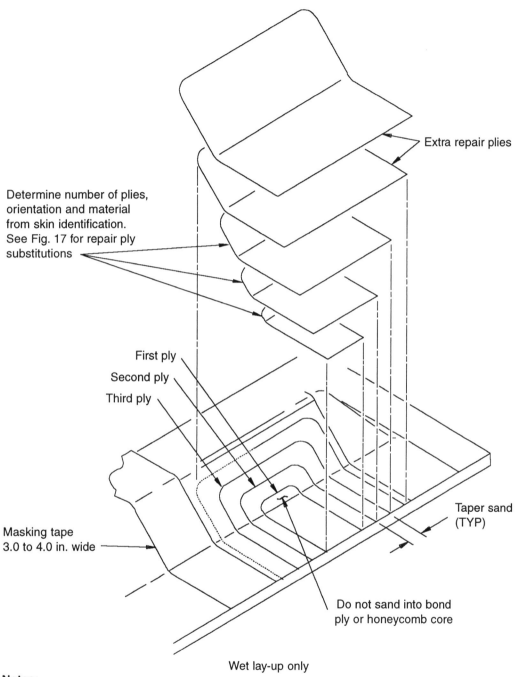

Extra repair plies

Determine number of plies,
orientation and material
from skin identification.
See Fig. 17 for repair ply
substitutions

First ply
Second ply
Third ply

Taper sand
(TYP)

Masking tape
3.0 to 4.0 in. wide

Do not sand into bond
ply or honeycomb core

Wet lay-up only

Notes:

- Refer to paragraph 3.G for the repair instructions.

*Fig. 10.11 Repair of damaged skin plies on a panel edge. (From Boeing SRM 51-70-17 Fig. 15.
Courtesy of Boeing Commercial Airplane Group.)*

10.2.4 Hybrid Repairs

Hybrid composites are normally produced in two forms:

1. Composites in which separate layers of carbon, glass, or aramid fabric or tape are laid up to achieve the required properties. These are the most common type in aeronautical applications.

2. Composites in which layers of fabric are used, with the fabric layers having different fibers woven into the same cloth (e.g., fabrics made from glass and aramid fibers, or carbon and glass, or carbon and aramid). In these fabrics, the proportion of each

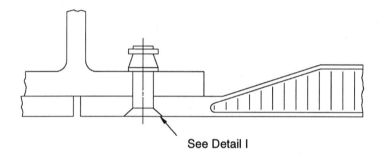

See Detail I

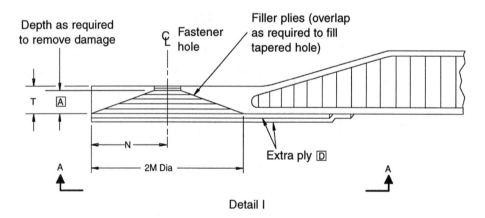

Detail I

Notes:

• Refer to paragraph 3.L for the repair instructions.

Fig. 10.12 Repair of damaged panel attachment holes. (From Boeing 757 SRM 51-70-17 Fig. 16, Sheets 1 and 2. Courtesy of Boeing Commercial Airplane Group.)

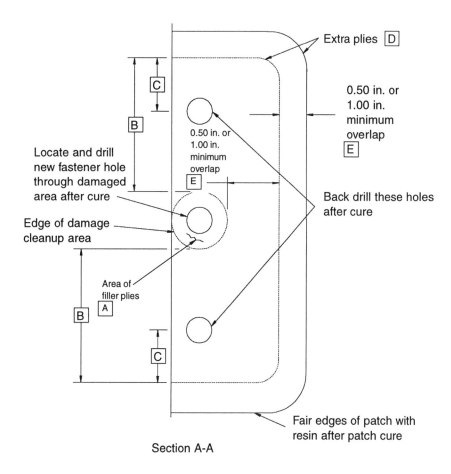

Section A-A

Notes:

- D = Fastener diameter.

- M = 5T maximum as shown, where T is the thickness of the edge band. Do not cut into core.

- This repair applies only where no more than two attach holes are damaged in any ten consecutive attach holes.

[A] Apply filler plies as required to fill the damaged area.

[B] Extend first extra ply far enough so that it extends at least 0.50 or 1.00 inch past edge of damaged area [E] .

[C] Extended first extra ply far enough to provide at least 2D edge margin.

[D] Orient extra repair plies in the same direction as the original outer layer (see Fig. 15).

[E] Some 350°F (177°C) original cure temperature components in heat-affected zones require 1.00 minimum overlap for each repair ply and are identified in the component repair charts. All other components require 0.50 minimum overlap.

Fig. 10.12 (continued)

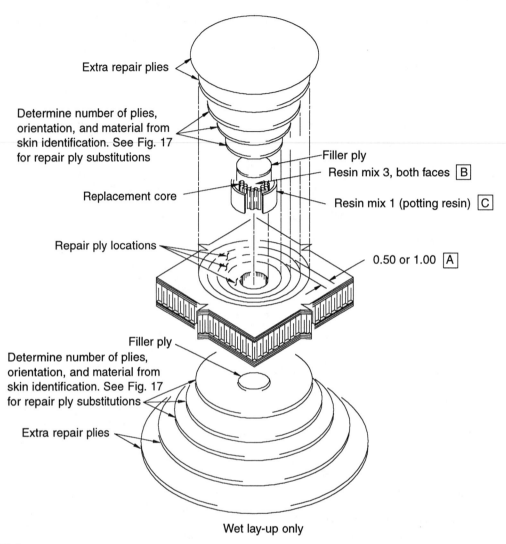

Extra repair plies

Determine number of plies, orientation, and material from skin identification. See Fig. 17 for repair ply substitutions

Filler ply

Resin mix 3, both faces [B]

Replacement core

Resin mix 1 (potting resin) [C]

Repair ply locations

0.50 or 1.00 [A]

Filler ply

Determine number of plies, orientation, and material from skin identification. See Fig. 17 for repair ply substitutions

Extra repair plies

Wet lay-up only

Notes:

[A] Some 350°F (177°C) original cure temperature components in heat-affected zones require 1.00 minimum overlap for each repair ply and are identified in component repair charts. All other components require 0.50 minimum overlap.

[B] Apply resin mix 3 to honeycomb core faces just prior to repair ply application.

[C] Resin mix 1 may be applied to periphery of existing core instead of on core plug.

Fig. 10.13 Repair to honeycomb panels with access from both sides. (From Boeing 757 SRM 51-70-17 Fig. 12. Courtesy of Boeing Commercial Airplane Group.)

fiber employed can be varied to suit the design requirements. When hybrids must be repaired, ensure that each layer of cloth is of the correct fiber type, weight, and orientation and that it is laid up in the sequence specified on the drawing. Hybrid repairs are also covered in each aircraft type SRM.

10.2.5 Blind Repairs

This type of repair is usually described in the SRM as, "Repair of damage to external and internal skins with access limited to one side." See Figure 10.14.

10.2.6 Injected Repair

Injected repair is sometimes permitted but is not recommended unless there is reason to believe that the cause of disbond is not some form of contamination. Where contamination exists, the process is unlikely to be effective. Two forms are possible:

1. Injection to repair delamination near the center of a solid laminate. This may be effective because, if the surface is undamaged, no contaminant is likely to have reached the surfaces that require bonding. In this case, two holes can be drilled at the edges of the mapped delamination and adhesive or resin can be injected into one hole until it flows out from the other. The method can be tried for edge delamination; however, in this case, the faces to be bonded are likely to be contaminated, and it is impossible to gain access to clean them except with a fine solvent spray. See Figure 10.15.

    For metal panels with thin edge doublers, it is possible to remove corrosion at delaminated edges with an acid etch (after lifting out the disbonded film adhesive layer), if it is thoroughly rinsed with distilled water and dried before injecting a suitable adhesive.

2. Some OEMs allow a similar repair to disbonded honeycomb panels for small areas. The repair may be successful if the disbond is not a result of moisture ingress or surface breaking damage that may have allowed water or other contamination to enter. The process is not recommended if contamination is suspected. The method should be used for radomes (only if permitted by the OEM) or if the disbond is outside the scanning area because it will affect the strength of the radar signal by increasing signal transmission losses.

10.2.7 Solid Laminate Repairs

Repairs to solid laminates take two forms:

1. Resin-bonded repairs of various types, as shown in Figure 10.15
2. Mechanically fastened repairs, with or without adhesive bonding

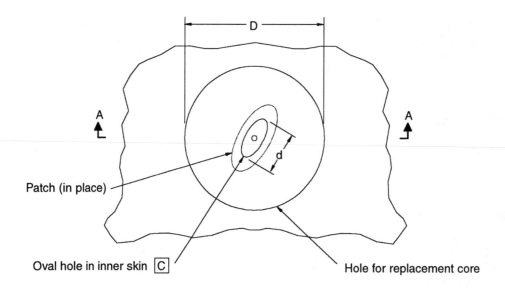

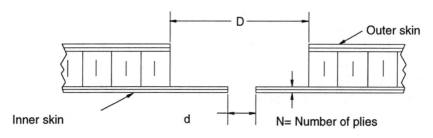

Section A-A
(Patch not shown)

Detail I

Notes:

- This repair illustrates the use of a spring steel clamp. However, any suitable retaining device may be used.

- D = d + N + 1
  D = Major diameter of oval hole in inner skin
  N = Number of plies

  For example:
     If d = 0.50 inch
     Then,   D = 0.50 + 2 (plies) + 1
              D = 3.50 inch diameter

    A Resin mix 3 applied to core plug or original core.

    B Make taper and overlap per Fig. 14.

    C Major diameter d of oval hole in inner skin is limited to 1.5 inch for this repair.

Fig. 10.14 Blind repair, access from only one side. (From Boeing 757 SRM 51-70-03 Fig. 8, Sheets 1 to 3. Courtesy of Boeing Commercial Airplane Group.)

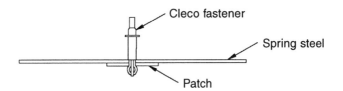

Assemble patch and spring steel
Detail II

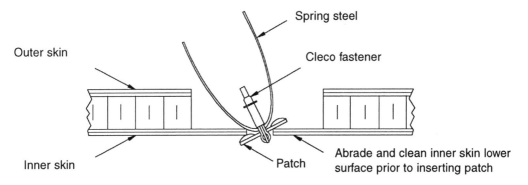

Insert patch into oval hole
Detail III

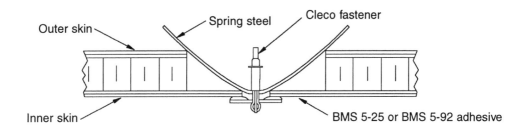

Hold patch in place while curing
Detail IV

Fig. 10.14 (continued)

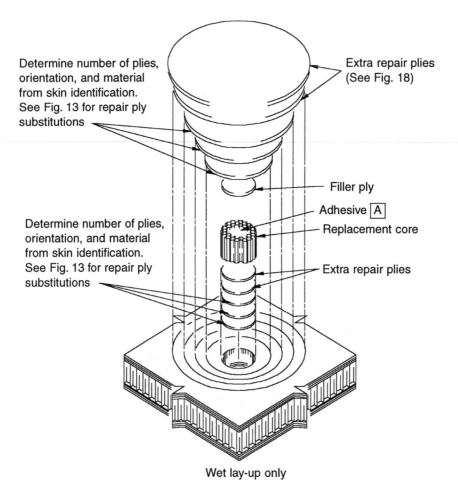

Determine number of plies, orientation, and material from skin identification. See Fig. 13 for repair ply substitutions

Extra repair plies (See Fig. 18)

Filler ply

Adhesive [A]

Replacement core

Determine number of plies, orientation, and material from skin identification. See Fig. 13 for repair ply substitutions

Extra repair plies

Wet lay-up only

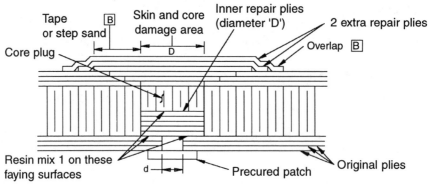

Tape or step sand [B]

Skin and core damage area

Inner repair plies (diameter 'D')

2 extra repair plies

Core plug

Overlap [B]

Resin mix 1 on these faying surfaces

Precured patch

Original plies

Secttion through repair (wet lay-up only)

Fig. 10.14 (continued)

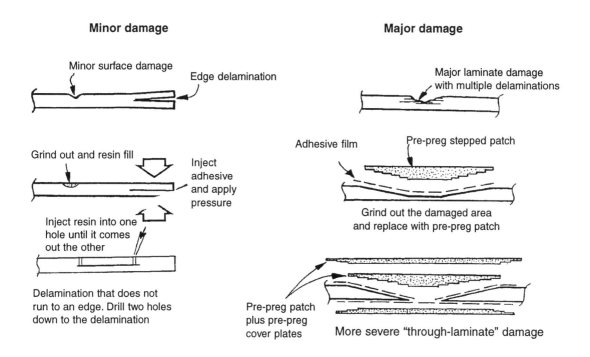

Minor damage

Minor surface damage

Edge delamination

Grind out and resin fill

Inject adhesive and apply pressure

Inject resin into one hole until it comes out the other

Delamination that does not run to an edge. Drill two holes down to the delamination

Major damage

Major laminate damage with multiple delaminations

Adhesive film

Pre-preg stepped patch

Grind out the damaged area and replace with pre-preg patch

Pre-preg patch plus pre-preg cover plates

More severe "through-laminate" damage

*Fig. 10.15 Basic types of solid laminate repair showing injection repairs.
(Courtesy of Aero Consultants (United Kingdom) Ltd.)*

These repairs appear similar to sheet metal repairs, but the parts used for repairs must be of the correct pre-preg type and have the required composite ply orientation. The specified fasteners must be used to ensure adequate load transmission and the avoidance of corrosion in the fasteners. Some fasteners used in composites require small interference fits to improve the fatigue life of joints, and special sleeved fasteners may be used to protect the holes during installation. Special drills and techniques may be required to produce good holes. (See Chapter 11.) The specified torque loading must be carefully observed with fasteners in composites, and large washers may also be required. In both cases, consult the SRM for details on the type and size of repair permitted in each component and for the materials and procedures required. It must be emphasized that solid laminate repairs must be made carefully because they are used in the load-carrying structural parts. As heavier structures are made of composites, this type of repair will become more common. Training in this type of repair will increase in importance because solid laminate repairs require special drills, techniques, and fasteners that are different from those used for metals.

10.2.8 Potted Repair

Potted repair is generally recommended only for small areas of core damage. The SRM calls for damage in excess of 12.5 mm (0.5 in.) diameter to be repaired with new core material. Potting compound may be used for very small areas of core damage and when permitted by the SRM. Climbing drum peel tests to ASTM D-1781 have shown that fully potted honeycomb core results in a lower peel strength than a direct bond between the skin and honeycomb. This is believed to occur because a good fillet is produced when the skin and honeycomb are directly bonded and also because potting produces a continuous glue line which, once cracked, continues to crack. Potting compound itself often is less tough than a good adhesive. It is instinctive to think that the filled honeycomb will be better because of the greater bond area, but test results do not support this conclusion. If in doubt, run some tests because test results that you have obtained yourself are most convincing.

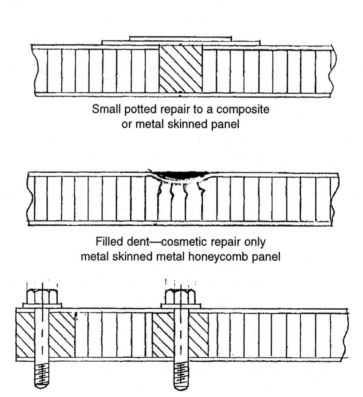

Small potted repair to a composite
or metal skinned panel

Filled dent—cosmetic repair only
metal skinned metal honeycomb panel

Potting around bolt holes in lightly
loaded interior trim panels

Fig. 10.16 Potted repairs.

Another form of potted repair is to fill small dents in metal skinned honeycomb panels caused by heavy hail or other impact. If the honeycomb core is metallic, the core will be buckled but not fractured and normally will retain approximately 50% of its compression and shear strength. This type of repair is only cosmetic and should be conducted within SRM limits. A suitable chemical etch treatment should be given to the skins before potting to ensure that the filler remains bonded. Another example of the use of potting is when honeycomb core around bolt holes is filled with potting compound to increase the compression strength of the honeycomb when the bolts are tightened. If panel edges are repaired then this potting must also be replaced to drawing requirements. Figure 10.16 illustrates all three methods.

10.2.9 Metal-to-Metal Bonding

This process should be performed according to the SRM after treating the metal surfaces as previously mentioned. See Chapter 15 for more details.

10.2.10 Plastic Welding (Solvent or Heat)

Some plastics may be joined by the use of suitable solvents to soften the mating areas and to allow the adjacent surfaces to bond together by interdiffusion of the molecules. This can be done with acrylic plastics commonly known as Plexiglas, Oroglas, and other trade names and also with polycarbonate, polystyrene, polyamides (nylons), and acrylonitrile-butadiene-styrene (ABS). The correct choice of solvent for each type of plastic is critical to avoid crazing and cracking. Thermoplastics can be bonded by raising them to their melting temperature and applying pressure, or by ultrasonic, hot tool, or hot gas welding. If interior or exterior items made from these materials require repair, refer to the SRM, Component Maintenance Manual (CMM), or the OEM for specialist advice on specific repair techniques. Thermoplastics can be bonded using adhesives if specialized surface treatments are given. For more details, see Ref. 10.5. Thermosetting plastic materials such as epoxies, polyesters, and phenol-formaldehyde are relatively easy to bond with a range of adhesives. Check the SRM for the specific materials and processes required.

10.3 Adhesive Usage

Paste Adhesives: For details of paste adhesives, see Chapter 2, Section 2.5.1.

Foaming Adhesives: For details of foaming adhesives, see Chapter 2, Section 2.5.2.

Film Adhesives: For details of film adhesives, see Chapter 2, Section 2.5.3.

10.4 Bagging Materials, Release Sheets, Peel Plies, Breather Cloths, and Application Techniques

Correct bagging materials must be used to achieve the required vacuum pressure. The details given in Section 10.4.2 should be studied carefully. Figures 10.13 and 10.14 show the lay-up sequence of the various materials, which is important to ensure good air and volatile bleed, to achieve the required vacuum pressure, and to prevent resins and adhesives from adhering to heater blankets.

This section includes not only the bagging material but also release sheets, perforated release sheets, breather and bleeder cloths, peel plies, and other materials required for a complete bagging operation.

10.4.1 Terminology

The following list of terminology is taken primarily from SAE AIR 4844 Composites and Metal Bonding Glossary:

- **Bagging:** Applying an impermeable layer of film over an uncured part and sealing edges so that a vacuum can be drawn.

- **Bagging Film Sealant Tape:** This is a soft mastic type of tape which is slightly tacky and is used to seal bagging film to the repair area or to join parts of the bagging film if two sheets are used.

- **Bag Side:** The side of the part that is cured against the vacuum bag.

- **Breather (Breather Cloth):** A loosely woven material such as a glass fabric or Osnaburg cloth that will serve as a continuous vacuum path over a part or the repair area, but it is not in direct contact with the part or the repair area. Its purpose is to allow the air inside the vacuum bag to be removed, thereby enabling atmospheric pressure to be applied to the part or repair area.

- **Bleeder Cloth:** A nonstructural layer of material used in the manufacture or repair of composite parts to allow the escape of excess gas and resin during the cure. The bleeder cloth is removed after curing and is not part of the final composite.

- **Bridging:** 1) A condition in which fibers do not move into or conform to radii and corners during molding, resulting in voids and dimensional control problems. 2) A condition in which one or more plies of a pre-preg span a radius step of the fluted core of a radome without full contact. 3) A condition where part of a vacuum bag does not go down into a radius and thus no pressure can be applied at that point. When making up a vacuum bag, it is important to avoid this.

- **Debulking:** Compacting of a thick laminate under moderate heat and pressure (i.e., noncuring conditions and/or vacuum to remove most of the air) to ensure seating on the tool and to prevent wrinkles. This process may be carried out by debulking a few layers at a time, in a series of debulking operations, rather than by debulking the whole lay-up in one operation.

- **Mold Release Agent:** A lubricant, liquid, or powder (often silicone oils and waxes) used to prevent sticking of molded articles in the cavity.

- **Parting Agent:** A material, liquid, or solid film used on the tool surface to ease removal of the assembly. See "Mold Release Agent."

- **Peel Ply:** A layer of open-weave material, usually fiberglass or heat-set nylon, applied directly to the surface of a pre-preg lay-up. The peel ply is removed from the cured laminate immediately before bonding operations, leaving a clean, resin-rich surface that needs no further preparation for bonding, other than application of a primer where one is required. Note: For many purposes this is acceptable, but research has shown that the release agent used on the peel ply sometimes leaves contamination on the surface of the composite, and stronger and more durable bonds are achieved if sanding is used to remove this contamination.

- **Release Agent:** A material that is applied in a thin film to the surface of a mold to keep the resin from bonding to the mold. Also called parting agent.

- **Release Film:** An impermeable layer of film that does not bond to the resin being cured.

- **Separator:** A permeable layer that also acts as a release film. Porous Teflon-coated fiberglass is an example. It is often placed between lay-up and bleeder to facilitate bleeder system removal from a laminate after cure. Also called separator cloth. This item is sometimes reusable.

- **Vacuum Bag:** The plastic or rubber layer used to cover the part to enable a vacuum to be drawn.

10.4.2 Selection Criteria

Vacuum Bags: Several important criteria must be considered when selecting a vacuum bag, as follows:

1. Will it withstand the required cure temperature?

2. Is it flexible enough to be draped over the shape of the part?

3. Does it have good tear resistance?

4. Is it nonporous?

5. Will it stretch enough to prevent bridging at corners?

6. Will it adhere well to the tape sealant to provide good vacuum?

7. How much does it cost? This is important because, to avoid contamination of the next job, vacuum bags should not be reused. See SAE ARP 5143 Draft—Vacuum Bagging of Thermosetting Composite Repairs for materials and techniques.

Questions 1 and 7 also apply to release sheets, breather and bleeder cloths, and peel ply materials.

Note that all vacuum bagging fabrics, films, and tapes must remain dry to prevent adding moisture to the repair. The storage environment should be noncontaminating and free of dust and oil. Materials should be protected from exhaust fumes, soot, oils, sprayed silicone, mist, rain, or other obvious particulate contaminants.

Release Sheets: Release sheets may be made from a range of materials. They may be release sheets of plastic film, known as "release film," in which case they may be perforated to allow the escape of volatiles and surplus resin for the vertical bleed-out technique. Likewise, they may be nonperforated for the squeeze-out or edge bleed-out technique. Release films are made from nonstick materials, the most common being modified halohydrocarbons such as polyvinyl fluoride (PVF), fluorinated ethylene propylene (FEP), polyester, and nylon. FEP films are used with all types of epoxy resins and are lightweight and conformable. When using perforated release film, note that the size of the perforations selected in combination with the bleeder cloth can have a direct effect on the resin content of the cured component. Too much bleed-out can result in a resin-starved component; too little can result in one that is resin rich. Nonperforated release film, also known as nonporous parting film or separator, is placed over the bleeder plies and is intended to stop resin flow from bleeder plies into breather plies. The film should be spliced or otherwise overlapped by a sufficient amount to prevent bleed through of the resin or adhesive to the vacuum bag. Keep resin out of the breather plies to ensure a good vacuum over the entire repair area.

Release sheets also may be made from coated fabrics known as release fabrics. In this case, the weave may be open or completely sealed, depending on the type. For example, PTFE (Teflon) coated fiberglass may be sealed, or it may allow the escape of volatiles and bleed out of resin, depending on the type supplied.

Airtech International Inc. and many other companies supply a range of materials and a good catalog illustrating the use of these materials. As with all materials involved in composite curing, the release films/fabrics must be able to operate at the curing temperature required. They also must release from the resin being used, and the choice of release material must be suitable for the resin system. Always check with the supplier to ensure that release materials are suitable for the job.

Breather Cloths: These are thick, loose layers of woven cotton cloth, polyester nonwoven random mats, or similar noncontaminating materials that are placed inside the vacuum bag to provide a continuous vacuum path for air and volatiles to escape, enabling a good vacuum to be drawn. Sisal or polyamide nylon rope can be placed around the periphery of the vacuum bag to provide a low-air-resistance ventilation channel to minimize vacuum drop across the vacuum bag.

Bleeder Cloths: Bleeder cloths are the same materials as breather cloths, but they are located so that surplus resin, coming through the perforated release film from the wet lay-up or pre-preg being cured under vacuum or autoclave pressure, can be absorbed, thereby making the job cleaner and neater when the vacuum bag, release film, and breather cloths are removed. Again, these materials must withstand the curing temperature.

Peel Plies: Peel plies are used to provide a surface that can be bonded directly without cleaning or other surface treatment. The idea is good, and it is difficult to prepare large areas in any other way. However, some of these systems leave contaminants on the surface, and the bond strength and durability is not as good as with a sandpaper-abraded or grit-blasted surface. Some peel plies are made from heat set and scoured polyester. Some are made from heat set and scoured nylon, and others are made from fiberglass with a release coating. Use a peel ply that does not leave unwanted deposits on the surface at the cure temperature required. Nylon should not be used above 180°C (350°F). Fluorinated ethylene propylene (FEP) peel plies tend to leave fluorine contamination on the surface, which is not good for bonding. Careful checks should be made before using any peel ply that is not approved by an OEM. It is difficult to remove overheated peel plies; therefore, use good-quality materials and ensure that the selected type is adequate for the cure temperature to be used at the highest temperature likely to be reached. Some allowance should be made for the possibility that areas not being measured by a thermocouple may become slightly hotter than anticipated. Peel plies should not be used at the limit of their capability. SAE ARP 5143 Draft states that further surface preparation may be required prior to subsequent bonding operations. If additional preparation is specified, it must be used. Use a water break test before bonding. (See Section 10.1.6 for discussion of the water break test.) If the surface does not pass this test, then cleaning followed by abrasion should be used. Then, repeat the water break test until the surface is acceptable for bonding. The bonding surfaces will require drying after this test.

10.4.3 Surface Versus Envelope Bagging

See also SAE ARP 5143 Draft—Vacuum Bagging of Thermosetting Composite Repairs and the SRM for the aircraft component under repair.

Surface Bagging: In this method, the high-temperature bagging film is applied to the tool face or the surface of the component in the case of repairs and is sealed around the perimeter of the tool or the repair area. See Figure 10.17.

This method should be treated as the normal method because it does not apply vacuum pressure to the structure of the tool. It presses the part down to the tool face, which has atmospheric or autoclave pressure behind it. The tool must be made significantly larger than the part to allow room all around its edges for any tapered edging blocks and breather cloths plus space for the bagging film sealant tape. This extra tool surface area must be specified when ordering the tool because a tool of the same size as the part can be envelope bagged only if it is sealed at the sides. Note that if surface bagging is used, the tool face must be completely nonporous and hence vacuum tight. The use of rivets or bolts in the construction should be avoided because each of these will have to be sealed and made vacuum tight.

The maximum acceptable pressure drop (leak rate) from a vacuum bag is less than 127 mm (5 in.) of mercury (2.5 psi or 0.2 bar) in five minutes. Any leak rate higher than this must be investigated and corrected.

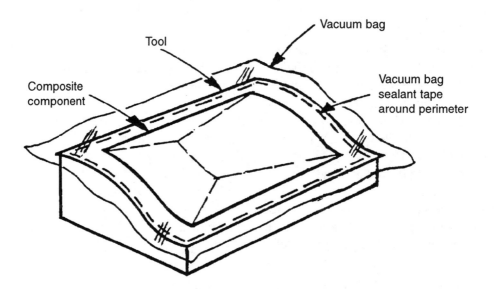

Fig. 10.17 Surface bagging. (Courtesy of Aero Consultants (United Kingdom) Ltd.)

Envelope Bagging: This technique is ideally suited to the repair or remanufacture of small parts in which the whole tool and composite assembly can be inserted into a length of tubular high-temperature bagging film. See Figure 10.18.

In this case, full vacuum pressure is applied over the whole area of the tool. It must be made stronger to avoid creep distortion at the cure temperature. This can be a particular problem with large welded aluminum alloy tooling. These alloys have poor creep resistance, and the tool will suffer creep distortion at the cure temperature unless its construction is very heavy. It is necessary to warn against the use of this method for certain repairs. A problem occurs with large control surfaces, such as rudders, in which composite sandwich panels are used with spars and ribs. If such a large assembly was envelope bagged, the pressure would collapse the panels. Each panel is autoclave bonded by itself before assembly, and the assembled design cannot withstand the forces involved. Any repairs must either be locally surface bagged or the panel must be removed from the assembly and envelope bagged by itself. SRMs usually contain this warning.

10.4.4 Pleating

Where necessary, the vacuum bag should be carefully folded and pleated to fit into any recesses, thereby avoiding the problem of bridging.

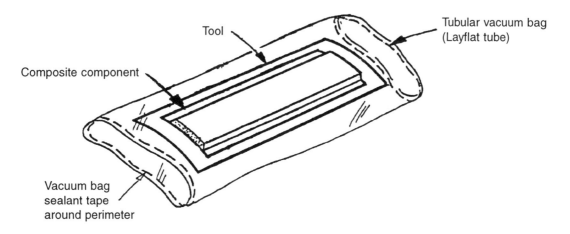

Fig. 10.18 Envelope bagging. (Courtesy of Aero Consultants (United Kingdom) Ltd.)

10.4.5 Bagging Sequence

The sequence of bagging operations is important, and each piece of release sheet, bleeder, or breather must be assembled in the correct order. This can differ slightly, depending on whether the vertical bleed or squeeze-out (edge bleed) method is used. For more details, see the SRMs or SAE ARP 5143 Draft.

10.4.6 Vacuum Requirements and Principles

These are worth a significant discussion here because vacuum pressure is the most convenient, cheapest, and most easily applied method for repair work. It is simply a matter of making good use of atmospheric pressure by understanding the techniques involved. The standard atmospheric pressure at sea level is 1013.2 mb or 14.7 psi.

10.4.6.1 Vacuum Bonding

This is the simplest and most common method for repairs and is sometimes used for the production of large parts if the technique developed by Armstrong (Ref. 10.6) is used. This method has been successfully adopted for the remanufacture of undercarriage doors for the Concorde and Boeing 747 and numerous other parts. Vacuum pressure alone is often used to form decorative panels and other nonstructural items. It has the advantage that it can be used over large areas and is easier to handle than bags of sand, steel, or lead shot. The depth of these materials that is necessary to achieve a useful pressure is remarkably large and especially so when compared with the amount of effort required to achieve similar pressure using vacuum. Although an apparently simple process, it is not always sufficiently understood to ensure good results. For this reason, the subject will be covered here at length.

The method consists of surrounding the parts to be bonded with a plastic bag and then extracting the air from the bag as completely as possible with a vacuum pump, thereby employing the pressure of the earth's atmosphere as a means of holding parts together. This pressure must be maintained while the adhesive or resin is allowed to cure either at room temperature or as the result of heating with an electrically heated mat, hot air blowers, lamps, or an oven. It follows that the vacuum bag must be well sealed to avoid loss of pressure due to leakage and that the pump must run continuously during the curing process to cope with the small amount of leakage that always occurs. Recognize that even when the best possible vacuum is achieved, we can achieve only something less than the pressure of the atmosphere itself.

The International Standard atmospheric pressure at sea level is 1013.2 mb, or 14.7 psi. This is the pressure used by aircraft captains to set altimeters to give the required separation when using allocated flight levels. The range of variation in the United Kingdom, monitored over a ten-year period in the London area, is from approximately 965 to 1045 mb. That is a variation from 48.2 mb below to 31.8 mb above the standard value. (In American units, the figures are -0.70 psi to +0.46 psi about the standard; that is, from 13.8 psi to 14.96 psi.) Careful use of vacuum gauges while pumping with an ordinary rotary vacuum pump has shown that approximately 11 psi (or 22 in. of mercury) is the best that can be obtained and that 10 psi (or 20 in. of mercury) is the normal maximum likely to be achieved. These pressures require good breather cloths and well-sealed vacuum bags. Most vacuum pumps are fitted with gauges, and these will show pressures down to 26 to 28 in. of mercury (13 to 14 psi). These figures indicate the condition of the pump, but it is important to recognize that they do not indicate the actual vacuum pressure achieved on the job itself. The pressure on the job must be measured with a gauge located on the opposite side of the work from the vacuum outlet to the pump. This will indicate the true pressure available to hold the work together while it cures.

The pressure due to one standard atmosphere is one bar.

1 bar = 106 dynes/cm^2
1 bar = 750.1 mm of mercury
1 bar = 29.53 in. of mercury

All values apply at 0°C (32°F) in a latitude of 45°.

10.4.6.2 Application of Vacuum Pressure to Plates and Assemblies Using Nonperforated Honeycomb

The diagrams first presented in this section illustrate what happens when a vacuum is produced between two plates; the later diagrams show how vacuum can be used to apply pressure during the bonding of nonperforated honeycomb. Where perforated honeycomb is used, no great difficulties are associated with the production of adequate vacuum pressures. However, perforated honeycomb allows water vapor to enter the cells after the bonding process is completed, unless sealing is extremely good. This results in corrosion in the case of aluminum honeycomb and disbonding. Bond strength also is reduced. Serious weight gain can result from water entry. The water can then freeze at altitude and expand, which can force the skins away from the honeycomb. To avoid these problems, perforated honeycomb is used rarely in aeronautical applications. This raises the problem of how to achieve adequate bonding pressure for film adhesives with vacuum alone when 14.7 psi is already in the honeycomb. This pressure can be increased by heating, especially if moisture becomes steam, and the best vacuum pressure is only approximately 10 psi. A solution will be offered later after further discussion of vacuum principles. The use of the vacuum technique for holding parts together consists of applying the pressure of the atmosphere to a chosen set of surfaces and removing it from others. Aircraft operators at, say, Nairobi (7000 ft above sea level) or Denver (4000 ft above sea level) would have greater difficulty

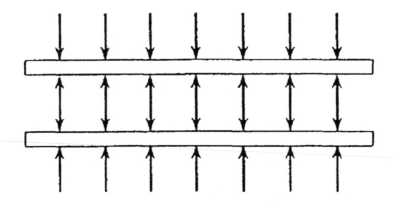

Fig. 10.19 Air pressure around two plates before the application of a vacuum.

than others using this technique because, at the higher altitudes, they have less air pressure with which to work. For the same reason, they require longer runways and high-lift devices. See Figures 10.19 and 10.20.

These plates have atmospheric pressure on the outside and reduced pressure on the inside. Consequently, the two plates are held together (as long as the vacuum pump keeps running at the same efficiency and the bag does not leak) with a pressure that equals the difference between the atmospheric pressure of the day and the pressure on the inside. This difference is the pressure that is available for vacuum bonding. It has the advantage of simplicity, can be produced at low cost and maintained fairly easily, and can cover any area because the vacuum bags can be joined with a strip of sealant. These bagging materials and sealants are available at a range of prices corresponding with the maximum temperature required in service. Large areas would require several vacuum pumps with several extract points.

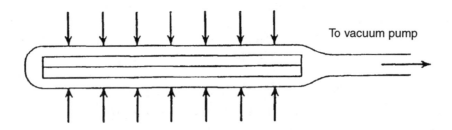

Fig. 10.20 Air pressure around two plates after air has been evacuated from the vacuum bag.

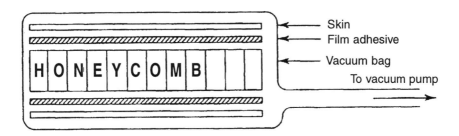

Fig. 10.21 Vacuum bag around typical honeycomb structure.

Difficulty arises when trying to bond nonperforated honeycomb. (See Figure 10.21.) The adhesive film is slightly tacky and relatively soft. Consequently, as the vacuum pump begins to extract air from the bag, some air is removed from the honeycomb cells, especially near the edge of the panel. However, as the vacuum pressure increases, the film adhesive forms a seal over the ends of the honeycomb cells, and air is no longer extracted. Because of this, the principle of vacuum bonding is defeated. See Figure 10.22.

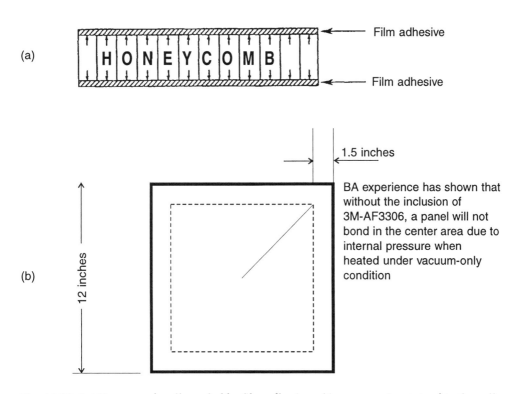

Fig. 10.22 (a) Honeycomb cells sealed by film adhesive. Air pressure is retained in the cells.
(b) 300 mm (12 in.) square test panel, illustrating good bonding around only the edge.

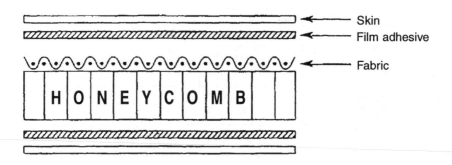

Fig. 10.23 Use of fabric for extraction of air from nonperforated honeycomb.

Pressure is retained inside the cells, thereby reducing the pressure difference on which the vacuum bonding process relies. The problem is aggravated by the high temperature at which the film adhesive must be cured (120 or 180°C [250 or 350°F]). In the case of the 120°C (250°F) cure, the pressure increases by a factor of 1.5 inside the cells; at 180°C (350°F), it increases by even more. Instead of having a pressure difference in favor of holding the parts together, the situation could be reversed if no air was removed from the honeycomb cells during the cure. For example, if the pressure in the middle cells was reduced to 9.85 psi before heating, then the increase in temperature to 120°C (250°F) would restore it to 14.7 psi. Ensuring a bonding pressure difference of only 3 psi would require an initial pressure in the cells of no more than 7.5 psi. To ensure an actual pressure of 6 psi, the pressure in the cells would have to be reduced to 5.8 psi before heat is applied. The technique for doing this with nonperforated honeycomb is to use a suitable fabric on one side of the sandwich between the adhesive and the honeycomb. See Ref. 10.6 and Figure 10.23.

Experiments became essential because fabric manufacturers could supply no data on air-flow through the fabric in the x direction, which governs behavior for vacuum bonding purposes. See Figure 10.24.

Fig. 10.24 Airflow direction through fabric.

The fabric selected from a range tested was 3M-AF 3306, which is a nonwoven polyester, monofilament fabric, sometimes called a positioning cloth. It has a low moisture uptake and is used as a carrier cloth in some film adhesives. Unfortunately, it is supplied in two thicknesses with the same part number.

Two factors are important:

1. The fabric should not be used at a thickness greater than 0.003 in. (Two layers of the original version of AF-3306 gave a total thickness of 0.0015 in.)

2. One layer of film adhesive of 0.085 lb/ft^2 weight or two layers of 0.06 lb/ft^2 should always be used, as a minimum, to ensure the following:

   - An adequate amount of adhesive to absorb the fabric
   - An adequate adhesive fillet between the skin and the honeycomb cell ends

10.4.6.3 Practical Techniques to Ensure Adequate Applied Vacuum Pressure to Parts Under Repair

The application of a sufficient vacuum pressure evenly over the work depends on the ability of the vacuum pump to draw air easily from all parts of the repair area. It depends on the use and correct location of suitable and sufficient breather or bleeder cloths and on breather blocks, as shown in Figure 10.25. See also Figures 10.26 and 10.27.

These edging blocks prevent edge crushing and help to ensure a good flow of air from the vacuum extract pipe (usually 12.5 mm [0.5 in.] bore aluminum tube generously perforated with 4.5 mm [3/16 in.] holes) to the breather cloths. Breather cloths are necessary if the air is extracted at the top from a fitting similar to that used to mount the vacuum gauge.

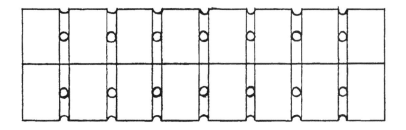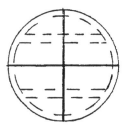

Fig. 10.25 Vacuum breather blocks for panel edges. Cut into quarters after turning and drilling. Make from beech wood.

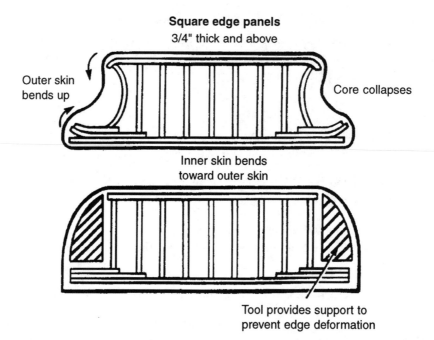

Fig. 10.26 *Square edge panels—one use of edge tooling.*

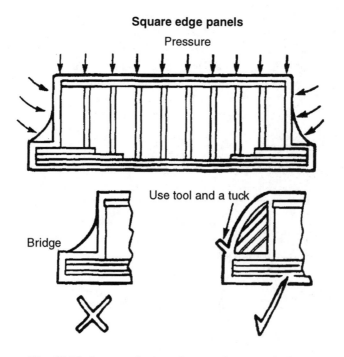

Fig. 10.27 *Square edge panels—another use of edge tooling.*

In principle, the problem is simple. To obtain a good vacuum, certain requirements must be met:

- Leaks must be avoided.

- The container (vacuum bag) and the base plate or mold must not be porous.

- The vacuum pump must be large enough to cope with the volume of the container.

- The extract pipe and breather cloths must have an adequate area through which the air can flow and should, ideally, be fairly short. The longer the pipe and other airflow pathways, the lower the pumping efficiency and the lower the vacuum pressure that will be achieved on the job.

- The breather cloths must be correctly located to ensure that the bag cannot seal off at the edge of the job.

- A vacuum gauge must be used on the repair to check that the above factors have been addressed and that the required vacuum has been achieved where needed.

Because the vacuum is being used solely as a means of applying atmospheric pressure to a repair and the vacuum produced is poor compared to a more scientific usage, it is really only a significantly reduced pressure. Nevertheless, it is extremely useful, convenient, and valuable as a tool if correctly applied. Calculations can be done for viscous flow because the pressure is too high for molecular flow. It is sufficient to use a rotary pump and a Bourdon tube type of vacuum gauge. Oil diffusion pumps and Pirani or Penning type vacuum gauges are irrelevant at these pressures. After the vacuum bag has been evacuated, the pumped volume is small in the case of honeycomb parts and almost nil when doubler plates are being bonded. Therefore, a small pump can be used. Calculations have shown that, at the low volumes involved, doubling the pumping speed has little effect; however, doubling the diameter of the extract pipe has a large effect. Thus, the use of a sufficient number of layers of breather cloth is critical because they are an essential part of the air extraction system. Together with the breather blocks, they prevent the vacuum bag from sealing off at the edge of the part. They also constitute the smallest, and therefore most critical, part of the extract pipe system and must provide a large enough area through which the air in the bag can be pumped.

Experiments have shown that four layers of coarse weave glasscloth, or their equivalent in breather cloth, are a necessary minimum for achieving a vacuum pressure of 10 psi at a gauge on the job. Many companies now supply special fabrics for use as breather/bleeder cloths. Porous bleeders are also available and are made from fiberglass fabric coated with PTFE (Teflon) and other release agents. They are listed as release fabrics. These are useful bleeder cloths because they do not adhere to any adhesive or resin and can be reused many times. They are similar to but thicker than perforated release films. Fine weave release fabrics are now produced, which leave no marks on the finished part. The same companies also supply vacuum bags, valves, vacuum gauges, release films (both perforated and nonperforated), and breather and resin absorption cloths of various weights. Use a vacuum gauge on the far side of the work from the vacuum extract pipe to ensure that the pressure is sufficient. When using vacuum pressure for repairs, remember to apply heat for approximately 50 mm (2 in.) all around the entire repair to ensure that the edge of the repair patch does not fall below the required temperature. This means that the air sealed into the honeycomb around the repair will be heated, together with any water vapor that may be present, to a temperature approaching that of the heater blanket. The pressure will increase inside the cells from atmospheric to a higher value. Figure 10.28 shows the pressure increase without allowance for water vapor or steam pressure above 100°C (212°F).

The original adhesive will soften; however, because it is cross linked, it cannot melt a second time. Repairs performed at the original cure temperature, using only vacuum pressure, often lead to the skin being blown off the honeycomb around the repair. Because this has occurred many times at different repair bases worldwide, it is strongly recommended that repairs to honeycomb structures should be made at temperatures at least 25°C (45°F) and preferably 50°C (90°F) below the original cure temperature of the part, unless positive pressure above vacuum can be applied. Most people who ignore this recommendation find that doing so is expensive.

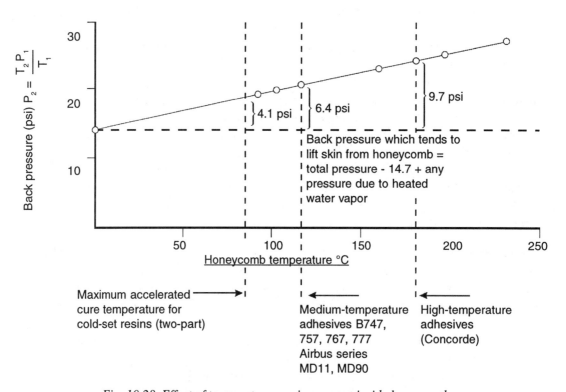

Fig. 10.28 Effect of temperature on air pressure inside honeycomb.

All vacuum bonded repairs using film adhesive will need 3M-AF 3306 polyester fabric across the full width of the repair patch. Without this, adequate pressure will not be achieved over the honeycomb core. Even on test panels with honeycomb to the edge (see Figure 10.22), only the first 38 mm (1.5 in.) around the edge were found to be well bonded.

10.4.7 Caul Plate and Dam Usage

Use of Caul Plates: Caul plates are defined in SAE ARP 5143 Draft—Vacuum Bagging of Thermosetting Composite Repairs as, "Smooth support plates, usually metal, cured composite, or silicone sheet, free of surface defects, the same size and shape as a composite lay-up, used immediately in contact with the lay-up during the curing process to transmit normal pressure and temperature and to provide a smooth surface on the finished laminate." They can assist uniformity of temperature if a heater blanket is used. For more details, see SAE ARP 5143 Draft. See also SAE ARP 5144 Draft—Heat Application for the Local Repair of Thermosetting Composite Materials.

In repair work, caul plates usually are the same size as the heater blanket. The heater blanket should be 50 mm (2 in.) larger than the repair. SRMs specify caul plates made from thin gauge aluminum alloy (i.e., 0.4 mm [0.016 in.]), because it has a high thermal conductivity, evens the temperature, and flexes and conforms to the surface profile. Copper sheet may also be used because it conducts heat well. Airbus and Boeing SRM pages state that the use of caul plates is optional. Caul plates also may be made from silicone rubber sheet approximately 1.5 mm (0.0625 in.) thick. Metal caul plates should conform to the surface profile under moderate finger pressure to confirm that they will follow the profile under vacuum pressure. Airline experience suggests that 0.8 mm (0.032 in.) 2024-T3 aluminum alloy plates are better because thinner caul plates can allow the thermocouple to mark the work surface. Caul plates must be flexible enough to follow the major surface contours of the part. If significant double-curvature is involved, the aluminum alloy type must be wheeled to profile to a high degree of accuracy such that vacuum pressure will be sufficient to cause the caul plates to adopt the required profile during the curing process. Metal caul plates normally are used on exterior convex surfaces; silicone rubber caul plates are used on interior concave surfaces. In both cases, the thermocouples can be placed on top of the caul plates to avoid leaving an imprint in the work surface. However, experiments must be run to determine the temperature difference between the top of the caul plate and the actual work surface. A solid release film is required between the caul plate and the work surface. As was mentioned previously, the glue-line or pre-preg surface temperature is critical, and every effort must be made to ensure that the adhesive or resin being cured remains at the correct temperature. The Boeing 757 SRM 51-70-17 Figure 4, Sheet 1, shows the thermocouple buried in the repair, and there does not appear to be a common view on where they should be located. What is certain is that the correct cure temperature must be achieved at the bond line. Caul plates also can serve to reduce telegraphing on composite skins bonded to honeycomb cores. This descriptive term means that the skin will not sag into the honeycomb cells in the same way as telephone wires sag between posts; skin tends to do this under vacuum pressure. The SRM should be consulted regarding thermocouple location. If a decision is made to use another location than that recommended in the SRM, then the suitability of the alternative position must be checked by experiment before the repair cure cycle begins. SAE ARP 5143 Draft also states, "Caul plates may be used to bridge over an area during drying. A caul plate may be used over areas where the skin has been removed. The caul plate prevents pressure from damaging the honeycomb cells by bridging over an area where pressure is not desired." In this case, a stiffer caul plate would have to be used, and its thickness would depend on the size of the honeycomb section being covered. Thin plates deflect considerably under vacuum pressure, and experimentation is usually necessary.

Dams: A dam is defined as, "A boundary support or ridge used to prevent excessive edge bleeding or resin runout of a laminate." The bagging tape used to seal vacuum bags is normally used for this purpose.

10.4.8 Vertical Bleed Method

This method is commonly used both in manufacture with pre-pregs and in repair with both pre-pregs and wet lay-up. See SAE ARP 5143 Draft. In the vertical bleed method, the first layer of release sheet is of the perforated type covered with bleeder cloth, and resin can bleed out vertically over the entire repair area. See Figure 10.29. In this method, the breather cloths must be carried to the edge to ensure good air extraction and hence a good vacuum over the entire repair area. Experience is required because the size and number of perforations in the release sheet, the thickness of the breather cloth, the resin viscosity, the

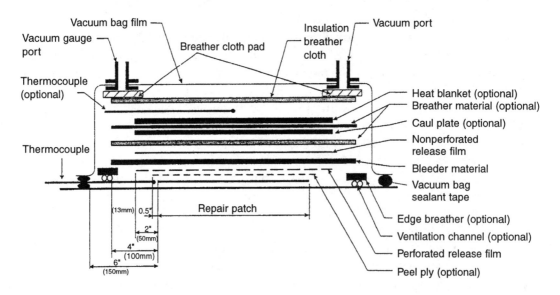

Fig. 10.29 Vertical bleed method. (From SAE ARP 5143 Draft, Fig. 9.)

cure temperature, and the vacuum pressure achieved significantly affect the amount of resin bled from the lay-up and hence the final resin content of the composite or repair. SAE AMS 2980/1 Carbon Fiber Fabric and Epoxy recommends 2-mm (0.08-in.) diameter perforations and 5% of the surface area perforated. Check the SRM for the method of resin bleed and bagging lay-up required for each job.

10.4.9 Squeeze-Out (Edge Bleed-Out) Method

In this method, which is recommended for small wet lay-up repairs, no release sheet is used because the vacuum bag also serves this purpose. All the resin that is bled away must be squeezed out at the edges of the repair—hence, the name of the squeeze-out method. Note that breather cloth is used only around the edges. The fabric layers to be used for the repair plies, with a layer of peel ply on top, are first impregnated with resin between two cover sheets of polyethylene or vacuum bag foil. The bottom cover sheet is removed before the patch, complete with peel ply, is applied to the repair area. The top cover sheet is then removed, and the vacuum bag is laid up. After laying up the vacuum bag, excess resin is removed by pushing a spatula carefully over the outside surface of the bag to ensure that any air and excess resin is squeezed out to an edge bleeder. Spatulas are made of PTFE or polyamide (nylon) with rounded edges to avoid damage to the vacuum bag. The use of a layer of peel ply helps in detecting any remaining entrapped air. See SAE AMS 2980/1. The peel ply is removed after the cure is complete. Care is required to avoid distortion of the fabric by the spatula because the correct orientation of the fibers must be maintained, and the fibers should not be damaged. On large components, extra care is required to avoid resin entrapment, which can result in high-resin-content areas and thickness-tolerance

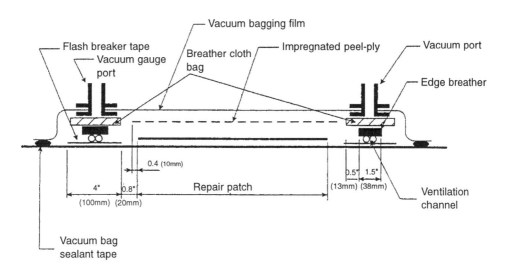

Fig. 10.30 Squeeze-out (edge bleed-out) method. (From SAE ARP 5143 Draft, Fig. 11.)

problems. This method must be used when one or both sides of a laminate are covered with Tedlar or a foil. In this case, no squeezing is done, apart from that used to apply the Tedlar or foil to the skin. The vacuum pressure is relied on to remove any excess which, in the case of pre-pregs, should be small. See Figure 10.30. Check with the SRM to ensure that the specified method of resin bleed and vacuum bagging is used.

The Repair Techniques Task Group of the Commercial Aircraft Composite Repair Committee (CACRC) found that this method gives better results in terms of repair strength; however, it is difficult to see how large an area is practical, and this may need to be determined by further experiments. The value of this method is that it helps to remove bubbles (voids) which would otherwise reduce the strength of the laminate. If performed carefully, it produces a laminate with fewer voids and more uniform thickness than the vertical bleed method, particularly on more viscous resins where the air is not drawn through all the plies and the bubbles are expanded by the vacuum. (See SAE ARP 5143 Draft.) In this method, several strips of bleeder cloth are placed around the periphery of the lay-up and are overlapped approximately 12 mm (0.5 in.) to create a continuous border or frame around the repair area.

10.4.10 Zero-Bleed Method

This method is used, mainly in production, with zero-bleed pre-pregs designed to achieve high-fiber-content composites. They are cheaper to use and result in less resin wastage and less consumption of consumables such as bleeder cloth. However, great care is required to ensure that no further resin loss occurs or resin-starved composites may be produced. These pre-pregs generally have a resin content of only 30 to 34%; therefore, further loss is unacceptable. Careful processing is needed to achieve a low void content. Check the SRM for the resin bleed-out method and bagging lay-up required and ensure that the correct method is employed. See Figure 10.31.

10.4.11 Ply Compaction and Debulking

This process is used when laying up pre-pregs to minimize the amount of air entrapment and the potential for voids or delamination. It is particularly important for thick laminates. For tooling pre-pregs, a vacuum debulk/bleed procedure is recommended after the second ply on top of the gel coat and then after the addition of every fourth subsequent ply. Perforated release sheet is used to allow the air to escape freely. Check the SRM or tooling pre-preg literature for specific requirements.

10.4.12 Debagging Precautions and Typical Problems

Debagging should not occur until the repair has cooled to room temperature. Even then, debagging should be done carefully. Bagging sealant tape should be peeled from the component surface slowly and at 180° or parallel to the skin. See Figure 10.32.

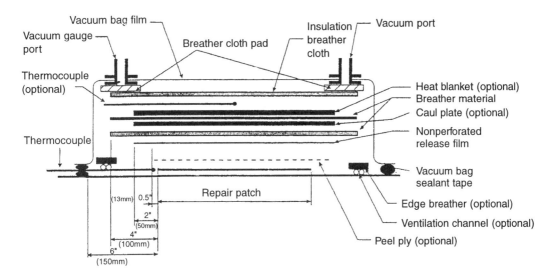

Fig. 10.31 Zero-bleed method. (From SAE ARP 5143 Draft, Fig. 10.)

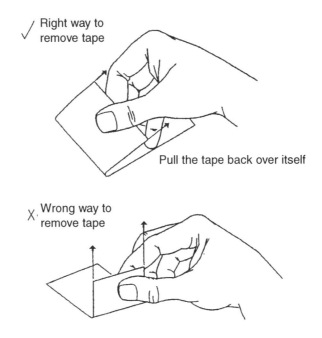

Fig. 10.32 Peel angle when removing bagging or other tapes. (From SAE ARP 4916.)

Rapid peeling at 90° to the skin may cause delamination of the outer layer of fabric. Removal of release sheets and breather/bleeder cloths should also be done carefully and in the same way. It is better to remove any surplus resin or film adhesive that has migrated to the wrong place by careful sanding than to try to rip these materials from the surface with the release sheets or cloths to save a little time. Attempts to save a few minutes often generate several hours of additional repair work.

10.5 Curing Stages and Temperatures—Heating Techniques

See SAE ARP 5144 Draft—Heat Application for the Local Repair of Thermosetting Composite Materials for details of recommended methods.

10.5.1 Curing Stage Definitions

These are covered in more detail in Chapter 2 but are repeated briefly here for convenience.

- **"A" Stage:** The initial state of the resin as produced by the manufacturer.

- **"B" Stage:** An intermediate stage in the reaction of certain thermosetting resins in which the material softens when heated and is plastic and fusible but may not entirely dissolve or fuse. Resin in an uncured pre-preg or film adhesive is in this stage.

- **"C" Stage:** The final stage in the reaction of certain thermosetting resins and adhesives in which the material is practically insoluble and infusible (i.e., it is fully cured and hardened).

- **Gelation:** This important stage falls between the "B" and "C" stages. Time-expired material can reach this stage without being used if it remains at room temperature for enough time. Gelation is defined as, "The point in a resin cure when the resin viscosity has increased to a point such that it barely moves when probed with a sharp instrument." When curing thermosetting film adhesives and pre-preg resins, the heat-up rate must be within the required limits because the adhesive/resin must melt, flow, and wet the bonding faces before the point of gelation is reached. If the rate of heating is too slow, it can gel before it can melt or flow and thus prevent good bonding. When using two-part resins as matrix resins for composite skins, use the mixed resins within 75% of their stated pot life; otherwise, air from inside the honeycomb cells will not come through the resin. Air must pass from the cells and through the resin to ensure a positive bonding pressure.

For more information, see Ref. 10.7 and Section 2.4.4 of Chapter 2.

10.5.2 Low-Temperature Cure

Two versions of low-temperature cure are possible:

1. Room-temperature cure (i.e., cure at whatever the room temperature happens to be). This should fall within the range 15 to 25°C (59 to 77°F) but may be higher in summer. Many two-part paste adhesives and resins are typically cured in this way.

2. Warm-cured two-part systems, a term used synonymously with intermediate-temperature curing adhesives, in which mild heat in the range 50 to 80°C (122 to 176°F) is used either to hasten the curing process or to slightly improve mechanical and other properties.

10.5.3 Elevated-Temperature Cure

Likewise, elevated-temperature cure includes two temperature levels:

1. Elevated-temperature repair, which is defined as, "A repair using a resin system that will be cured above 70°C (160°F)."

2. Hot-cured repair, which is defined as, "A repair that requires a temperature at or above 100°C (212°F) to set."

Both systems are almost certain to require a fairly high temperature to initiate cure. They include one-part paste adhesives, one-part resin systems, film adhesives, and pre-pregs. All of these contain a curing agent that must be heated above a specified temperature before cure can occur. Unlike two-part room-temperature systems, they will not cure properly unless they are held within a specified band of temperature for a specified time. (This information is given on the data sheet for each material.) They also require storage in a freezer, and a record of out-time must be maintained.

10.5.4 Direct Versus Indirect Heating

Direct heating (i.e., steam-heated platens) is a preferred method for the manufacture of flat honeycomb panels, and heater blankets are convenient for small repairs and, in some cases, large ones in which the structure has a fairly uniform thickness. Heater blankets can be a problem near hinge fittings, ribs, and spars where heat can be soaked away faster than it can be added. Indirect heating, such as hot air or radiant lamps can be beneficial, especially if the heat can be contained in some sort of cover. Heat soaked away from a point can be replaced more rapidly because the air is continually moving. Lamps without some form of heat circulation can create their own hot spots. The most suitable method should be used for each application. In some cases, a combination of methods may be used, such as hot air or lamps directed onto a cold spot to even the temperature when a heater blanket alone cannot produce a uniform temperature. Glass fiber or other insulation can be used to reduce heat loss during hot bonding or to compensate for heat sinks, to some extent.

10.5.5 Ramp Rates and Soak Cycles

The material data sheet will also give the minimum and maximum heat-up rates (ramp rates) for each material. The minimum and maximum heat-up rates must be programmed to remain within the specified limits. They must be slow enough to allow the escape of volatiles and fast enough to ensure that melting, flowing, and wetting can occur before the gelation point is reached and the material is unable to flow, as previously described. The cure cycle for some special materials may require that a particular temperature must be held for a period of time (soak cycle) and then increased to a higher level for another period. The data sheet will specify any special requirements. For typical cycles, see Figure 10.33.

10.5.6 Temperature Control and Monitoring

Temperature control and monitoring is important for all hot-curing materials, and it is normal to keep a chart record with the job card as evidence of correct cure. See SAE ARP 5144 Draft—Heat Application for the Local Repair of Thermosetting Composite Materials.

Hot-bonding controllers normally have a minicomputer program to give the correct ramp rate for increase of temperature up to the cure temperature, followed by a dwell period at the cure temperature, and then a controlled decrease to room temperature. Forced cooling is not normally used. Therefore, the cool down is usually uncontrolled if the work cools too slowly, but it will be controlled if it attempts to cool too rapidly. This will help to prevent severe residual stresses, which could develop if cooling occurred too quickly.

Thermal mass can create a problem, especially when a repair requires the use of tooling. Tooling must be of low thermal mass and of a thermal conductivity equal to or better than that of the part under repair. It is useless to program a particular heat-up rate if the thermal mass is too high, or the thermal conductivity too low, for the autoclave, vacuum press, heater blanket, or other heat source to achieve the temperature demanded in the time called for by the program. The tooling material had to be changed on one occasion because of this.

10.5.7 Thermocouple Placement

It is important to use a number of thermocouples, preferably six or more for large repairs. If hot-bonded repairs are to reach the standard required, the pre-preg or film adhesive must be cured at the correct temperature, or at least within the permitted band of variation, and

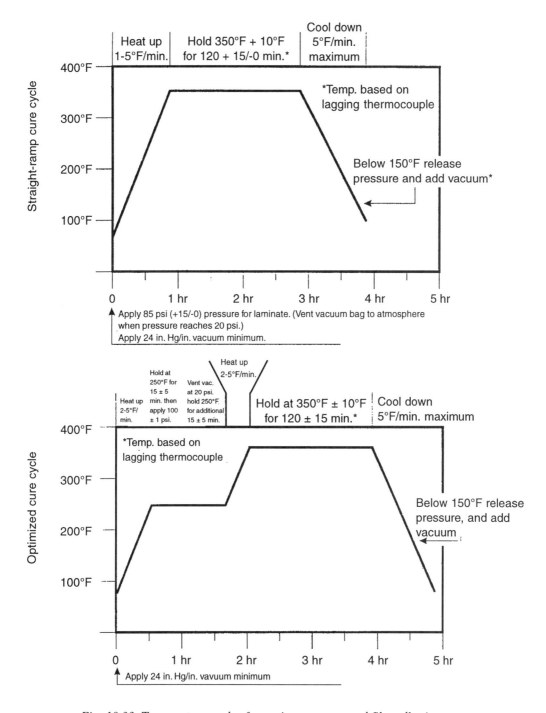

Fig. 10.33 Temperature cycles for curing pre-preg and film adhesives.
(Courtesy of Abaris Training Resources Inc.)

for the correct length of time at the specified pressure. To achieve this, it is necessary to know the temperatures at various points on the repair to establish that they are correct and within limits. The hot bonder should be designed to enable switching from one thermocouple to another. Thus the temperature at a low point could be raised by using that particular thermocouple as the control. It may be necessary to add fiberglass blankets to some areas to raise the temperature if there is a danger of other areas becoming too hot. If the temperature at one point is too low, heater lamps may have to be added. If too hot, air cooling may be helpful. Many techniques are available to adjust the temperature if enough thermocouples are used to indicate the actual temperature at sufficient points on the part. Thermocouples should be placed at several points around the edge, with one or more at the center of the part and extras at any heat-sink points such as large metal fittings, spars, and ribs. When curing film adhesives or pre-pregs, recognize that the glue-line temperature is what matters because it is the adhesive or matrix resin that is being cured. Most people involved in bonded repairs have, at some time, made the mistake of heating an oven or a blanket to the right temperature, only to find that the adhesive itself was not fully cured because the glue line did not reach the required temperature. Therefore, thermocouples must be placed at or in the glue line to ensure that the adhesive or matrix resin reaches the cure temperature. See Section 10.4.7 of this chapter for a discussion of caul plates. Also see SAE ARP 5144 Draft.

Leading and Lagging Thermocouples: When several thermocouples are used around a repair, one usually reads higher than all the others and is known as the leading thermocouple. Another thermocouple usually reads lowest, and this is known as the lagging thermocouple. The remaining thermocouples all may read slightly differently between these two extremes. As mentioned elsewhere, the lowest reading must not fall below the bottom limit for proper curing of the pre-preg or film adhesive. Likewise, the highest reading should not exceed the maximum recommended curing temperature. If this problem occurs, then insulation should be used in the cold areas to reduce the spread of temperature so that the upper temperature can be brought within limits.

Temperature Profile: Most adhesives and resins must be heated within specific rate limits and cooled at suitable rates. Therefore, the temperature controller on the oven, autoclave, or portable hot bonder must be programmed to operate at the correct heat-up and cool-down rates and to provide the required length of dwell at the cure temperature. Today, this is usually done by programming a microprocessor having a memory that allows several programs to be set. Thus, the average repair shop simply has to select the appropriate program number, and the heat-up, dwell, and cool-down rates are set automatically for the chosen resin system. See Figure 10.33 for a typical example.

10.5.8 Temperature Control Problems

Several problems are possible with temperature control, and some have already been mentioned. Other likely problems are:

- Drafts of cold air if work is done near a doorway or in a hangar. A cold draft as the work reaches cure temperature can be serious. If possible, do not perform hot bonding in areas where doors can be opened during a repair. When this is impossible, draft excluder screens should be erected around the repair area, and the temperatures should be monitored while the doors are open.

- Occasionally, a thermocouple may break, and this alone is a good reason for using more than one. Also, use at least two in any known critical areas. In the event of failure, it is a simple matter to switch to another thermocouple.

- Areas of known heat sink (such as metal fittings, spars, and ribs) should be monitored for some time to ensure that temperatures remain within limits. All repairs are "one-off" jobs; thus, there is no opportunity to conduct a few trials to make adjustments. Every repair is a first-time exercise, with only one chance to complete it correctly. The only adjustments possible are those that can be made while the work is in progress. Therefore, progress must be monitored either for the whole time or at least until temperatures have stabilized and the likelihood of problems is low.

Thermal Runaway/Overheat Protection: Select equipment for hot bonding that includes overheat detection and protection. The hot bonder can be designed to switch off in the event of overheating or underheating, or to give an audible warning so that corrective action can be taken before the repair is damaged and must be repeated.

10.5.9 Hot Bonder and Heater Blanket Usage

Hot Bonders: The use of hot bonders is probably the most popular method for performing hot-cured repairs. Today, modern hot bonders are lighter in weight and are therefore more portable than earlier versions. They are made by many different companies, and their design is improving all the time. The latest versions are being made much simpler to operate. They can be programmed to retain several heating cycles in the microprocessor memory, and only the program number for the cycle required must be selected. See Figure 10.34 for two typical hot bonders. Similar products are made by Heatcon Composite Systems and ATACS, among many. See also Ref. 10.8.

The latest hot bonder from Aeroform has 10 thermocouples per zone, all independently scanned and selectable to control from the hottest, coldest, average, or any of the 20 thermocouples by selection. All hot bonders undergo constant improvement. Most hot bonders include a vacuum pump of the venturi type, a six-channel temperature recorder,

(a)

(b)

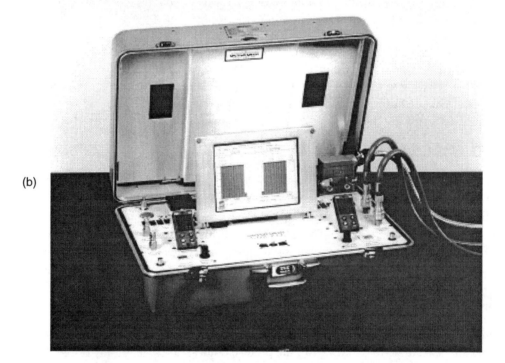

Fig. 10.34 (a) GMI hot bonder. (Courtesy of GMI.) (b) Aeroform hot bonder.
(Courtesy of Aeroform.)

six thermocouples, and the ability to switch to any thermocouple for use as the control couple. Low-temperature and low-vacuum alarms are usually provided. The temperature recorder prints a chart, providing documentary evidence that the correct temperature cycle was used. Some hot bonders have two heating circuits to enable two blankets to be heated and controlled independently at the same time. These are rated at 30 amps. Check the specification for each type. Ensure that the type purchased meets all local electrical safety regulations and that a flameproof type to the required specification is purchased for use in hangars or on or near aircraft. Safety circuits also are required for heater blankets. See Section 6.11 of Chapter 6. See also Figure 10.35 for a typical hot bonder equipment setup.

Heater Blankets: Heater blankets are usually supplied in a range of standard sizes but can be made to special order. See Figure 10.36. Molded blankets can also be made to suit the shape of radomes or other curved parts if necessary. Most repairs are limited to a maximum of 180°C (350°F), with the majority made at 120°C (250°F) or less. However, heater blankets can be made to a normal maximum operating temperature of 260°C (500°F) and to 350°C (662°F) for special purposes.

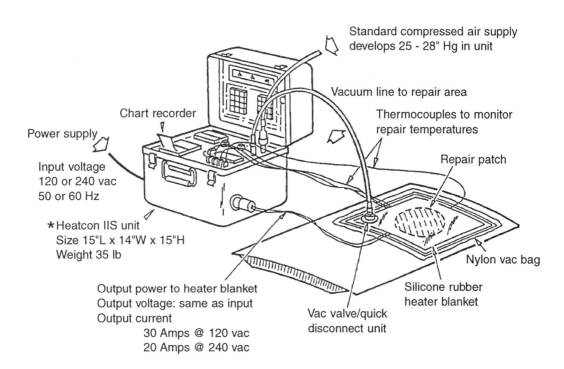

Fig. 10.35 Hot-bonding equipment setup. (Courtesy of Aero Consultants (United Kingdom) Ltd.)

If normal 13 amp electrical sockets are used at 220 to 240 volts, the maximum available power is approximately 3000 watts. The size of blanket that can be supplied is limited by the power rating required (i.e., watts/in.^{2}, watts/ft^2, or watts/m^2). For most repairs, 5 watts/in.^{2} is sufficient; low-temperature repairs can use a rating approximately half of this. At 220 to 240 volts, this means a maximum blanket size of 600 x 600 mm (24 x 24 in.) at the higher rating; however, most repairs are smaller than this. Large repairs will require more than one power socket. Note that a blanket should be larger than the repair area. It has been found that 75 mm (3 in.) all around is a good figure for thin skins of 0.635 mm (0.025 in.) thickness and 100 to 125 mm (4 to 5 in.) all round on thicker skins. The blanket must be significantly larger than the repair area. On some occasions, two blankets may be necessary, supplied by two separate sockets.

10.5.10 Heat Sinks

When performing hot-bonded repairs to large composite panels attached to relatively heavy substructure such as hinge fittings, operating jack fittings, spars, or ribs, these items may be expected to act as heat sinks and draw heat away from the repair area.

If the panel to be repaired cannot be removed from the structure, as is often the case, either add thermal insulation over these areas to reduce heat losses or add extra heat using lamps or hot air blowers to minimize the temperature differences from one part of the repair to another. Film adhesives and pre-pregs must be cured within the specified temperature

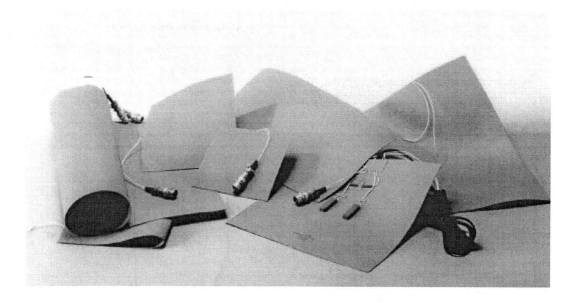

Fig. 10.36 Heater blankets. (Courtesy of GMI.)

limits given on their data sheets. Ensure that the maximum variation over the repair area does not exceed these limits. Curing time varies with temperature, and the curing time may have to be increased to compensate for a part of the repair area being at a lower temperature than another part. Use thermocouples at several positions around a repair and at the center. If it is obvious from a study of the part that a particular area has a heat sink, then a thermocouple should be placed in this area to monitor the effect. Then, either heat or insulation should be added to raise the temperature to the required value. If necessary, conduct a trial run without any adhesive or pre-preg to assess temperature differences, and try various heating or insulation methods before a proper repair is attempted. This could significantly reduce problems with the actual repair because, after the resin has begun to flow and cure, it is difficult to abandon a repair halfway and to remove the partially cured materials without causing further damage.

10.5.11 Thermocouples

These are defined in SAE AIR 4844 Composites and Metal Bonding Glossary as, "A device which uses a circuit of two wires of dissimilar metals or alloys, the two junctions of which are at different temperatures. A net electromotive force (emf), or current, occurs as a result of this temperature difference. The minute electromotive force, or current, is sufficient to drive a galvanometer or potentiometer." See also SAE ARP 5144 Draft—Heat Application for the Local Repair of Thermosetting Composite Materials.

In special laboratory cases, where great accuracy was required, it was common many years ago to place one junction (the cold junction) in melting ice and the other at the point of temperature measurement. In portable equipment, such as hot bonders, a cold junction compensation circuit is used, and only the hot junction employed for temperature measurement is visible to the user.

Thermocouples may be made from several different combinations of metals and alloys, depending on the temperature range to be covered and the accuracy of temperature measurement required. For high-temperature purposes, platinum is used with an alloy of platinum and rhodium. Inevitably, these are expensive; therefore, cheaper materials are used for low-temperature applications.

Two types of thermocouples are likely to be used with hot bonders.

- **Type "J" (Iron/Constantan):** In this type of thermocouple, the positive lead is iron, and the negative lead is an alloy called Constantan, which consists of 60% copper and 40% nickel. Type "J" thermocouples are used in the temperature range of 20 to 700°C (68 to 1292°F) for continuous use and -180 to 750°C (-292 to 1382°F) for short-term use. Type "J" is often, but not always, associated with calibration in degrees Fahrenheit. Color codings have changed recently. The old British coding was outer sheath black, positive lead yellow, and negative lead blue. The new color code is the

International Electrical Code (IEC 584.3) outer sheath black, positive lead black, and negative lead white. In the United States, American National Standards Institute (ANSI) codes may continue to be used.

- **Type "K":** In this type of thermocouple, the positive lead is a nickel/chromium alloy, and the negative lead is a nickel/aluminum alloy. Type "K" thermocouples are used in the temperature range of 0 to 1100°C (32 to 2012°F) for continuous use, and -180 to 1350°C (-292 to 2462°F) for short-term use. Color codings have also changed recently. The old British code was outer sheath red, positive lead brown, and negative lead blue. The new color code is the (IEC 584.3 code) outer sheath green, positive lead green, and negative lead white. In the United States, ANSI codes may continue to be used.

Both types can be supplied in two classes of temperature measurement accuracy. Class I have an accuracy of ±1.5°C (±2.7°F), and Class II have an accuracy of ±2.5°C (±4.5°F). The emf or current generated depends on the alloy combination in use; thus, the temperature-measuring instrument employed must be calibrated to account for this. Therefore, only the correct type of thermocouple must be used with a given design of temperature-measuring instrument, and it should be used only within its permitted temperature range. Remember this when ordering spares. Good contact is required at the junctions of the two materials, and welding is the normal method of joining. Twisting the wires together is not adequate and does not give the measurement accuracy necessary. Simple welders are supplied by hot-bonder companies and should always be used.

Thermocouples that break in service can be repaired by welding with a suitable thermocouple welder supplied by many companies. The damaged ends can be cut off and the new ends welded again. GMI supplies a suitable welder. Note any instructions regarding thermocouples given in the temperature-controller handbook. In some cases, thermocouples are required to be within a specified range of length to maintain their resistance within specified limits. Check the minimum permitted length stated in the manual, and do not cut off more than is necessary for each repair.

10.5.12 Distortion of Parts During Heating or Cooling

Different materials have different thermal conductivities and different thermal expansion coefficients. If different materials are bonded together (e.g., steel and aluminum or aluminum to a composite part), then slow heating and slow cooling is recommended, together with firm tooling. Some local panting or distortion is inevitable if dissimilar materials are bonded and the construction is not balanced in a way similar to a composite lay-up. Also residual stresses will remain after cooling; slow cooling can help to minimize these. The nearer to room temperature dissimilar materials are bonded, the less the distortion that will occur. The use of tough adhesives with a moderate modulus and a good strain to failure is also helpful. Where practical and permissible, bond different materials together with room-temperature curing adhesives to minimize distortion and residual stress.

10.6 Post-Repair Inspection

In addition to the following information, this subject is covered in more detail in Chapter 7.

Nondestructive Inspection (NDI) Requirements: For lightly loaded and nonstructural items, a tap test often is sufficient. Structural items require ultrasonic inspection, as detailed in Chapter 7. For metallic components, common practice is to perform one or more of the destructive tests listed below, using samples made by the same processes and cured at the same time as the repair. These tests are similar to those used by the OEM to confirm the quality of the original parts.

Visual Inspection: This inspection confirms that nothing is obviously wrong with the part. It also confirms that any parts that have been removed before the repair have been refitted (e.g., metal hinge fittings, bearings, edge seals, or static dischargers). See Section 7.1 of Chapter 7.

Tap Test: The tap test is covered in detail in Section 7.2 of Chapter 7 and is often sufficient for minor parts.

Coupon Testing: When confirmation of bond quality is required by destructive methods, test coupons are usually made using the same surface preparation and materials and then they are cured at the same time in the same oven or by the same method as the repair itself. In production of some items such as floor panels, it is possible to take an offcut from the sheet from which they are made; then the tests are regarded as truly representative. Otherwise, great discipline and integrity is required to make the tests as valid as possible. They must be made using the same materials from the same batch and processed in the same way as the repair. With metal bonding, the test coupon and the repair parts must be given the same surface treatment. If treatments are done at different times or the adhesive or prepreg is from a different batch, then the results may be meaningless. Two results are possible: (1) either the test coupon passes the test when the repair is inadequate, which permits an unsafe part to fly; or (2) the test coupon fails when the repair is actually good, which could mean that a good repair is removed and repeated at great cost and to no purpose.

Neither result is acceptable. The following are the most common destructive tests.

- **Climbing Drum Peel Test to ASTM D-1781:** This is a long-established test to check the quality of the bond between the skin and the honeycomb or other core material in a sandwich panel. See Figure 10.37.

- **Wedge Test to ASTM D-3762:** This test was originally developed by Boeing as a quality control test on its phosphoric acid anodizing line. The test is good for that purpose, and its use has been extended to composite adherends. It is a good check on the quality of surface preparation, especially if the wedge test pieces are immersed in water with the wedges in place. See Figure 10.38.

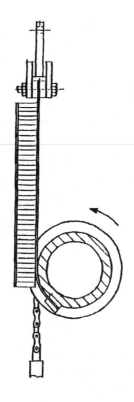

Fig. 10.37 Climbing drum peel test to ASTM D-1781.

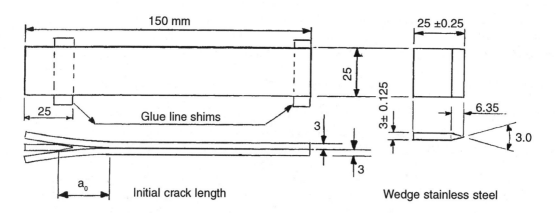

Fig. 10.38 Wedge test to ASTM D-3762. Make from 2024-T3 clad aluminum alloy or as specified, 3 mm (0.12 in.) nominal thickness or 10 SWG (0.128 in.). See notes for shim thicknesses. Use 5X to 30X magnification to locate crack tip.

- **Lap Shear Test to ASTM D-1002:** This is the standard test for lap shear strength, which is the most common method of testing adhesives. The test can indicate if the adhesive has been properly cured and, in the case of two-part adhesives, will show if the mix ratio was seriously in error and/or the cure was incomplete for some reason (i.e., the material being time expired). Although not as sensitive as the wedge test, this test can also show interfacial failure if the surface preparation is incorrect. See Figure 10.39.

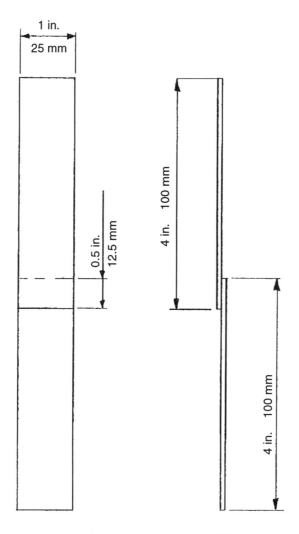

Fig. 10.39 Lap shear test to ASTM D-1002.

- **Peel Test ("T"-Peel) to ASTM D-1876:** This test can be used if peel strength is important. If used with water in the crack, it can also be used in lieu of the wedge test to check surface preparation quality. Cohesive failure in the adhesive will be achieved only if the surface preparation is very good. See Figure 10.40.

Typical Inspection Problems: For general inspection purposes, an inspector often will have difficulty in finding the time or the drawings to study the detailed construction of an assembly. The following are some problems that can arise:

- If ultrasonic or other instrumental methods are used and calibration of test blocks is necessary, then a knowledge of the structure is essential to perform the calibration on samples of the right type and thickness.

- In the case of purely visual examination, the difference between manufacturing defects and barely visible impact damage (BVID) must be known. This requires experience and training.

- With the tap test, know that thick skins do not respond to this test because they give a good sound alone. This is true for thick metal skins and for thick composite skins; it is effective only if delamination is located in the first few layers.

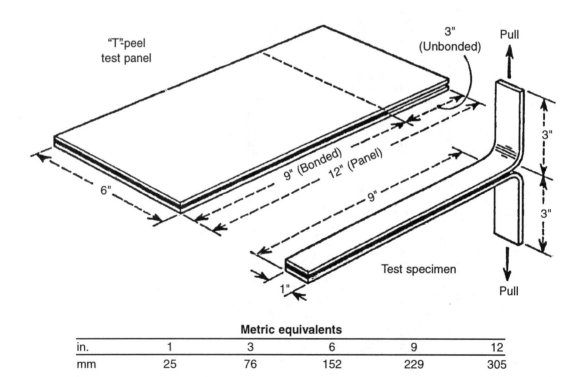

Metric equivalents				
in.	1	3	6	9	12
mm	25	76	152	229	305

Fig. 10.40 "T"-peel test to ASTM D-1876, test panel and specimen. (Courtesy of ASTM.)

10.7 Surface Restoration

This topic must be considered after all repairs are completed and the appropriate restoration must be carried out. In every case, establish from the SRM or drawings the standard of finish or protection required. The finish may be needed for protection from weather, for appearance, for lightning protection, or for some other reason; however, it is seldom an optional choice. In the case of radomes, a conductive band of special paint may be required in a specific area. Elsewhere, on a radome only nonconductive paints may be used, and they too must not exceed the specified thickness because this can affect the radar signal transmission. In every case, the surface finish is a part of the design, and its replacement must be done as correctly as every other job.

Filler Usage: In this application, surface fillers for the improvement of surface smoothness and finish are discussed. These are usually epoxy or polyester compounds produced specially for the purpose. They are normally of fairly high viscosity and are thixotropic so that they remain where they are placed to fill undulations and do not flow to areas where they are unwanted. Another meaning of the word is the fillers used in epoxy and other resins to modify properties, reduce weight, etc. Suitable fillers can be used to reduce the thermal expansion coefficient of epoxy filler compounds and to bring them close to aluminum alloys and nearer to carbon and other composites. This type is covered in Chapter 2, Section 2.7.1. See also Ref. 10.9.

Surface fillers are used to fill dents in metal skinned metal honeycomb structures. For composites, they are used to fill small pits and pinholes in the surface where the surface gel coat has been omitted or has worn away or when the surface is somewhat resin starved. In the last case, apply a coat of low-viscosity gel coat resin to thoroughly wet the fibers. A filler can be applied afterward if the surface is not yet smooth enough. The filler can then be rubbed down with abrasive paper to achieve the finish required, and then the surface can be painted and a good appearance produced. Good quality filler compounds with a high strain to failure must be selected. Unfortunately, fillers often can be rather brittle. If used on surfaces subject to large temperature variations, they frequently crack and defeat the purpose for which they were employed. The problem can occur in two ways:

1. If the surface is, say, carbon fiber, and expands very little, then a surface filler with a high thermal expansion coefficient will try to expand or contract as a result of temperature changes while the surface moves very little. The filler may thus generate its own fatigue cracks.

2. If the filler has a low thermal expansion coefficient and/or a low strain to failure and the surface being filled is aluminum alloy, then the reverse can occur and the metal surface can expand more than the filler.

Thus, choose fillers that match the materials being filled, in terms of thermal expansion coefficient and/or a high strain to failure to ensure that any movement due to temperature change or loading can be accommodated for an acceptable lifetime.

Sealant Usage: Sealants are covered in detail in Chapter 2, Section 2.8.1.

Erosion-Resistant Coatings: These are covered in Chapter 2, Section 2.8.5.

Conductive Coatings and Metal Wire Cloths: These topics are discussed in Chapter 2, Section 2.8.4 to the end of Section 2.8.

Primers and Topcoats: These topics are covered in Chapter 2, Sections 2.8.2 and 2.8.3.

10.8 References

10.1   Horton, R.E., McCarty, J.E., *et al.*, *Adhesive Bonded Aerospace Structures Standardized Repair Handbook*, AFML-TR-77-206/AFFDL-TR-77-139, U.S. Air Force Materials and Flight Dynamics Laboratories, U.S. Air Force Systems Command, Wright-Patterson AFB, OH 45433, 1977.

10.2   Baker, A.A. and Jones, R. [eds.], *Bonded Repair of Aircraft Structures*, Martinus Nijhoff Publishers, P.O. Box 163, 3300 AD Dordrecht, The Netherlands, ISBN 90-247-3606-4, 1988.

10.3   Armstrong, K.B., "Effect of Absorbed Water in CFRP Composites on Adhesive Bonding," *International Journal of Adhesion and Adhesives*, Vol. 16, No. 1, 1996, pp. 21–28.

10.4   De Bruyne, Norman, *My Life,* Midsummer Books, Cambridge, ISBN 1-85183-080-4, 1996.

10.5   Allen, K.W., *Joining of Plastics,* The City University, London, RAPRA Review Report 57, RAPRA Technology Ltd., Shawbury, Shrewsbury, Shropshire SY4-4NR, United Kingdom, 1992.

10.6   Armstrong, K.B., *The Repair of Adhesively Bonded Aircraft Structures Using Vacuum Pressure*, Society for the Advancement of Material and Process Engineering, Series 24, Book 2, 1979, pp. 1140–1187.

10.7   Kelly, A. [ed.], *Concise Encyclopedia of Composite Materials*, Pergamon, Oxford, UK, ISBN 0-08-042300-0, 1994.

10.8   Wegman, R.F. and Tullos, R.T., *Handbook of Adhesive Bonded Structural Repair*, Noyes Publications, Park Ridge, NJ, ISBN 0-8155-1293-7, 1992.

10.9   Lee, Henry and Neville, Kris, *Handbook of Epoxy Resins*, McGraw-Hill Book Company, New York, 1967.

Mechanical Fastening Systems

11.1 Introduction

Advances in bonding technology have greatly reduced the need for mechanical fasteners in applications where parts are assembled permanently. However, many installations must be removable for access or maintenance, and others use thick composite laminates for which mechanical fastening is the only practical solution. Mechanical fastening systems are widely used today for these applications. New fasteners are continually being specially designed and introduced for use with advanced composite materials; indeed, some of these new fasteners are made from composites. (See Section 11.9.) Mechanical fasteners sometimes are used for temporary repairs to composite parts when the speed of repair is especially important or if facilities for bonded repairs are unavailable. They are also used for permanent repairs to thicker solid laminate structures such as wing, tailplane, and fin skin panels, and stringers and spar webs, where the overlap lengths required for a sound bonded joint would be heavier than for a mechanically fastened joint. Although the amount of interference is less than that for metals, fasteners are sometimes used with interference fit holes in composites to improve the fatigue life of joints. This means that close tolerance fasteners must be used in holes drilled accurately to provide the required degree of interference. Also, these fasteners must be installed with great care using the correct equipment. Repairs to solid laminate composites must be taken seriously because these are the more structural and heavily loaded parts of the aircraft. As more primary structures are being made of composites, more training will be necessary in this type of composite repair. In many ways, this composite repair is similar to sheet metal repair except that more care is required to drill good holes, and special drills and drilling techniques are required. The correct fasteners must be properly installed. Symbol systems are widely used to identify fasteners in metals and composite materials on drawings, and the National Aerospace

Standards NAS 523 standard is commonly used as a basis by the aerospace industry. Each manufacturer may use a slightly different system. A gunsight cross is used to pinpoint the location of each fastener, and each quadrant has an alpha and/or numeric code that provides details such as the hole size, fastener required, head location (near side or far side), and whether a countersink is required. See Figure 11.1.

For information on joint design, see Ref. 11.1.

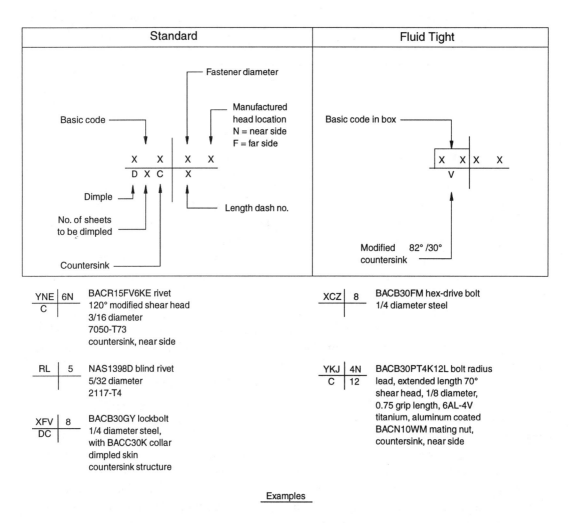

Fig. 11.1 Fastener symbols and interpretation. (From Boeing B.757 SRM 51-40-01. Courtesy of Boeing Commercial Airplane Group.)

11.2 Fastener Types

Fastener types vary considerably, and some of those used in metals and most of the types commonly used in composites at the time of this writing are described briefly below. The total range of fasteners is so vast that if all fasteners used with metals were included here, a separate book would be necessary to adequately cover the subject. Ref. 11.2 is a small but helpful and well-illustrated book on the subject. The types mentioned below are mainly those recommended for use with composites. Those used for metals can also be found in the Structural Repair Manual (SRM) 51-40-00. Manufacturers of these items produce helpful technical and installation data for these fasteners. The subject is well covered in SRM Chapter 51-40-00 for fasteners for metals and composites.

11.2.1 Fastener Standards

National Aerospace Standards: The National Aerospace Standards (NAS) have been in use in the United States for many years and are continually adding to the list of new fasteners, anchor nuts, and other small and frequently used components. Obsolete parts are periodically removed. The items listed as NAS parts are often commercial items required by the aircraft industry. The NAS specification establishes quality control and purchasing standards, and the part is allocated an NAS number. Some examples are NAS 679 for hexagon nuts, NAS 1738 for universal head bulbed Cherrylock rivets, and NAS 1739 for countersunk head Cherrylock rivets. These rivets conform to Procurement Specification NAS 1740. After the NAS number is known, full details can be obtained by reference to the standard. This information can also be obtained from manufacturers' catalogs.

American Military Standards: These Military Standards (MS) are those developed originally for use on military aircraft and equipment in the United States. They are presented in a way similar to NAS standards and should be consulted when fasteners made to these specifications are used.

Other Fastener Standards: Similar aerospace standards exist in the United Kingdom, France, Germany, Japan, China, and elsewhere. These standards should be consulted when working on Airbus, British Aerospace, Fokker, and other types of aircraft. Always consult the SRM chapter on fasteners (SRM 51-40-00) for specific details. For marine, automobile, and other applications, the standards for these industries should be used. The Society of Automotive Engineers (SAE) provides many fastener standards for automobile and general engineering work in the United States. See SAE J429 for bolts, J1199 for metric bolts, and J1200 for blind rivets. In Europe and elsewhere, the use of International Organization for Standardization (ISO) metric standards is increasing.

11.2.2 Fastener Compatibility

Fastener compatibility has always been a concern, whether the problem was the choice of material for screws in wooden aircraft, bolts, and screws in aluminum alloys and steel, or titanium components or fasteners for composite constructions. In all cases, the possibility of corrosion must be considered, and the correct materials and protective coatings must be used. See the galvanic series in Table 5.3 of Chapter 5 and also aircraft SRMs. In the case of composites, fiberglass and aramid fibers are compatible with most materials; however, in marine applications, the materials chosen for fasteners and their use in a salt spray environment must be considered. Automobile applications on the underside of the vehicle will have to consider the type of salt used on roads in winter for de-icing purposes. Carbon fiber and certain hybrid components can react galvanically with some fasteners, causing corrosion. Avoid the use of aluminum or magnesium fasteners in these components. The Boeing B.757 SRM states, "Fasteners installed in graphite composite structure must be bare or aluminum-coated titanium or corrosion-resistant steel (CRES). Cadmium-plated corrosion-resistant steel may also be used. Aluminum or alloy steel fasteners are not allowed in graphite composite structures." In all aircraft applications, use only the fastener type approved by the SRM or manufacturing drawing. In marine, automobile, railway, and other applications, use only approved materials suitable for the operating conditions in which the item is used. See Ref. 11.3. For other useful information on fasteners, see Ref. 11.4.

11.2.3 Rivets

Rivets are made in a variety of types for a range of purposes and are described in detail below.

11.2.3.1 Solid Rivets

Solid rivets are used almost exclusively in metal sheets. Occasionally, the softer types may be used in composites. Solid rivets are fasteners made from soft metal which are squeezed or hammered to form a large upset head on one or both sides of the sheets they are fastening. The soft metal hardens as a result of cold working during installation. The strong aluminum alloys used for rivets may require normalizing (also known as solution treatment) to soften them sufficiently to allow the deformation required for head formation during the riveting process. The precise temperatures and times for this process vary slightly with each alloy, and the correct treatment must be given according to the material specification. They are kept frozen and stored at a temperature of -23°C (-10°F) or lower after water quenching following normalizing to slow the rate of natural age-hardening. These rivets must be used quickly (i.e., within 15 minutes of removal from the freezer), or they become too hard for the riveting operation. See Table 11.1, which is from Boeing B.757 SRM 51-40-02.

Table 11.1
Solid Rivets

Material	Suffix	Head Marking	End Marking
2017	D	Raised dot	None
2117	AD	Dimple	None
5056	B	Raised cross	None
2024	DD	Two raised dashes	None
7050	KE	None	Indented circle
Monel	M	None	None

The 2017, 2117, 5056, and 7050 aluminum alloy rivets and monel rivets may be stored and used at room temperature. Rivets made from 2024 aluminum alloy must be heat-treated and then stored at a temperature of -23°C (-10°F) or lower. After removal from cold storage, they must be completely driven within 15 minutes. These rivets must not be returned to cold storage after they have been removed. The same treatment is given before bending and forming these alloys. These rivets must be bucked to form the inside head in one burst of percussion riveting. Three to four seconds driving duration is the optimum; seven seconds should be the maximum. The rivets work harden rapidly, and a second burst has no effect. Bucking bars should be smooth, especially when driving the hard 2017 aluminum alloy rivets. Modern high-strength shear pins, such as the Cherrybuck and the Cherry E-Z Buck, are bucked by squeeze riveting. Both of these can be squeezed or riveted more than once. The former is a bimetal, one-piece fastener, which combines a 95 ksi shear 6Al-4V titanium shank with a ductile Ti/Cb tail to provide weight savings over two-piece, pin and collar, shear pin fasteners and to reduce installation costs through one-piece automated riveting equipment. The latter is a 50 ksi ductile Ti/Cb solid rivet, which requires special tooling to flare the hollow end. Its ductility eliminates the sheet distortion found when using monel rivets, and it is excellent in fatigue critical applications. All of these are more suitable for metal parts than for composite parts. In all rivet installations, if the head produced is not satisfactory, the rivet must be drilled out and a new one must be installed. The alloys most commonly used for rivets in aircraft structures are aluminum, magnesium, and monel. Other industries may use copper or steel rivets. Most rivets have a manufactured head on one side and a shank of smaller diameter which is inserted into the hole. The most common head styles are the protruding or universal head and a flush countersunk head. Depending on the type of rivet and its end use, the included angle of the countersunk head may be 70°, 90°, 100°, 120°, and, for some fasteners for composites, 130°. Replacement rivets in existing holes must have the correct angle; thus, the angle required must be checked. Many high-strength aircraft rivets are made to close tolerances, and therefore accurate holes are required. See the SRM for hole quality required.

11.2.3.2 Blind Rivets

The advantage of blind rivets, and the reason for using them, is that they can be installed from one side with special tools. They are ideal for shear applications in which production time or backside access is limited. Some blind rivets have hole fill capability and can be used on wing fuel tank structures and pressure bulkheads, giving improved sealing and fatigue properties. They also are available in several different alloys and can have either universal or countersunk head styles. The most common angle for countersunk heads is 100°, but the actual angle used should be checked for each type that may be called up. Versions vary. For example, in some cases, a stem with a bulbed end is pulled partially through a hollow shank, which expands on the far side to retain the fastener in the hole (e.g., CherryMax). Others rely on expansion of the sleeve by a larger plug diameter on the stem (e.g., CherryMax "A"). The stem is retained either by friction or a locking ring. If the component is subject to vibration or if retention is critical, the positive locking system is recommended. An example that uses the positive locking system, in which the locking collar is mechanically attached to the stem, is the CherryMax blind rivet. Earlier types relied on friction alone for locking. The original blind rivets used in aviation and elsewhere were available only in oversize diameters because their strength was poor, and engineers were reluctant to use them for structural applications. Not all blind fasteners are strong enough for important structural applications, and they should not be substituted for solid rivets without the permission of the SRM or an approved design authority. As in many other areas, advances have been made in the design of blind fasteners, and some of them are actually stronger than some solid rivets. Many types are available, and each requires its own special tools, which can present problems to maintenance bases when many types of aircraft or other vehicles are repaired. These are used mainly in metal sheets and should be used in composites only if specified in the SRM. See Figure 11.2.

11.2.3.3 Hollow End Rivets

The hollow end solid rivet is specially designed to allow the swelling of the end without damaging expansion within double countersunk holes in composites. Stronger alloys can be used for higher shear strength. These special titanium/columbium alloy rivets are ideal for use in composite materials in which side loads may fracture the composite matrix. Close tolerance holes are needed to give the best results with these fasteners. They must be installed with squeeze riveting equipment. Special forming dies are required to flare the hollow end of the rivet into the countersink. See Figure 11.3 and Boeing B.757 SRM 51-40-02. Cherry (now part of Textron Aerospace Fasteners) supplies its Hollow End E-Z Buck for this market. This type is used in composites.

Pop Standard blind rivet with domed head

Break head   **Break stem**

Aircraft-Chobert blind rivet with countersunk head

Without sealing pin   **With sealing pin**

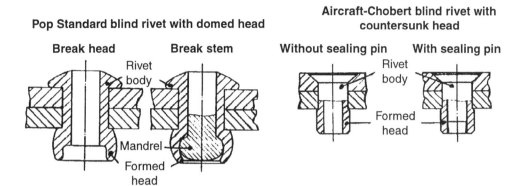

Bulbed Cherrylock blind rivet with countersunk head

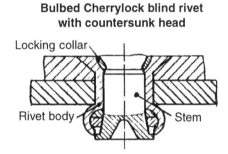

Cherry Max blind fastener with countersunk head

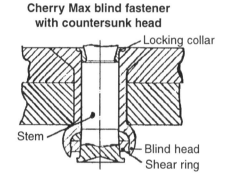

Visu-Loc blind fastener with flush head

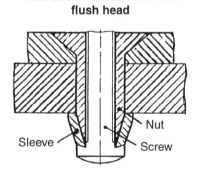

MBC blind rivet with countersunk head

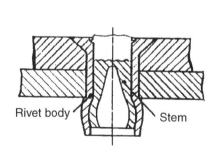

Fig. 11.2 Types of blind rivets and fasteners.

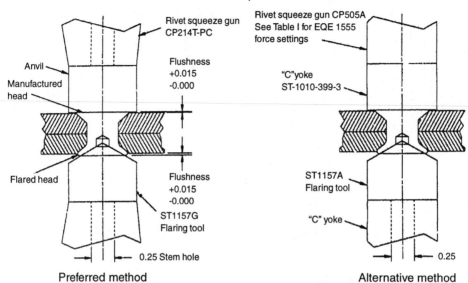

757-200
Structural Repair Manual

Preferred method

Anvil
Manufactured head
Rivet squeeze gun CP214T-PC
Flushness +0.015 -0.000
Flared head
Flushness +0.015 -0.000
ST1157G Flaring tool
0.25 Stem hole

Alternative method

Rivet squeeze gun CP505A
See Table I for EQE 1555 force settings
"C"yoke ST-1010-399-3
ST1157A Flaring tool
"C" yoke
0.25

Tool and flushness requirements for hollow-ended (BACR15GA) rivets
Detail I

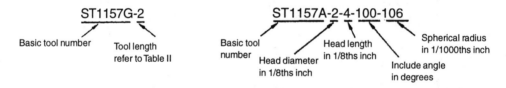

ST1157G-2

Basic tool number

Tool length refer to Table II

ST1157A-2-4-100-106

Basic tool number
Head diameter in 1/8ths inch
Head length in 1/8ths inch
Include angle in degrees
Spherical radius in 1/1000ths inch

Flaring Tool Description

Nominal Rivet Diameter	Squeeze Force (lb) ±200 lb
5/32	3000
3/16	3500
7/32	4000

Force settings for power supply unit EQE 1555 [A]
Table I

Tool Length Number	Tool Length "L" (inches)
-1	0.150
-2	0.300
-3	0.500
-4	0.750

Notes:

[A] Power supply unit is available from
E.F. Bailey Co.
5160 4th Avenue South
Seattle, WA 98108
Specify Boeing standard pump SPL 1555-7-1

*Fig. 11.3 Hollow end rivet and installation. (From Boeing B.757 SRM 51-40-02,
Fig. 17. Courtesy of Boeing Commercial Airplane Group.)*

11.3 Screws, Nuts, and Bolts

Fasteners designed to withstand both shear and/or tensile loads and that are retained with some type of nut or collar can be categorized as bolts. The most common style is the hexagon head, which is turned or held with a wrench or socket. Internal wrenching bolts use an Allen-type wrench to install. A range of recesses for screwdriver-type bits is also available. The most common are Torq-Set, Phillips, and Tri-Wing. Tri-Wing recesses have not been used for new design since 1988 but remain in use on old aircraft. A problem with all of these is that the correct size and type of bit must be used. All bits are different, and the use of the wrong bit will usually result in slippage, rotation of the bit in the recess instead of rotation of the bolt, and finally scrappage of the bolt because of damage to the recess and possibly also scrappage of the bit. With all of these recess types, the screwdriver must be held close to 90° to the panel under firm pressure. Otherwise, the bit may slip in the recess and fail to rotate the bolt (which may damage the recess), or the bit may slip and scratch or score the panel on which work is being done. The problem occurs when trying to remove these types of fasteners after they have been installed for a long period of time. When removing these fasteners, it is even more important to hold the screwdriver precisely square to the surface and also to apply firm pressure to give the best chance of rotating the screw easily. Use of a new bit for the removal process is also helpful.

Some specialty bolts are not turned at all but use threaded or swaged collars to retain them. Small-diameter bolts, which permit the use of screwdrivers, are referred to as machine screws. All have a solid shank and threads or serrations at the end. Some provide excellent sheet take-up capability. See Figure 11.4.

11.3.1 Specialty Fasteners (Special Bolts)

Many specialty fasteners are available, but not all are suitable for use with composites. Early trials showed that some types required clamping loads that caused crushing of composite materials, and new developments or modifications were required to provide a larger bearing area on the rear face to solve this problem. In general, the larger the bearing area, the greater the preload that can be applied to the structure, resulting in improved joint strength.

11.3.2 Lockbolts

Lockbolts are special factory fasteners used for quickly bolting parts together. Lockbolts have annular grooves, not helical threads. Once swaged in place, they cannot back off. Their nonthreaded design also offers the advantage of superior vibration resistance. They have excellent sheet take-up capability (i.e., in the process of being tightened, they automatically clamp the sheets of metal or composite together). Not all fasteners do this, and sheet take-up capability is a useful feature. They are inserted from one side, and then a

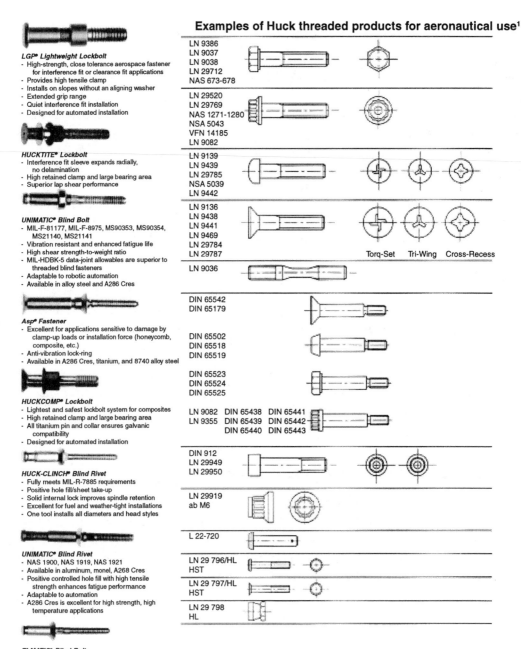

LGP® Lightweight Lockbolt
- High-strength, close tolerance aerospace fastener for interference fit or clearance fit applications
- Provides high tensile clamp
- Installs on slopes without an aligning washer
- Extended grip range
- Quiet interference fit installation
- Designed for automated installation

HUCKTITE® Lockbolt
- Interference fit sleeve expands radially, no delamination
- High retained clamp and large bearing area
- Superior lap shear performance

UNIMATIC® Blind Bolt
- MIL-F-81177, MIL-F-8975, MS90353, MS90354, MS21140, MS21141
- Vibration resistant and enhanced fatigue life
- High shear strength-to-weight ratio
- MIL-HDBK-5 data-joint allowables are superior to threaded blind fasteners
- Adaptable to robotic automation
- Available in alloy steel and A286 Cres

Asp® Fastener
- Excellent for applications sensitive to damage by clamp-up loads or installation force (honeycomb, composite, etc.)
- Anti-vibration lock-ring
- Available in A286 Cres, titanium, and 8740 alloy steel

HUCKCOMP® Lockbolt
- Lightest and safest lockbolt system for composites
- High retained clamp and large bearing area
- All titanium pin and collar ensures galvanic compatibility
- Designed for automated installation

HUCK-CLINCH® Blind Rivet
- Fully meets MIL-R-7885 requirements
- Positive hole fill/sheet take-up
- Solid internal lock improves spindle retention
- Excellent for fuel and weather-tight installations
- One tool installs all diameters and head styles

UNIMATIC® Blind Rivet
- NAS 1900, NAS 1919, NAS 1921
- Available in aluminum, monel, A268 Cres
- Positive controlled hole fill with high tensile strength enhances fatigue performance
- Adaptable to automation
- A286 Cres is excellent for high strength, high temperature applications

TI-MATIC® Blind Bolt
- MIL-F-8975
- Highest shear strength-to-weight ratio of any aerospace blind bolt
- All titanium construction for metal to metal, metal to composite, or composite to composite structures
- Adaptable to automation
- Wide grip range with a flush pin break
- Vibration resistant and enhanced fatigue life
- Large blind side bearing surface provides high sustained clamp load to ensure joint integrity

Examples of Huck threaded products for aeronautical use[1]

LN 9386
LN 9037
LN 9038
LN 29712
NAS 673-678

LN 29520
LN 29769
NAS 1271-1280
NSA 5043
VFN 14185
LN 9082

LN 9139
LN 9439
LN 29785
NSA 5039
LN 9442

LN 9136
LN 9438
LN 9441
LN 9469
LN 29784
LN 29787

Torq-Set Tri-Wing Cross-Recess

LN 9036

DIN 65542
DIN 65179

DIN 65502
DIN 65518
DIN 65519

DIN 65523
DIN 65524
DIN 65525

LN 9082 DIN 65438 DIN 65441
LN 9355 DIN 65439 DIN 65442
         DIN 65440 DIN 65443

DIN 912
LN 29949
LN 29950

LN 29919
ab M6

L 22-720

LN 29 796/HL
HST

LN 29 797/HL
HST

LN 29 798
HL

[1]Apart from the German "Luftfahrt-Norm" LN all future DIN standards and numbers from 9000 to 9499, 29500 to 29999 and 65000 to 65999 are reserved for the air-industry. DIN-EN air- and aerospace standards are registered under DIN-EN 2000-9999.

Fig. 11.4 Huck threaded products showing head recess types. (Courtesy of Huck International Inc.)

special tool swages a collar on the back side before breaking the extra stem material. Because installation equipment is expensive, these are usually substituted with other bolts such as Hi-Loks at repair stations. When used in large quantities, the speed of production outweighs the cost of special production tooling. Huck has developed low-cost pneumatic tooling designed specifically for small-volume repair applications. Cherry and Huck type lockbolts are available in sizes ranging from 3 to 10 mm (1/8 to 3/8 in.) and are made in nominal and oversize versions in aluminum, steel, and CRES materials. Monogram Aerospace Fasteners Inc. manufactures the Huck LGPL® series of lockbolts designed for use in composite materials. Lockbolts are also produced under the trade names Hy-Per and Hy-Per-Lite by Hi-Shear Corporation, and by other suppliers under their own names.

1. **The HuckComp® Lockbolt:** This special version of the Huck lightweight titanium lockbolt system is designed specifically for carbon fiber composite applications and is an all-titanium system with a Ti-6Al-4V pin and a flanged, commercially pure titanium swage-on collar. The HuckComp lockbolt can be supplied in either a pull stem version for installation with common pulling-type tools or a stump version for installation with portable squeezers or automatic drill/rivet machines such as a DRIVMATIC machine. The bearing areas under the head and the collar have been increased to avoid any damage problems during installation or use.

2. **Cherry® Lockbolts:** These are available in pull or stump type.

3. **HuckTite® Sleeved Lockbolt:** HuckTite is a sleeved version of the HuckComp. This sleeved fastener can be used to give interference fits in carbon fiber composite structures without causing installation damage, while offering performance gains in terms of improved structural fatigue, electrical continuity, and fuel tightness. To achieve this, a thin-walled 0.2 mm (0.008 in.) A-286 sleeve is pre-assembled on the pintail/lockgroove section of the lockbolt. The pintail/lockgroove section with the sleeve is then inserted into a clearance hole, with the sleeve acting as a protective bushing for the hole during final installation.

4. **Monogram Radial-Lok:** This is also a sleeved lockbolt, which performs similarly to the HuckTite bolt described above. It provides a tighter joint from an interference fit and a better fatigue life, and it can be fitted into an 0.08 mm (0.003-in.) clearance fit hole, which is completely filled after installation. It is a sleeved version of the Composi-Lok fastener.

5. **Hi-Loks®:** These specialty fasteners developed for the aerospace industry offer several advantages over standard bolts. They install quickly with uniform torque and provide significant weight savings. Numerous styles and alloy types are available for specific needs. The distinguishing feature common to all is a special breakaway nut that severs at a predetermined torque, and a recess in the end for an Allen key. The Allen key is

mounted in the wrench tool and prevents rotation of the pin while the nut is tightened to a sufficient torque to shear the collar. See Figure 11.5. Hi-Loks are normally used when the following features are required:

- High shear strength
- Impact riveting of solid rivets is prohibited, and they cannot be squeezed
- Expansion of the shank would damage the materials being joined
- High clamp-up is required for sealing or other purposes

Hi-Loks are most commonly used to join metal parts. For use in carbon fiber composites, they must be titanium or CRES to avoid corrosion of the fastener. They are supplied by Hi-Shear Corporation (which holds the patent) and by SPS Company and Voi-Shan Manufacturing Company, which manufacture these fasteners under license.

6. **Hi-Lite® Fasteners:** Hi-Loks are being superseded by the Hi-Lite fastening system, which is also excellent for joining composite materials. This series offers useful weight reductions on the earlier type. The Hi-Lite ST pin is designed for clearance or interference fit applications. See Figure 11.6.

7. **Hi-Lex® Fasteners:** Hi-Shear Corporation also produces the Hi-Lex fastening system specifically for composite materials, as shown in Figure 11.7. This system uses a small interference fit and a counterclockwise shallow grooved thread on the shank with a normal clockwise thread for the collar. These fasteners use a 130° countersunk head, and the interference fit ensures good electrical contact in the event of lightning strike.

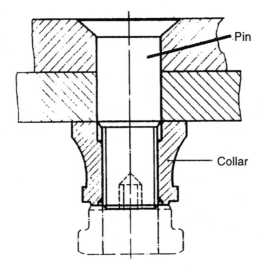

Fig. 11.5 Hi-Lok fastener with flush head. (Courtesy of Hi-Shear Corporation.)

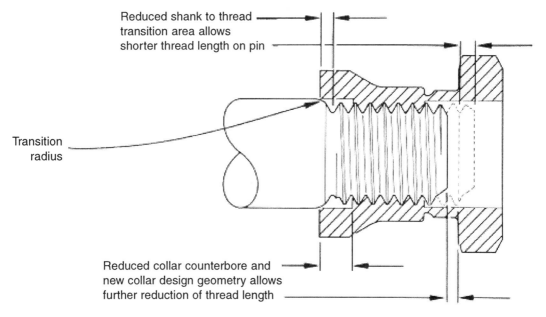

Reduced shank to thread transition area allows shorter thread length on pin

Transition radius

Reduced collar counterbore and new collar design geometry allows further reduction of thread length

Fig. 11.6 Hi-Lite fasteners. (Courtesy of Hi-Shear Corporation.)

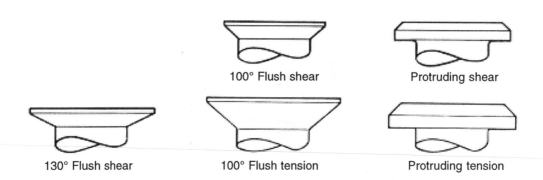

100° Flush shear	Protruding shear
130° Flush shear	100° Flush tension	Protruding tension

Matching Collar Styles

A choice of collars is available for the Hi-Lite Multi-Purpose ST Pin. They can be used interchangeably. This choice gives the Hi-Lite Fastening Systems flexibility. Each collar has been engineered to match the ST pin, and each pin/collar pair forms a balanced unit.

Hi-Lite® ST™ Collar

A threaded collar featuring automatic torque control to guarantee precise minimum/maximum preload range without need for calibrated torque-control tools. Offers the optimum strength-to-weight balance. For general use with high-volume Drivmatic automatic installation equipment or with hand tools. Sealing collars are available.

Hi-lite® ST™ Swaged Collar

A swaged collar specialized for Drivmatic-type equipment already tooled for swaged collars. The Hi-Lite ST Swaged Collar has strength and reliability. It installs with existing swaged collar tooling with modified insert.

Hi-Lite® ST™ Forged Collar

A strong threaded collar with reduced profile to permit easy installation in restricted access areas. It has the automatic preload control and visual inspection advantage of all the threaded Hi-Lite ST collars. Uses existing installation tooling. Sealing collars are available.

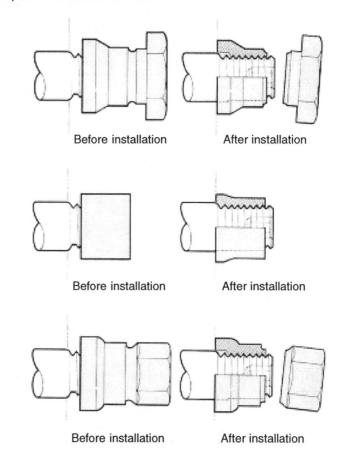

Before installation   After installation

Before installation   After installation

Before installation   After installation

Fig. 11.6 (continued)

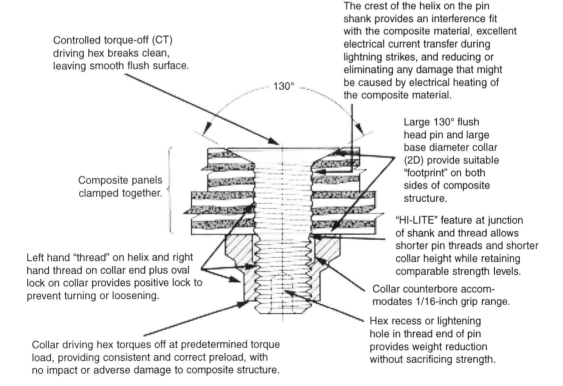

The crest of the helix on the pin shank provides an interference fit with the composite material, excellent electrical current transfer during lightning strikes, and reducing or eliminating any damage that might be caused by electrical heating of the composite material.

Controlled torque-off (CT) driving hex breaks clean, leaving smooth flush surface.

130°

Large 130° flush head pin and large base diameter collar (2D) provide suitable "footprint" on both sides of composite structure.

Composite panels clamped together.

"HI-LITE" feature at junction of shank and thread allows shorter pin threads and shorter collar height while retaining comparable strength levels.

Left hand "thread" on helix and right hand thread on collar end plus oval lock on collar provides positive lock to prevent turning or loosening.

Collar counterbore accommodates 1/16-inch grip range.

Hex recess or lightening hole in thread end of pin provides weight reduction without sacrificing strength.

Collar driving hex torques off at predetermined torque load, providing consistent and correct preload, with no impact or adverse damage to composite structure.

Fig. 11.7 Hi-Lex fasteners. (Courtesy of Hi-Shear Corporation.)

11.3.3 Hex Drive Bolts

Hex drive bolts, known as "Eddie" bolts, are made by Voi-Shan Manufacturing Company, now part of the Fairchild Aerospace Fastener Division. They are used on the Boeing 777 series of aircraft and in many other applications. These bolts are shear-type bolts. When used in composite materials, they have a large washer on the nut side to spread the load due to tightening. This special design requires a modified ratchet or box wrench with a hexagonal Allen-type key through its center. The Allen key prevents rotation of the bolt while the ratchet tightens the nut. A wrenching device at the top of the nut shears at a predetermined torque, thus eliminating the need for a torque wrench. The collar (nut) is deformed at the top to provide a swage locking function. See Figure 11.8, which is taken from Sheets 1 and 2 of Boeing B.777-200 SRM 51-40-02, pages 38 and 39. Sheets 3 and 4 following the two sheets shown in Figure 11.8 provide details of the types of washers required for prevention of installation damage, corrosion protection, counterbore substitution, and grip adjustment. Sheet 5 gives head contour limits and assembly details, and Sheets 6 and 7 give pin protrusion limits. These and all other fasteners must be assembled correctly, and use of the SRM is essential.

11.3.4 Specialty Blind Bolts

Of the many specialty blind bolts, some of those specially designed for use with composites have sleeves to protect the composite hole during installation. Those used with metals do not need a sleeve. In some cases, close tolerance holes are required.

1. **Cherry MaxiBolt® Blind Bolts:** These bolts are used for bolting metal sheets. They use single-action-type installation tools. A range of titanium blind bolts are also supplied by Cherry.

2. **Composi-Lok® Blind Bolts:** Composi-Lok blind bolts from Monogram Aerospace Fasteners Inc. are specifically designed for use with composites and offer a blind side bearing area 70% greater than that afforded by other systems. This large bearing area allows clamping forces or preloads more than twice those of pull-type systems, resulting in superior joint integrity and fatigue strength. The fasteners are available in A-286 CRES and titanium for galvanic compatibility, as well as low-profile protruding hex, 100° flush, and 130° shear head styles. The Composi-Lok II version of the fastener incorporates the manufacturers drive nut concept which simplifies installation, eliminates cam-outs, and reduces the nose pieces required to install 4 to 10 mm (5/32 to 3/8 in.) diameter fasteners to only two. Low-cost hand tools are also available, further supporting maintenance base repairs.

3. **Radial-Lok® Blind Bolts:** Radial-Lok blind bolts, also manufactured by Monogram Aerospace Fasteners Inc., incorporate the blind (bearing) sleeve design of the Composi-Lok and therefore provide the same beneficial bearing area, high preload, and sheet take-up characteristics. Unique to this blind fastener is its ability to achieve full radial

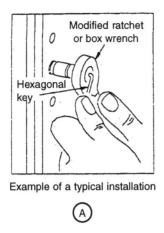

Example of a typical installation

(A)

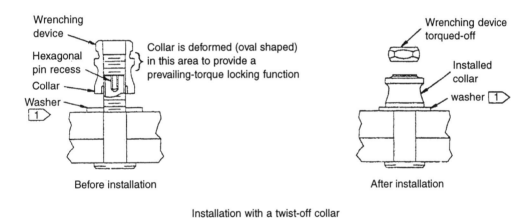

Fig. 11.8 Hex drive bolt installation. (From Boeing B.777-200 SRM 51-40-02, Sheets 1 and 2, pages 38 and 39. Courtesy of Boeing Commercial Airplane Group.)

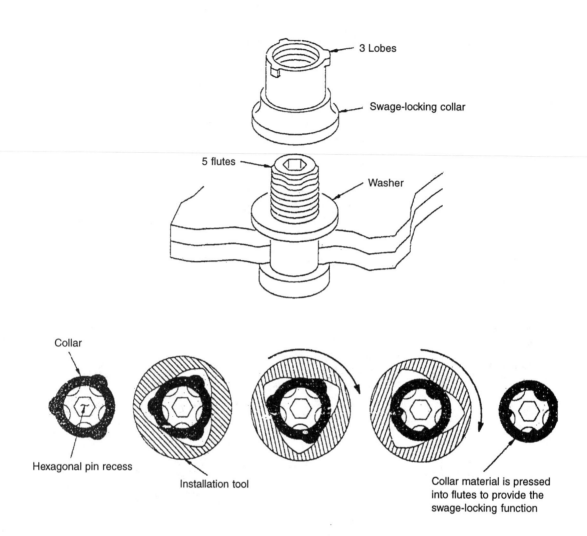

3 Lobes

Swage-locking collar

5 flutes

Washer

Collar

Hexagonal pin recess

Installation tool

Collar material is pressed
into flutes to provide the
swage-locking function

Installation with swage-locking collar and
swage-locking (fluted shank) bolt

Fig. 11.8 (continued)

expansion uniformly throughout the entire structure thickness. This expansion is achieved in a clearance fit (0.08 mm [0.003 in.] tolerance) hole, eliminating costly close tolerance hole preparation. This hole-filling capability increases the fatigue life of the fastener and ensures structural integrity under severe cyclic loading conditions in both aluminum and composite structure. Radial hole filling is also advantageous to enhanced electrical continuity and fuel sealing characteristics. The fastener is available in 100° flush and low-profile protruding head styles in 5 to 10 mm (3/16 to 3/8 in.) diameters, including oversizes for repair work.

4. **Huck UniMatic® Blind Bolts:** This type of blind bolt is an all A-286 stainless steel blind bolt with a pull-type, single-action, flush break, and a single bulbing action on the blind side. It can be supplied with 120° included angle shear-type flush head or with 100° tension head. It was not designed for composites and has a smaller blind side footprint than a Composi-Lok. Therefore, it is considered adequate only for applications in which the higher preload afforded by a larger footprint is undesirable.

5. **Huck Ti-Matic® Blind Bolts:** This titanium version of the Huck UniMatic blind bolt is available with a commercially pure (CP) titanium sleeve and an A-286 pin, with a CP titanium sleeve and an A-286 pin, or with a titanium sleeve and a 15-3-3-3 titanium pin. This fastener was designed specifically for use with composites with weight and galvanic compatibility in mind. The CP titanium sleeve provides an excellent, wide-bearing-area footprint on the blind side. However, care must be exercised to remove all gaps prior to installing this type of fastener in order to protect against possible delamination resulting from a high preload spike.

As the above demonstrates, one of the problems of repair is to check the type of fastener used in the original construction and to use the same type, possibly in an oversize version. If the correct type cannot be obtained, then a suitable alternative must be approved by a recognized design authority. See Ref. 11.5.

11.4 Spacers, Bolt Inserts, and Grommets

Spacers, also called inserts, bolt inserts, or grommets, are designed to protect sandwich panels from crushing loads when bolts are tightened to attach them to other structure. Shur-Lok spacers are used extensively in aircraft floor board installations. Young Engineers Inc. and many other companies make similar items. See their catalogs for the variety available. These two-piece inserts (Shur-Lok SL 5128) and similar (Figure 11.9) require a hole to be drilled through the panel and are pressed together to form a solid spacer through which a bolt or screw can be safely tightened. In many aircraft, the inserts for floor boards are more complex because each insert is required to carry a 1000-lb (455-kg) shear load to enable the floor panel to contribute to structural strength in the crash case. These special high-load-carrying inserts are usually two-part, interference fit type SL 5094 (similar to SL 5128, but with a larger flange top and bottom) that are adhesively bonded to the floor panel

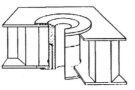

Plug and sleeve assembly

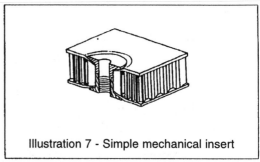

Illustration 5 - Shur-Tab potting operation

Occasionally, to maximize strength of a molded-in insert, it may be necessary to undercut the core to provide a larger volume of potting compound to be injected within the panel corespace resulting in a larger installed fastener "footprint" (see Illustration 6).

Illustration 7 - Simple mechanical insert

When a higher load-carrying capability is required in combination with metallic sandwich structure, Shur-Lok has developed and produced high-strength mechanical inserts that trap both top and bottom face skins, thereby making them integral with the structure (see Illustration 8).

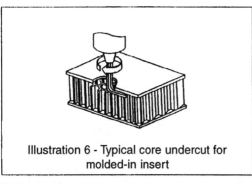

Illustration 6 - Typical core undercut for molded-in insert

Mechanical inserts

For metallic sandwich structure and light loading conditions, a simple mechanical insert such as the Shur-Lok SL101 series is recommended. These inserts require only a through hole for installation, and when flush mounting is required, the head style provides the advantage of automatic dimpling of the face skins during installation (see Illustration 7).

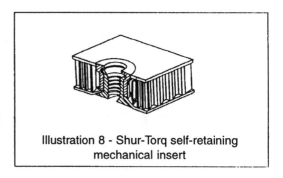

Illustration 8 - Shur-Torq self-retaining mechanical insert

Fig. 11.9 Examples of Shur-Lok inserts. (Courtesy of Shur-Lok International.)

skins on both top and bottom flanges and also in the hole itself. This means that they must be anodized and primed, preferably with a phenolic primer such as Redux 101 (if they are made from aluminum alloy) to achieve a strong and durable bond. They also must be bonded with a tough adhesive such as Redux 420 or similar. Prior to use, these inserts must be stored in clean, dry conditions. The bonding surfaces must not be handled before the adhesive is applied, or contamination will reduce the bond strength. Threaded inserts for mounting various items require a hole to be drilled either through the panel if they are of the two-part interference fit type, or down to (or near to) the bottom skin if they are of the bonded type. The two-part type should also be bonded to prevent rotation during bolt installation and removal. See also Chapter 16 and Figure 11.9. The bonded type is first attached to a tab coated with a pressure-sensitive adhesive. The insert is then positioned and retained by this tab, which has two small holes in it. The permanent adhesive, usually a two-part mix epoxy, is injected into one hole until it flows out the other. It is then injected into the second hole until it flows out the first. This procedure avoids or minimizes the risk of air entrapment. Shur-Lok International supplies this type of insert. For metal-skinned honeycomb panels in which only moderate loads must be carried, several two-part, interference fit type inserts are supplied by many manufacturers. These do not require adhesive bonding.

Asp® Fastener, 0.002-Inch Close Tolerance Shank, Adjustable Preload, Self-Sustaining Positive Mechanical Lock: This special fastener from Huck is designed to attach parts to sandwich panels without crushing the soft honeycomb, foam, or other core material. It is available in a range of materials and lengths to suit a variety of panel thicknesses. An equivalent product is manufactured by Monogram Aerospace Fasteners Inc. under the trade name of MAF.

11.5 Hole Preparation

11.5.1 Drilling

Most mechanical fasteners require holes through the materials to be connected. Machining holes in composite materials requires tools and techniques different from those traditionally used in metals. Thorough coverage is given in Boeing B.757 SRM 51-70-16 and probably in most other SRMs for aircraft using composites. Fiber breakout and/or delamination can occur during drilling, trimming, and countersinking if proper procedures are not followed. Aramids require special care.

Caution: Machining of composite materials produces dust, which may be hazardous to humans and machinery. Eyes, lungs, and skin must be protected. Electrical machinery should be protected from carbon fiber dust. The use of vacuum collection equipment is highly recommended. See also Chapter 6, Safety and Environment.

Delamination may occur between plies, whereas fiber breakout is caused when the drill breaks through the last (external) ply on the far side. Both are caused by excessive pressure. Generally, a high drill rotation speed and a slow and controlled feed drilling technique is recommended. See SRM 51-70-16. Fiber breakout can be further minimized by using a suitable backup such as a block of wood or plastic on the far side. Too much pressure and/or speed can quickly raise the temperature of the component and the drill bit, which can damage the surrounding resin and reduce the life of the drill bit. To minimize this, the drill bit can be lubricated with approved materials and methods. Use only approved lubricants such as cetyl alcohol ($C_{16}H_{34}O$), also known as Hexadecanol or n-Hexadecyl alcohol. This is a waxy solid of melting point 49°C (120°F). The composite should not exceed a temperature of 60°C (140°F) during drilling or machining. If this waxy lubricant melts, it is good indicator that the temperature limit has been reached. Boeing has developed the BOELUBE range of lubricants using blends of this and similar materials. The B.757 SRM permits filtered air, CO_2, non-oil-containing freon, BOELUBE, and cetyl alcohol as lubricants. (See Chapter 13.) Vacuum dust extraction can also aid cooling by providing an air flow. Special drills have been developed for machining advanced composite materials. See the SRM.

Aramid fibers may stretch and then break during the drilling or cutting process. The fibers may also split into a larger number of smaller fibers, resulting in a fuzzy hole or edge. Aramids have a greater tendency to heat up if high drill speeds are used. They are also more likely to cause self-feeding of the drill if ordinary twist drills are used. The Klenk drill was designed to reduce this problem. Applying a little extra resin over the area to be machined can help to further reduce this effect.

Machining a hole in carbon fiber is similar to drilling glass and is particularly grueling on bits. Carbide- or cobalt-tipped spade bits are good, but diamond polycrystalline coated tools are better. For carbon fiber, Boeing recommends drilling dry if possible. If these special drills are unavailable, a 135° split-point carbide or cobalt drill may be substituted for drilling most composite materials.

Most SRMs give useful information on the drilling, countersinking, and machining of composites. In view of the complexity of tool design and the use of specific techniques for different materials such as aramids, it is recommended that SRM 51-70-16 be studied prior to each drilling operation. Drill and cutter types are listed together with speeds and feed rates. Hole quality requirements are specified; scorching is not permitted, and a limit is placed on the amount of breakout damage allowed. No delamination is permitted on the outer ply. This section of the SRM also covers edge trimming and deburring. Use of 60- to 80-grit diamond-coated bandsaw blades and cutting wheels is recommended. The Commercial Aircraft Composite Repair Committee (CACRC) Document 95AM-1 states, "Equipment for sawing or machining shall be capable of achieving the dimensional tolerances defined for the specimens. Diamond tools shall be used. The machining speed shall be such that the local temperature of the specimen shall not exceed 60°C (140°F) during the cutting operation. The process may be performed in dry or wet condition. Wet condition is recommended to lower the cutting temperature and to avoid carbon dust. To cool the specimen, water without additives shall be used."

Warning: When appropriate, use a drill stop to limit penetration, to prevent injury to personnel on the far side of the structure, and to prevent damage to structure, components, wiring, or pipelines beyond the panel or the skin being worked. See Figure 11.10.

11.5.2 Hole Sizes and Tolerances

Most fasteners have a fixed diameter that requires careful hole preparation to prevent damage. Hole diameters must be chosen and machined carefully to ensure a proper fit. An interference fit hole in a composite component will result in fiber breakout when the fastener is inserted, unless the interference is carefully limited and preferably a sleeved fastener is used. Delamination also will eventually result from excessive movement of a fastener because of a loose fit in a sloppy hole. High local bearing stresses can occur due to tipping of a bolt or rivet in a loose hole. Reinforcing holes with grommets can alleviate this problem. If this cannot be accomplished, a close tolerance hole is preferred. Hole tolerance is not as critical when installing soft solid rivets because of fastener expansion within the hole. For hard special rivets with a hollow end or soft end, it is more important to produce a close tolerance hole. Always use the hole tolerance called for by the drawing or the SRM. A table of hole sizes for solid rivets and another for hollow ended titanium rivets (BACR15GA) may be found in Boeing B.757 SRM 51-70-16.

Pilot drilling, followed by undersize drilling and then reaming, is sometimes recommended. For more detail on fasteners, see Boeing B.757 SRM 51-40-00 through 51-41-09 and similar data for other types.

11.5.3 Edge Distance and Fastener Pitch

If the component is constructed of only unidirectional tapes placed in parallel, no holes can be allowed. Because only warp fibers exist, a fastener would easily break the resin bond and cause separation, regardless of distance from the edge. Adequate fiber reinforcement in several directions will prevent fastener pull out near the edge. A popular design is several plies of fabric reinforced by unidirectional tape running parallel to the edge. Grommeting holes will also help. See the SRM or drawing, and use the original design. Generally, an edge distance of four times the fastener diameter (4D) is recommended as the minimum edge distance for use with composites. When determining fastener size and quantity, fewer holes and larger fasteners are preferable to a large quantity of small ones. The minimum fastener pitch should be eight times the diameter (8D). The maximum pitch can be as high as thirty times the fastener diameter (30D) in sandwich panels. Check the SRM for the specific requirements for each repair.

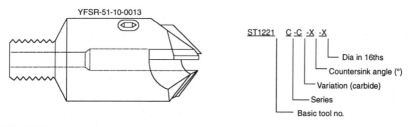

YFSR-51-10-0013

ST1221 C -C -X -X
- Dia in 16ths
- Countersink angle (°)
- Variation (carbide)
- Series
- Basic tool no.

Ⓐ ST1221C-C Cutter-Microstop, countersink 3-flute-carbide
Detail I

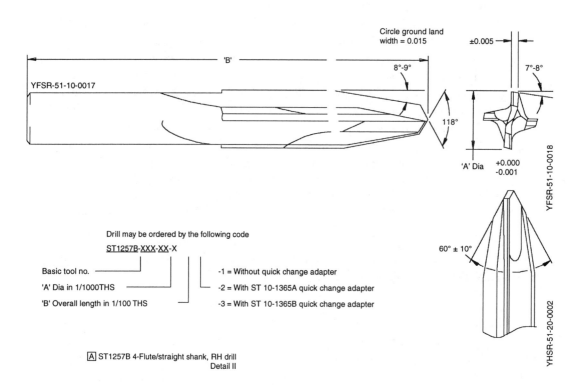

Circle ground land
width = 0.015

±0.005

'B'

8°-9°

7°-8°

YFSR-51-10-0017

118°

'A' Dia +0.000
         -0.001

YFSR-51-10-0018

Drill may be ordered by the following code

ST1257B-XXX-XX-X

Basic tool no.

'A' Dia in 1/1000THS

'B' Overall length in 1/100 THS

-1 = Without quick change adapter

-2 = With ST 10-1365A quick change adapter

-3 = With ST 10-1365B quick change adapter

60° ± 10°

YHSR-51-20-0002

Ⓐ ST1257B 4-Flute/straight shank, RH drill
Detail II

Fig. 11.10 Cutters for hole drilling in graphite and aramid structures. (From Boeing B.757 SRM 51-70-16, Sheets 1 and 2, Figs. 1 and 2. Courtesy of Boeing Commercial Airplane Group.)

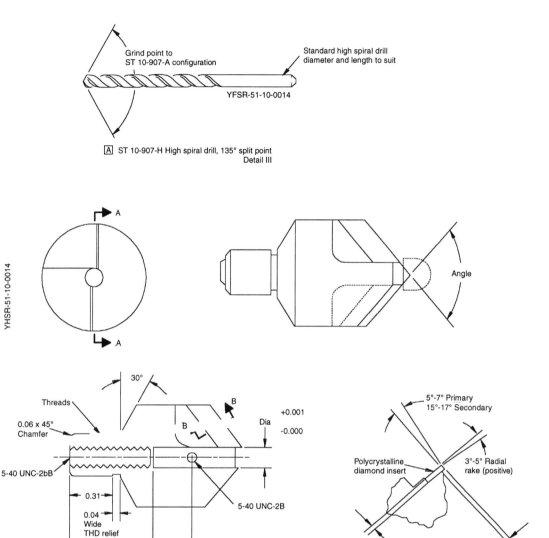

Grind point to
ST 10-907-A configuration

Standard high spiral drill
diameter and length to suit

YFSR-51-10-0014

A ST 10-907-H High spiral drill, 135° split point
Detail III

YHSR-51-10-0014

A

A

Angle

30°

B

Threads

0.06 x 45°
Chamfer

5-40 UNC-2bB

Dia
+0.001
-0.000

B

5°-7° Primary
15°-17° Secondary

Polycrystalline
diamond insert

3°-5° Radial
rake (positive)

0.31

0.04
Wide
THD relief

0.53

0.70

5-40 UNC-2B

Section A-A

0.060
+0.003
-0.000

0.06 Marking

Section B-B

A ST 1223C-D Cutter-2-flute, polycrystalline diamond
Detail IV

Notes
A Boeing standard tool number

Fig. 11.10 (continued)

375

11.5.4 Hole Protection

Touching up the exposed hole with resin, after drilling, followed by reaming after the resin has cured, will reduce moisture ingress and provide a more durable hole. Prime all holes and countersinks, and allow them to dry before installing fasteners. Apply sealant if specified on the drawing, in the SRM, or in modification kit instructions before installing fasteners. To reduce moisture ingress and to avoid crushing in honeycomb sandwich components, the hole must be sealed and reinforced with potting compound and/or grommets. If potting compound is used to reinforce the hole in lieu of grommets, twist an Allen key (wrench) to clean out a slightly larger diameter of honeycomb, then fill all voids and the hole with potting compound. All around the hole, 10 mm (3/8 in.) should be potted using a compound with a good compression strength (i.e., 35 MPa [5000 psi]). Redrill the fastener hole when the potting compound is fully cured. Some special fasteners use sleeves to protect the hole during the installation of interference fit fasteners. Fatigue Technology Inc. (FTI) supplies metal grommets (known as its GromEx system) as fastener hole reinforcements for composites. These grommets are expanded in their holes by FTI cold expansion methods. Advantages claimed are superior hole protection from lightning strikes, improved service life of mechanically fastened joints in composites, and protection of holes in removable access panels.

11.6 Fastener Installation and Removal

11.6.1 Fastener Installation

Temporary Clamping: Many fasteners do not have sheet take-up capability (i.e., the ability to pull tightly together the sheets of material being joined). To avoid gaps between the sheets after assembly, the parts to be fastened should be clamped together before the fasteners are installed. For the temporary retention of parts before assembly, "C" clamps (often known as "G" clamps) will work well if the edges of both parts meet. In other installations, temporary fasteners must be used to clamp the parts prior to assembly. Temporary hole clamps are available in several types, depending on clamping needs. See Figure 11.11. Use these with care to avoid damage to holes. These clamps should be used only in pilot holes. As fasteners are fitted and the parts are held firmly together, the pins should be removed and the holes opened carefully to final size, preferably by reaming.

The plier-operated spring type (M series) Wedgelock is useful for assembling two light parts that have no preloads to overcome. The wing-nut type (WNX) provides additional retention force needed for assembling thick or multiple sheets having slight preloads. The nut type (HNX) provides for quick installation with additional retention strength, but

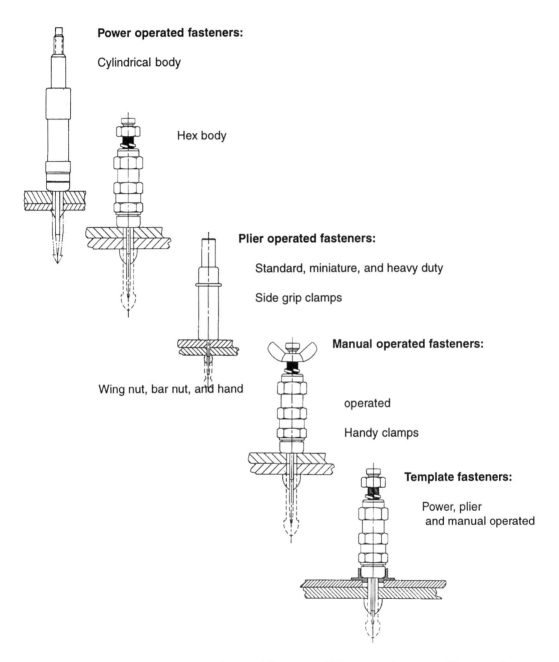

Fig. 11.11 Types of temporary clamps. (Courtesy of Monogram Aerospace Fasteners.)

Power Operated Cylindrical Body Fasteners

Description

The cylindrical configuration of this new temporary blind fastener eliminates tedious fastener-to-tool orientation problems. The result is a reduction of over 50% in installation and removal time as well as providing a "robot ready" unit.

The revolutionary Cylindrical Body Wedgelock design concept is complimented by the introduction of a captive nut, double lead thread, and larger geometry bearing wires. Utilization of a captive nut provides positive disengagement, allowing unobstructed removal of the fasteners even when used in wet sealant applications.

The double lead thread design further enhances overall installation and removal efficiency while the larger bearing area makes the fastener the optimum choice for both composite and metallic structures.

Features/Advantages

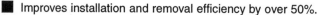

CBX

- Cylindrical design eliminates fastener-to-tool orientation problems.
- Improves installation and removal efficiency by over 50%.
- Installation tooling is compact and lightweight.
- One tool drives all diameters and grip accommodations.
- Reusable fastener life significantly increased.
- Larger blind side bearing area reduces potential material damage.
- Ideal for manual, automatic, and robotic operation.

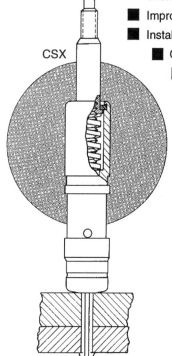

CSX

Now Available in Spring Loaded Design

Designed specifically for use in wet sealant or liquid shim applications.

The CSX utilizes an internal spring to retain a maximum of 100 lbs. of clamping pressure even with minor structure thickness variations due to sealant flow.

- Saves time and labor.
- Eliminates the need for re-torquing after installation.

Low Profile Versions

Monogram's Low Profile cylindrical body fastener allows for "closer proximity" to the skin surface of a structure, so that additional steps, such as drilling, may be easily completed without interference from fastener protrusion. Contact Monogram for additional information.

Fig. 11.11 (continued)

removal can be slow. A newly developed friction drive temporary fastener, the Cylindrical Body Wedgelock (CBX), provides good retention. It is both easily installed and removed. Use washers on the back side to prevent pull through when additional retention force is required.

The spring-loaded version (CSX) provides clamping pressure retention within structural thickness variations resulting from wet sealant or liquid shim flow. If none of these special clamps are available, use common bolts and wing nuts for temporary assembly.

Length Determination: To determine the proper fastener length, measure the combined thickness of the materials being fastened. This length is referred to as the grip length. See Figure 11.12.

If the hole is near an edge, simply measure the thickness with a caliper. However, many holes are not located so conveniently. A special grip gauge is available from the fastener manufacturers and can be inserted in the hole. It catches on the back side and then is read

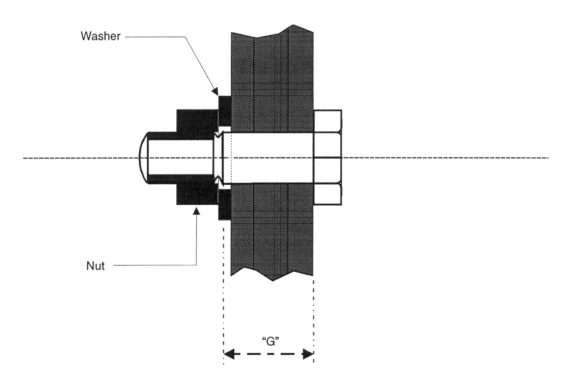

Fig. 11.12 The grip length "G" is the plain parallel portion of the bolt shank. No part of the thread is allowed to be in the hole. Up to three washers may be used to ensure that the nut can be fully torque loaded without bottoming on the shank.

by sighting a mark at the top surface. These gauges are different for blind bolts and blind rivets because of different grip length allowances. Ensure that the grip length gauge used is appropriate to the fastener involved. See Figure 11.13.

Solid Rivet Installation: All solid rivets, including hollow end, should be installed in composite parts using mechanical or pneumatic squeezers. Reciprocating guns (percussion riveters) can severely damage the matrix bond and should never be used on composite materials. The squeezer has interchangeable dies for different rivet types and sizes. Small 4.5-mm (3/16-in.) bolts, with the head markings ground off, will suffice when installing flush rivets. Choose a rivet length that extends through the total thickness of the parts to be joined plus 1.5 times the rivet shank diameter. Squeeze until the upset head is half as high as the original shank diameter and the diameter of the upset head is spread to 1.5 times the original shank diameter. Adjustments to the pneumatic squeezers can be made by simply adding or removing washers from beneath the interchangeable heads. To prevent hole expansion damage and to transfer load more effectively, use a washer under the upset head. To retain the washer during rivet installation, use a small neoprene sheet that is the same thickness as the rivet length visible from the back side. Use a leather punch to make a hole in the neoprene that is slightly smaller than the rivet diameter. Install as shown in Figure 11.14.

Blind Rivet Installation: Blind rivets are also installed using mechanical or pneumatic tools. All blind rivets can be further categorized as single- or dual-action fasteners. It is important to make this distinction because the tools are incompatible. The original locking stem type fasteners require a special tool that first pulls the stem, then seats a locking ring by pushing it in place. All inexpensive friction lock and new-generation locking stem fasteners are single action. Single-action fasteners may be distinguished in the following ways:

- A washer atop the manufactured head identifies CherryMax
- A shorter, fully serrated stem identifies Cherrylock "A" and CherryMax "A"
- A green identification at the end of the stem identifies a single-action blind bolt

Small diameters can be pulled using inexpensive, hand-held pop-rivet pullers. For large diameters and production installations, pneumatic tools are recommended. If the proper length is chosen and the tool is held perpendicular to the surface, no gaps should be present under the head. The stem and locking collar protrusion should be within the limits specified by the manufacturer. See Figure 11.15.

Caution: Pulling tools are not interchangeable. Damage can result if the wrong type of fastener is inserted in the pulling head.

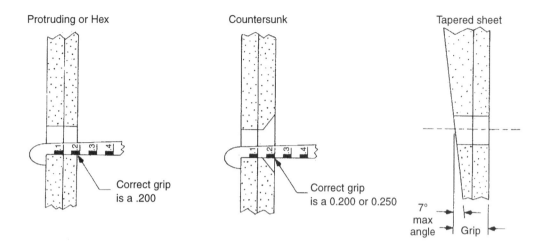

Measure the material thickness with a BFS-1A grip gage as shown. Insert gage into hole, draw gage back until hook contacts the blind sheet. Read where front skin surface coincides with numbers and lines on gage. If reading is directly on line, you may use either that grip or the next longer grip.

In those applications where a tapered sheet condition exsists on the blind side, the grip length must be determined by the depth at the centerline of the hole as illustrated above.

Fig. 11.13 Typical grip length gauge. (Courtesy of Monogram Aerospace Fasteners)

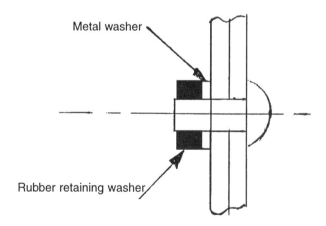

Fig. 11.14 Neoprene washer to retain metal washer during rivet installation.

Standard Bolt Installation: Standard bolts are inserted from one side and then are retained using a nut or collar. A washer should always be placed under the nut to prevent damage from rotation. Standard practice is to install bolts with the head up or forward when practical. Self-locking nuts are recommended if parts will be subject to vibration. Choose a bolt with a shank (grip length) long enough to ensure that no threads carry any bearing loads (i.e., the smooth shank of the bolt extends through the hole to ensure that no thread is against a bearing surface). Washer thickness can be adjusted to ensure a parallel

Installation sequence
with shifting washer

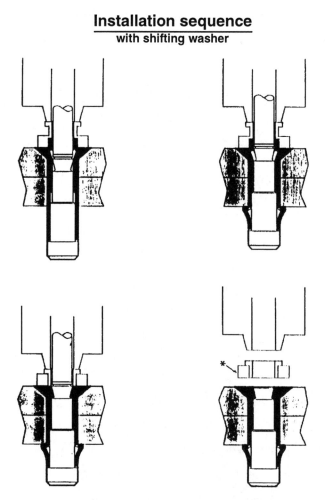

*During installation sequence, the Cherry shifting washer
collapses into itself, leaving a solid washer that is easily retrieved.

*Fig. 11.15 Cherry MaxiBolt showing identification washer.
(Courtesy of Textron Aerospace Fasteners Inc.)*

shank in the hole and also to prevent the nut from bottoming on the thread before the full clamping torque is reached. If the proper bolt length is chosen and the thread length is correct for the length of nut being used, one to three threads should be visible after tightening the nut. The bolt thread length may need to be longer if a self-locking nut is used. Considerable care is required to ensure that all these conditions are met. However, the conditions must be met. Torque values for bolting composites together must be checked and observed even more carefully than for metal parts. Check the SRM for the precise values required. The Boeing B.757 SRM 51-40-04 gives torque values for a variety of bolts in both metal and composite parts, and the whole of 51-40-00 on all types of fasteners is excellent. An important note in Boeing B.757 SRM 51-40-04 states, "The torque values quoted may be used only where they do not conflict with specific information contained within the Maintenance Manual, Component Maintenance Manual, Service Bulletins, or drawings of the Boeing Company." Generally, all bolts in a particular group should initially be torque loaded to a value somewhat below the maximum, and then they all should be tightened again to bring them to full torque. They should also be tightened in a sequence similar to a car cylinder head, starting from the center and then tightening bolts on opposite sides alternately and working outward. Where seals are involved, check the torque again after a few days and increase it to the correct value. Hi-Lok collars can be used for torque control, but the parts must be securely held together first. The collar has a frangible nut that breaks off at a predetermined torque, leaving a locking collar.

Lockbolt Installation: Two types of lockbolts are supplied:

- Pull type
- Stump type

The type selected depends on the type of installation tooling available. This tooling is expensive; therefore, not every shop will stock both types. The pull type can be installed with common pulling-type tools; the stump type can be installed with portable squeezers or automatic drill/rivet machines such as the DRIVMATIC machine in production situations.

Blind Bolt Installation: Blind bolds are installed with a heavy-duty pneumatic tool similar to that used with blind rivets. The original blind bolts used a special dual-action pneumatic puller that pulled the stem and then seated the lock ring by pushing it in place. New-generation fasteners are single action and are identified by a washer or frangible collar atop the manufactured head. Choose a proper diameter for the hole size, and determine the grip length with the correct grip gauge. If the proper length is chosen and the tool is held perpendicular to the surface, no gaps should be present under the head. The stem and locking collar protrusion should be within the limits specified by the manufacturer.

Caution: Pulling tools are not interchangeable. Damage can result if the wrong type of fastener is inserted in the pulling head.

Composi-Loks: Composi-Loks are supplied by Monogram Aerospace Fasteners Inc. They are blind bolts designed specifically for composite material applications and have a locking feature to resist loosening under vibration conditions. See Figure 11.16. Composi-Loks can be fitted using NAS 1675 type installation tooling. The use of Composi-Lok II Series fasteners will enhance the speed and efficiency with which the fastener is installed and will reduce total tooling requirements through utilization of the drive nut concept. The fasteners are prelubricated for proper performance as supplied and should not be subjected to cleaning or additional lubrication. Many special-purpose installation tools are available, including close-quarter power tools, 20° offset power tools, and close-quarter hand tools. The proper grip length fastener should be determined as specified in Section 11.6.1 and inserted in the prepared hole. After engagement of the fastener with the selected tool, the motor trigger is depressed and the fastener clamps the structure tightly together. Installation is complete when the corebolt component breaks torsionally in a predetermined break groove. The remaining corebolt protrusion is measured as a means of checking for proper installation and should fall within the limits specified by the manufacturer. Monogram supplies its RK 5000 removal kit designed specifically for removal of Visu-Lok and Composi-Lok fasteners. The company also supplies a corebolt break-off shaving tool known as the Little Shaver, designed for use with Visu-Lok and Composi-Lok fasteners.

Radial-Lok Installation: In general, Monogram's Radial-Lok fastener is installed in the same way as is the Composi-Lok. Because of the increased corebolt length and travel, however, the basic installation tool must be fitted with a longer housing than that found on the standard Composi-Lok installation tool. A longer master torque driver is required also. These components are available from the fastener manufacturer under part numbers MH550RL and MLTD550RL, respectively. After fitting these components, the tool will install Radial-Loks and Composi-Loks without modification. Installation of 8 to 10 mm (11/32 and 3/8 in.) diameter Radial-Loks may require a special high-torque-output motor, depending on fastener grip length and shop conditions. Inspection of the installed fastener can be performed by measurement of the corebolt protrusion. Dimensions should fall within the limits set by the manufacturer.

11.6.2 Fastener Removal

Solid Rivet Removal: Traditional use of a hammer and punch to produce a centering mark on solid rivets is discouraged. Use a spring-loaded, snap-type punch instead. Although this device aids in centering and starting the drilling process, it is possible that the drill will wander off center. Secondary damage to composite structure often results from improper handling of hand-held drills. Practice drilling rivet heads from a scrap part first, before attempting to do it on an expensive component. A skilled repairman will intermittently drill the rivet, making constant centering adjustments until the head pops off with a slight

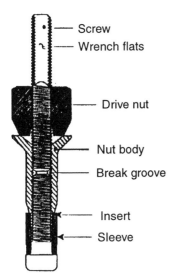

— Screw
— Wrench flats

—— Drive nut

—— Nut body
—— Break groove

—— Insert
—— Sleeve

The drive nut concept

Composi-Lok II differs visually and functionally from the original Composi-Lok in the following ways:

* An additional conponent - the DRIVE NUT.
* Only two nose pieces are required for installation of 5/32" through 3/8" diameter fasteners, regardless of head style. One nose piece drives 5/32" to 1/4" sizes while the second nose piece is utilized for 9/32" to 3/8" diameters. A separate wrench adapter is required for each diameter.
* Cam-outs are completely eliminated on flush and protruding head fasteners.
* Nose pieces never wear out.
* The DRIVE NUT stays on the discarded pintail after installation, minimizing the tendency of small pintails to find their way into F.O.D. critical areas. The steel DRIVE NUT permits easier retrieval (including magnetic pick-up).

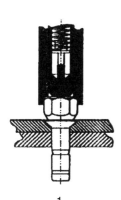
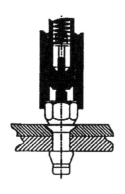
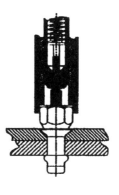
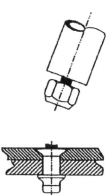

1	2	3	4
The Composi-Lok fastener is inserted into the prepared hole. The installation tool is placed over the screw simultaneously engaging the wrench flats and the DRIVE NUT.	Torque is applied to the screw while the DRIVE NUT is held stationary. If the DRIVE NUT is not already threaded down tight, then both the screw and the nut body rotate until the DRIVE NUT is "jammed" against the nut body. This restrains further restoration of the nut body. The screw continues to advance through the nut body, causing the sleeve to be drawn up over the tapered nose of the nut. Initial blind head formation is started.	Continued tightening removes sheet gap, completes the large blind head, and clamps the sheet tightly together.	When the sleeve forms tight against the blind side of the structure, the screw fractures at the break groove. The tool is pulled away and the pintail DRIVE NUT assembly is discarded.

Fig. 11.16 Composi-Lok II fastener installation. (Courtesy of Monogram Aerospace Fasteners.)

385

tilt of the drill. Extreme care must also be used when removing the remaining rivet shank. Forceful hammering will result in delamination and fiber breakout. If it is necessary to use a hammer and punch to remove the remaining shank, back up the component with a wooden, plastic, or metal block drilled with a hole in the center, which is large enough to allow the bucktail to fit in with the support. Tap lightly until the fastener is ejected. An alternative is precisely drilling the head dead center into the fastener until it pops through.

Blind Fastener Removal: Blind fasteners that have stem-retaining lock rings must be predrilled with a slightly smaller bit until the lock ring releases. Tapping out the stem should be done carefully, if at all, because there is usually no access to the back side to enable support to be given. Process manuals are available from manufacturers to explain precise details of removal. Careful drilling is preferred. Intermittently drill with slight pressure (to prevent spinning of the fastener in the hole), using the same diameter as the rivet shank. Be careful to remain dead center. The head should pop off with a slight tilt of the drill motor when the proper depth is attained. The remaining shank should offer little or no resistance to tap out. Some fastener manufacturers offer specialized equipment to aid the removal process.

Bolt Removal: When removing standard bolts, notice if the washer is under the head or under the nut, and then turn from that end. Be careful not to damage composite material with the socket or the wrench. Hi-Lok collars can be unscrewed with a small pair of channel-lock or vise-grip pliers. Swaged collars should be cut off or milled off. A special collar cutter is available, which splits the collar in two. Hollow end milling bits that have an inside diameter equal to the shank of the bolt can also be used in a drill motor. This method shaves off the collar. If obstructions prevent collar removal as recommended, drill off the head of the bolt and then tap it out the other way. Do not use a pneumatic or hand chisel. If special tools for fastener removal are supplied by a manufacturer, they should always be used.

11.7 Fastener Substitution

Fastener substitution is an important subject. Fasteners are made primarily from steels, stainless steels (CRES), or titanium alloys having a wide range of tensile strength. Titanium and CRES are used with composites. Tension bolts have thick heads; shear bolts have thin heads. The 130° countersunk head style has been developed specifically for thin composite sheets. Check the angle of each countersink before fitting a replacement fastener, unless it is of the original type. When replacing existing fasteners, always use the original type if possible. If that is impossible because of nonavailability or restricted access, then a fully equivalent fastener must be used. This is another fastener having all the properties of the original. An alternative fastener must be of the same or superior strength.

Other considerations are fluid tightness, temperature resistance, corrosion resistance, and magnetic properties. High-strength fasteners may create excessive stress in composites when squeezed. Fasteners designed for tension applications may be considered as an alternative to shear fasteners, but never replace a tensile fastener with a shear fastener. Replacement tensile fasteners should retain the same locking features as the original (i.e., cotter pin, self-locking nut). If a different diameter must be used, consider the quantity required to provide the same strength. Substituting with large-diameter fasteners may result in insufficient edge distance. Alternative fasteners must be approved by the company designated engineering representative (DER), the aircraft manufacturer, or other design authority. OEM-approved fastener substitution is covered in SRM 51-40-03.

11.8 Bonded Fasteners

Surface-mounted bonded fasteners, which have come into use fairly recently (Click Bond Inc. and similar), have found application in many lightly loaded situations.

Applications: Surface-mounted bonded fasteners supplied by Click Bond Inc. (Figure 11.17) are used to support fairly light loads such as electrical cable mountings, lightweight trim panels, light fittings, and many other low-load applications. Studs, standoffs, and cable tie mounts are supplied to MIL-I-45208; anchor nuts, nutplates, and dome nuts are supplied to MIL-N-25027. Further details can be obtained from these specifications or Click Bond Inc.

Installation: All bonded fasteners require similar surface preparations to those used for other bonded parts, if strength and long-term durability are required. An acrylic adhesive is often used to give a tough, durable bond for the Click Bond type. The fastener must be carefully positioned, and slight pressure must be applied. To retain the fastener while this adhesive sets, the manufacturer uses a clever retainer with a less aggressive adhesive. Other fasteners use pull-through rubber retainers.

Removal: Removal of bonded fasteners is likely to be a problem if the fasteners were well bonded. If the part is damaged, they can be cut away with the affected area and replaced after repair. The acrylic adhesives used for bonding them are tough down to low temperatures and are fairly strong up to the curing temperature of 120°C (250°F) commonly used for aircraft interior parts. This makes removal with dry ice or heat somewhat difficult. Soaking the bond area in warm water for several hours, followed by the careful use of a shallow metallic wedge, is suggested as a possible method. Unless an anodize was used on the part with a good primer, this should work.

Fig. 11.17 Examples of Click Bond bonded fasteners. (Courtesy of Click Bond Inc.)

11.9 Composite Fasteners

In this paragraph, the term "composite fasteners" actually means fasteners that are made from composite materials. They are a relatively new development and offer weight savings when their strength is sufficient for the loads involved. Examples from the Cherry Division of Textron are the ACP all-composite pins and nuts. See Figure 11.18. They are vibration resistant, and they eliminate potential corrosion problems and problems relating to coefficient of thermal expansion (CTE). Cherry ACP all-composite pins and nuts offer a significant weight reduction and ample strength for structural and nonstructural applications. They are available in flush, hexagon, and protruding head styles with head or point drive. Cherry ACP collars are internally threaded and are available in a variety of lightweight, high-strength, advanced composite materials. These self-locking collars are designed to be paired with titanium or A-286 CRES threaded shear pins and installed in metallic and composite structures. The unique torque-off element provides consistent joint preload and eliminates the need for costly torque wrench sets.

Introduction

Cherry ACP Pins and Nuts are an innovative, all composite fastening system providing significant weight savings and ample strength for structural applications.

Cherry ACP Pins are externally threaded and available in head or point drives, and flush, hex, and protruding head styles. Cherry ACP Nuts are internally threaded with 12 point nut configuration. An optional self-locking feature is available.

Cherry ACP Pins and Nuts are available in a variety of high-strength composite materials, which can provide resistance to extreme mechanical, electromagnetic, chemical, and environmental conditions.

Cherry ACP Threaded Shear Pins and Nuts are half the weight of aluminum.

Features

Reduced weight
Example: PEEK/LC is 1/3 the weight of 6Al-4V Titanium

High strength
PEEK/LC Shear strength: 53 KSI

Low radar signature

Galvanic compatibility

Sonic vibration resistance

Wide service temperature range
-453°F to 550°F, depending on material selected

Space compatibility

Cherry ACP Pins and Nuts are available in seven materials

Material	Code Letter
PEEK/LC	A
PEEK/C	B
ULTEM/GL	D
PEEK/LQ	J
ULTEM/LGL	K
PEEK/LGL	L
PMR 15/LC	P

Fig. 11.18 Cherry ACP pin and nut. (Courtesy of Textron Aerospace Fasteners Inc.)

Cherry ACP fasteners may be made from the following materials:

- Polyetherether ketone (PEEK)
- Polyetherimide (PEI)
- PMR15 polyimide

Five reinforcements may be used:

1. **Long Carbon:** Long carbon (LC) fibers are continuous strands of carbon that extend the length of the fastener from head to point. Fasteners reinforced with LC fibers are not compatible with aluminum structures because of galvanic corrosion caused by the electrolytic reaction of the carbon and aluminum when exposed to moisture.

2. **Long Glass:** Long glass (LGL) fibers are continuous strands of glass that extend the length of the fastener from head to point. LGL fasteners are compatible with aluminum structure and can be used in applications where LC fasteners cannot be used because of galvanic corrosion concerns.

3. **Long Quartz:** Long quartz (LQ) fibers are continuous strands of quartz that extend the length of the fastener from head to point.

4. **Short Carbon:** Short carbon (C) fibers are chopped strands of carbon that are randomly placed throughout the fastener during the injection molding process. Fasteners reinforced with C fibers are not galvanically compatible with aluminum structure.

5. **Short Glass:** Short glass (G) fibers are chopped strands of glass that are randomly placed throughout the fastener during the injection molding process. Fasteners reinforced with G fibers are galvanically compatible with aluminum structure.

Threaded shear pins should be torque loaded to their own specified values. See the SRM.

11.10 References

11.1   Hart-Smith, L.J., "An Engineer's Viewpoint on Design and Analysis of Aircraft Structural Joints," Proceedings of the Institute of Mechanical Engineers, Part G, *Journal of Aerospace Engineering*, Vol. 209, No. G2, Institute of Mechanical Engineers, London, UK, ISSN 0954-4100, 1995.

11.2   Hoffer, K., *Permanent Fasteners for Lightweight Structures*, Aluminum Verlag, Dusseldorf, ISBN 3-87017-177-4, 1984.

11.3   Cutler, John, *Understanding Aircraft Structures*, 2nd edition, Blackwell Scientific Publications, Oxford, UK, ISBN 0-632-03241-3, 1992.

11.4   Toor, Pir. M. [ed.], ASTM STP 1236 Structural Integrity of Fasteners, American Society for Testing and Materials, Philadelphia, 1995.

11.5   Phillips, Joseph L., *Fastening Composite Structures with Huck Fasteners,* Huck Manufacturing Company, Irvine, CA, 1986.

Chapter 12

Documentation

Aircraft work requires that a record be maintained of all maintenance performed. In the event of an incident, it should then be possible to trace back to the batch of material used and to the person performing the work. To achieve this, accurate and well-organized recordkeeping is essential. The following documents are used for composite repairs and metal bonding to provide a record that the work has been done correctly.

12.1 Process Control Documents

Repair processes with composites are important because the repair material is being manufactured at the same time as the component is being repaired or a patch manufactured. If a person working with sheet metal had to mix all the alloying elements, melt and roll the batch into sheet, and conduct the specified heat treatment, then that person could be said to have manufactured the material. In the case of composite and adhesive bonding, materials often must be mixed, applied, and then cured at the appropriate temperature and pressure to obtain the desired properties. Consequently, the quality of the end product is process dependent, and careful attention to detail is necessary. To produce the intended properties within the specified limits, the mixing (if required) and curing processes must be done correctly. Flow charts are often produced to assist in following the correct sequence of operations, both for the whole job and for critical subprocesses such as anodizing of aluminum alloys and surface preparation of composites including the water break test. (See Section 10.1.6 in Chapter 10 for a discussion of the water break test.) Technicians may be required to sign off each step to ensure that this is done, and critical steps should be monitored and cosigned by quality control personnel to ensure that no shortcuts are taken. This is necessary because no post-repair inspection exists today that can guarantee the integrity of a repair. Quality is dependent on the knowledge, training, and integrity of repair personnel. To show that procedures have been followed correctly, records are maintained of cure temperature and vacuum or autoclave pressure versus time. Current regulations require that these records be retained for at least five years.

12.1.1 Cure Chart/Data Strip

When a temperature controller is used to program the cure cycle for an adhesive, several thermocouples may be placed around the repair area to ensure that the entire curing area is heated, cured, and cooled according to the adhesive/resin manufacturer's requirements. These thermocouples, including the one used as the control, are normally connected to a chart recorder. Chart recorders usually have six or more channels, and the temperature is printed for each one on a paper roll at short time intervals to produce a graph of temperature versus time for all of them. This provides a permanent record, which can be torn from the roll and filed with the work cards for the job. Circular charts often are used to monitor room parameters such as temperature and humidity, and they must be replaced daily. Computers are rapidly replacing these and other paper records.

12.1.2 Routine and Nonroutine Work Documents

12.1.2.1 Routine Work Documents

Routine work documents are used for planned work. The inspection and/or repair process is predetermined, and instructions often are formally printed along with sign-offs for technicians and quality control personnel. The details given below relate to aircraft work, but other forms of transport and components may have their own requirements for work recording. These are an aeronautical example by courtesy of Aerobond (United Kingdom). Aircraft work to Federal Aviation Regulation FAR 145 (United States) and Joint Airworthiness Requirement/Joint Aviation Requirement JAR 145 (Europe) requires the following documentation to ensure the traceability of materials and personnel.

- **Component Master Worksheet, AB 101:** This sheet records the item of work, a description of the defect, and the repair actions taken, with a column for the mechanic's signature and authorization stamp. It also records the aircraft type, registration, part number, and serial number of the item under repair. See Figure 12.1.

- **Component Additional Worksheet, AB 102:** If several work items are required and they cannot all be recorded in the space on one form, these additional sheets must be used to provide a complete record of the work done. See Figure 12.2.

- **Materials Record Sheet, AB 103:** This sheet must be included in the total work pack for each job. It provides details of all materials used, including part numbers, serial numbers, and batch numbers. The material batch numbers allow full traceability of parts or materials used. This sheet must carry the job number for the repair and must be filed in the work pack. See Figure 12.3.

JAA Form 1S/No. _____
Release date _____
Job no. AB _____

UK

MASTER WORKSHEET

Date _____ Contracting company name _____ Sheet __ of __

A/C type _____ A/C reg _____ Component desc _____

Part/no. _____ Serial/no. _____ Component defect _____

Manual reference/revision date _____

Item no.	Desc of work/defect	Repair/rectify	Sig	Stamp

Form Number AB 101

Fig. 12.1 Component Master Worksheet. (Courtesy of Aerobond [United Kingdom].)

Job no. AB _____

ADDITIONAL WORKSHEET

Component desc _____ Part/no. _____ Serial/no. _____ Sheet __ of __

Item no.	Desc of work/defect	Repair/rectify	Sig	Stamp

Form Number AB 102

Fig. 12.2 Component Additional Worksheet. (Courtesy of Aerobond [United Kingdom].)

AEROBOND
UK

MATERIALS RECORD SHEET

Job no. AB _____

Component desc _____ Part/no. _____ Serial/no. _____ Sheet ___ of ___

Description	Part no.	Serial no.	Qty used	Batch no.

Form Number AB 103

Fig. 12.3 Materials Record Sheet. (Courtesy of Aerobond [United Kingdom].)

- **Component Record Card, AB 104:** This card records the part number, serial number, and other details of each component. This will enable future work on the same part to be related to previous repairs. Modification state also should be recorded. See Figure 12.4.

- **Inspection Report, AB 105:** This form provides for a report after a component has been inspected. The part may be new, secondhand, damaged, or repaired. The report may or may not recommend that work should be performed, depending on the condition of the part. For lifed items, the inspection includes a check that an acceptable amount of serviceable life remains for the intended use. See Figure 12.5.

- **Warranty/Investigation Report, AB 106:** This document provides for a formal investigation and report if an item is returned under warranty with a claim that the repairer has not performed satisfactory work. See Figure 12.6.

- **JAA Form 1 or Similar FAA Form 1:** This form is used to release a repaired part for service and certifies that all work carried out has been performed in accordance with the airworthiness regulations of the stated country and that, in respect to that work, the parts are considered ready for release to service. See Figure 12.7.

- **Quality Control Form 1, AB 108:** This form is used for periodic internal audit of the company to ensure that it meets the requirements of JAR 145 or FAR 145. See Figure 12.8.

- **Technical Instruction, AB 110:** This form is used to provide specific instructions for the performance of technical processes to ensure that adequate information is available to those performing the work. See Figure 12.9.

- **Three-Part Serviceable Label, AB 111:** This multiple label, consisting of three pages that are identical except for color, is shown in Figure 12.10.

  The first label (white) is attached to the incoming paperwork and is filed with it.

  The second label (blue) is filed under shelf-life records, if appropriate. If not required, this copy is discarded.

  The third copy (green) is attached to the part/material to ensure identification at all times.

- **Reject Note, AB 112:** This form is used to reject incoming parts or materials if they are unacceptable for any reason. Possible reasons include damage, the item being time expired, or the item being incorrect to order (i.e., the wrong part or material or the incorrect modification state). See Figure 12.11.

- **Stock Record Card, AB 113:** All incoming materials must have a stock record card. As material is used, this card must be updated to enable reorder at the appropriate time. The entry of the relevant job numbers against the material batch, as the required

COMPONENT RECORD CARD

Part no.
Serial no.

Customer		Order no.	Job no.
Component Desc		Defect	Qty
Work required		A/C type	A/C reg
Mechanic	Manual ref		Revision date
Work done		Release	JAA Form 1 no.
Date received	Start date	Target date	Release date

Materials record

Date req	Part no.	Desc	Qty req	Location	Date issued	Qty issued	Batch no.

Hours worked

Date	Name	Hours	Date	Name	Hours
				Total	

Form Number AB 104

Fig. 12.4 Component Record Card. (Courtesy of Aerobond [United Kingdom].)

Mead Cottage
Bran End, Stebbing,
Essex CM6 3RW England

INSPECTION REPORT

Telephone: 0279 814909
Fax: 0279 814903
24-Hour AOG Phone: 0850 351559

Component	Report ref
Part no.	Customer
Serial no.	Current J/no.
HRS/LDG/CYC since O/H	Current O/no.
Previous O/H repair by	W/sheet file ref
Date prev. O/H/repair	Date of report
A/C type	A/C regn.

General remarks

Detail report

Conclusions

Signed .... Authority .... Date ....

Quality audit engineer	Accountable manager

Approval authorities: CAA 00156

Form Number AB 105

Fig. 12.5 Inspection Report. (Courtesy of Aerobond [United Kingdom].)

Mead Cottage
Bran End, Stebbing,
Essex CM6 3RW England

WARRANTY/INVESTIGATION REPORT

Telephone: 0279 814909
Fax: 0279 814903
24-Hour AOG Phone: 0850 351559

Component	Report ref
Part no.	Customer
Serial no.	Current J/no.
HRS/LDG/CYC since O/H	Current O/no.
Previous O/H repair by	W/sheet file ref
Date prev. O/H/repair	Date of report
A/C type	A/C regn.

1. Reason for return

2. Condition on receipt

3. Tests prior to strip

4. Findings

5. Previous history (if known)

    Date .... Reported defect .... Findings ....

    Job no. ....

6. Conclusions   *Reported defect confirmed/not confirmed

    (a)   Reported/Present defect caused by or resulting from

    (b)   Reported/Present defect should/could not have been apparent at last Overhaul/Repair/Test

    (c)   Parts affected were/were not replaced/adjusted/removed for access/at previous Overhaul/Repair/Test

    (d)   Could any part of unit have been dismantled for system fault finding in the field (by customer) Yes/No/Not known

7. Action taken   *Major   *Minor   *Overhaul   *B.E.B. (if B.E.R. give reasons)

Signed .... Authority .... Date ....
          Quality Engineer

8. Liability	*Accepted/not accepted	
	*Labor/material charges accepted/not accepted	Accountable manager
	*Copy to customer/customer liaison yes/no	Date
Distribution	Accountable manager	Additional copies to
	Quality audit engineer	

Form Number AB 106

Fig. 12.6 Warranty/Investigation Report. (Courtesy of Aerobond [United Kingdom].)

Civil Aviation Authority
Authorized Release Certificate - Airworthiness Approval Tag

1. Country United Kingdom	2.		3. Certificate ref no.

4. Organization	Postal address Mead Cottage Bran End, Stebbing, Essex CM6 3RW	Repair facility address Unit 26, Golds Nursery Business Park, Elsenham, Nr. Bishops Stortford Hertfordshire CM 22 6JX	Telephone: 0279 814909 Fax: 0279 814903 24-Hour AOG Phone: 0850 351559	5. Work order/contract

6. Item	7. Description	8. Part no.	9. Eligibility (*)	10. Qty	11. Serial/batch no.	12. Status/work

13. Remarks

14. New parts
Certifies that the part(s) identified above except as otherwise specified in Block 13 was (were) manufactured/inspected in accordance with the airworthiness regulations of the stated country and/or in the case of parts to be exported with the approved design data and with the notified special requirements of the importing country.

15. Used parts
Certifies that the work specified above except as otherwise specified in Block 13 was carried out in accordance with the airworthiness of the stated country and in respect to that work, the part(s) is (are) considered ready for release to service (see below).

16. Signed	18. Date	19. Issued by on behalf of the Civil Aviation Authority under authorization reference: CAA 00156
17. Name		

Authorized Release Certificate-Airworthiness Approval Tag
User/Installer Responsibilities

20. (*) Cross check eligibility for more detail with parts catalog.

Notes: 1. It is important to understand that the existence of the document alone does not automatically constitute authority to install the part/component/assembly.
2. Where the user/installer works in accordance with the national regulations of an Airworthiness Authority different from the Airworthiness Authority specified in Block 2, it is essential that the user/installer ensures that his/her Airworthiness Authority accepts parts/components/assemblies from the Airworthiness Authority specified in Block 2.
3. Statements 14 and 15 do not constitute installation certification. In all cases the aircraft maintenence record must contain an installation certification issued in accordance with the national regulation by the user/installer before the aircraft may be flown.

Fig. 12.7 FAA/JAA Form 1, Authorized Release Certificate—Airworthiness Approval Tag. (Courtesy of Aerobond [United Kingdom].)

UK

QUALITY AUDIT FORM (QC FORM 1)

Premises

		Satisfactorily	Unsatisfactorily	NIA
1	Housekeeping	( )	( )	( )
2	Emergency exits	( )	( )	( )
3	Identification on no smoking signs	( )	( )	( )
4	Fire extinguishers	( )	( )	( )
5	First aid kits	( )	( )	( )
6	Surrounding area	( )	( )	( )

Technical manuals

7	A/C manuals	( )	( )	( )
8	Amendments	( )	( )	( )
9	Location	( )	( )	( )
10	Company exposition	( )	( )	( )
11	Engineering notices	( )	( )	( )

Technical records

12	Completed components workpacks	( )	( )	( )
13	Company copies of JAA Form 1	( )	( )	( )
14	Test equipment/calibration certs	( )	( )	( )
15	Closed component record cards	( )	( )	( )

Procedures

16	Goods inward	( )	( )	( )
17	Stores/storage	( )	( )	( )
18	Components workpacks	( )	( )	( )
19	Inspection	( )	( )	( )
20	Dispatch	( )	( )	( )

Tooling

21	Calibration standards	( )	( )	( )
22	General condition	( )	( )	( )
23	Storage when not in use	( )	( )	( )

Form Number AB 108 page 1

Fig. 12.8 Quality Audit Form (Quality Control Form 1).
(Courtesy of Aerobond [United Kingdom].)

AEROBOND
UK
QUALITY AUDIT FORM (QC FORM 1)

Comments	
Item no.	Comment

Action	
Item no.	Action

Outstanding items from previous audits

Completion

Name _____ Signature _____ Date _____

Form Number AB 108 page 2

Fig. 12.8 (continued)

AEROBOND UK AEROBOND UK **TECHNICAL INSTRUCTION**

ATA	No.	Issue	Date

Effectivity	Compiled by:	Approved by

Form Number AB 110

Fig. 12.9 Technical Instruction Form. (Courtesy of Aerobond [United Kingdom].)

SERVICEABLE LABEL

Shelf life expires
Description
Part no.   Serial no.
Condition/work carried out
JAA Form I/batch no.
Qty   Date   Stock location
Job no.   Ins stamp
Form Number AB 111

Fig. 12.10 Three-Part Serviceable Label. (Courtesy of Aerobond [United Kingdom].)

**AEROBOND UK
REJECT NOTE**

Reason for rejection _____

_____ Date _____

Order no. _____ Part no. _____ Serial no. _____

Description _____ Qty _____

Supplier _____

Address _____

Incoming certificate no. _____ Stamp _____

Action taken on rejection _____

Action taken by QA _____

Remarks _____

Form Number AB 112   (Top Copy Office, 2nd Copy Quality Assurance)

Fig. 12.11 Reject Note. (Courtesy of Aerobond [United Kingdom].)

Stock Record Card					
Form Number AB 113					
Date	Part number	Description	Batch number	Qty	Location

Fig. 12.12 Stock Record Card. (Courtesy of Aerobond [United Kingdom].)

amount is allocated to each job, allows complete traceability of the actual usage of each batch of material, even if the batch is used for many individual jobs. If a suspect batch of material is reported later, each part can be identified by job number and, if necessary, those parts can be recalled for rework. See Figure 12.12.

- **Unserviceable Label, AB 114:** This form/label is used to identify a part or component as unserviceable and to ensure that the part or component cannot be used until all necessary repair work has been done. In some cases, the item must be scrapped. See Figure 12.13.

12.1.2.2 Nonroutine Work Documents

Nonroutine work documents provide a forum to document problems or damages found during inspection. They should also carry job numbers for auditing purposes. These documents may be in-house documents for company use, and in more serious cases, one of the following. Some examples of serious-defect reporting are as follow:

- **Incident Reports:** These reports are raised by airlines if a problem has occurred and has been reported in flight. A report is made, and a reply is requested from engineering stating what action is to be taken. This corrective action may be a request to the OEM for a modification or a modification or repair scheme raised by the airline.

Description of part _____

Part no._____ Serial no._____Qty _____

Condition _____

_____

Job no._____ Inspector_____ Stamp _____

Form Number AB 114

Fig. 12.13 Unserviceable Label. (Courtesy of Aerobond [United Kingdom].)

- **Maintenance Occurrence Reports:** These Maintenance Occurrence Reports (MORs) are needed when a defect that was not previously known is found on the ground and is considered to be a hazard that should be reported. These are reported to the OEM of the aircraft and to the Federal Aviation Administration (FAA), Civil Aviation Authority (CAA), or other regulatory body.

12.1.3 Nondestructive and Destructive Inspection Data

Nondestructive (NDI) and destructive inspection results must be recorded and retained. Photographs, charts, and written observations are all admissible. Quality assurance personnel retain photographic records.

12.1.4 Coupon Test Results

Destructive test coupons such as lap shear, wedge tests, or climbing drum or "T"-peel tests may be called for by manufacturers, design authorities, or regulatory agencies. These must be made strictly as specified and tested to International Organization for Standardization (ISO), American Society for Testing and Materials (ASTM), or national standards specified by the relevant authority. Results obtained must meet the stated requirements and must be retained for the life of the component or as required by the local regulatory authority.

12.1.5 Clean-Room Temperature and Humidity

The temperature and humidity of the clean room is critical to the integrity of any repair and must be monitored 24 hours a day. A slow-moving circular chart recorder has been the traditional choice for this task. Computer control and monitoring is preferred. These measuring devices should be examined before lay-up begins to ensure that conditions are within acceptable limits. The charts should be monitored hourly during the lay-up process and should be retained as permanent records.

12.1.6 Return to Service/Log Book Sign-Offs

Most commercial air transport maintenance programs require the signature of a qualified technician or inspector to return the component or aircraft to service after maintenance has been performed. Log books are used by pilots to record service problems and are the logical place for maintenance personnel to summarize their work and to affirm airworthiness. When work has been done by an outside shop, that work must be certified on an FAA/JAA Form 1, or equivalent document, by an authorized and qualified person.

12.2 Calibration Records

Measuring devices are commonly categorized as either Class A or Class B. Simple devices (i.e., hand scales, plastic calipers, and gauges used for work that does not require close tolerances) are labeled Class B. Class A measuring devices must be accurate within specified limits; therefore, they require periodic calibration. All tools and equipment requiring periodic calibration must have a Calibration History Card. See Figure 12.14. The quality audit engineer and the accountable manager are responsible for ensuring that all such equipment is recalibrated at the appropriate time by a suitable company. This calibration work, which may be done on-site by a visiting engineer or at an approved company with suitable facilities, ensures that the equipment is always maintained in good condition and to the required standard of accuracy. If repairs are needed because the item has broken or is not performing correctly, the item should be recalibrated after repair, before it is accepted for return to service. Most items are recalibrated annually unless a shorter period is specified. Stickers are normally applied to the tool itself and are located in a conspicuous place on items requiring calibration; these serve as a reminder of the due date for calibration. The sticker should carry four pieces of information:

- Item name
- Item inventory number
- Initials or signature of the person who last calibrated the item
- Expiration date

In addition, any such item that is dropped or damaged should be returned to stores immediately for repair and recalibration.

Calibration History Card
Form Number AB 115

Description
Part no.
Serial no.
Company asset no.

Calibration details

(a)   Calibration date
Calibration due date

(b)

EXPIRATION DATE/ERROR CARD

Description _____

Part no. _____ Serial no. _____

Date calibrated _____

Date calibration due _____

Certificate no. _____ Stamp _____
Form Number AB 116

Fig. 12.14 (a) Calibration History Card; (b) Expiration Date/Error Card.
(Courtesy of Aerobond [United Kingdom].)

Chart Recorders: These items usually are recalibrated annually. A Calibration History Card should be maintained. A chart recorder is included on most thermal devices used for elevated temperature bonding and curing (e.g., hot bonders and ovens). A cash-register-sized paper roll is imprinted with parameter data during the cure cycle. Some rolls are preprinted similar to graph paper, and pens trace lines corresponding to cure temperatures sensed by the thermocouples. Others print dots at close intervals, giving the same effect. Add the component name, part number, serial number, job number, time, and date to this chart before filing it.

Temperature Controllers: Temperature controllers also should be recalibrated annually, and a Calibration History Card (Figure 12.14) should be maintained as previously described.

High-Temperature Cut-Out (High Limit): These are usually included in hot bonders but may be purchased separately for ovens, vacuum presses, and similar equipment. They are useful in preventing runaway overheating if thermocouples fail. They should also be checked annually, and a Calibration History Card (Figure 12.14) should be maintained as previously described.

Vacuum Gauges: These gauges require frequent calibration, and thus Calibration History Cards (Figure 12.14) should be maintained. The gauges are not always handled carefully; therefore, regular calibration is recommended because the cure pressure is dependent on accurate readings.

Other items that usually are calibrated periodically include torque wrenches, hot bonders, moisture meters, weighing scales, and pressure gauges.

12.3 Material Control Records

The following records refer to matrix resins, adhesives, pre-pregs, potting compounds, and sealants. These materials should be treated in the same way as perishable foods; that is, their shelf life should be carefully monitored and correct storage conditions should be ensured. The shelf life is the period for which the manufacturer warrants its product if the product is kept under controlled conditions. Many materials may be qualified for use beyond the shelf-life expiration date if subjected to an approved batch test. Some compounds require only an on-site performance test. Pre-preg materials require more sophisticated testing and must be sent to an OEM-approved testing facility. The facility will recertify the material, if it is acceptable, for a specified further period. This certification must be attached to the original material control record.

Receiving Inspection: Incoming adhesive-type materials should be checked for correct part number, general condition, and expiration date prior to acceptance. Damaged or contaminated materials must be rejected. Materials requiring refrigerated storage in transit should be checked to ensure that they are at the proper temperature and that they have been transported at the proper temperature (i.e., in dry ice). The packaging should contain a temperature-indicating strip to enable the receiving inspector to confirm that the required transit temperature has not been exceeded. The materials should be transferred immediately to a refrigerator and stored in their original containers, which specify the expiration date. If the transit temperature was exceeded by a small amount, the supplier should be contacted for advice and possible recertification. Any new certification should be placed in the material records. If in doubt, the material should be returned for replacement or scrapped. If the material must not be stored at low temperature, this condition must be carefully noted and suitable storage conditions must be provided.

Manufacturer Certifications: For matrix resins, adhesives, potting compounds, and pre-preg materials, it is critical to ensure that accompanying paperwork is sent to confirm that quality assurance testing has been done by the manufacturer and that the materials conform to specification. This is essential if in-house retesting is to be avoided. Manufacturers must be required to send Certificates of Conformance to the specification to which the materials were ordered. These should be retained on file for the period of time required by the local regulatory authorities; however, at minimum, they should be retained until the last of the material has been used or scrapped.

Freezer Log: A freezer log is an ongoing record of the internal temperature and humidity of the freezer. Periodic consultation of the gauges should be confirmed by notation on a form requesting the time, temperature, and humidity, as well as the initials of the person performing the check. Computer monitoring and logging is preferable. All materials in refrigerated storage must have a shelf-life expiration date firmly fixed to each roll, tin, or other type of pack to identify their expiration date under these conditions. This shelf life is applicable only if these materials are stored at the temperature specified. (See the out-time log in Figure 12.15.) Many materials have a shelf life of six months, and others have a shelf life of one year, but only if they are refrigerated at -18°C (0°F). However, some materials have shelf lives as short as three months. Check the data sheet for each material, and store it under the specified conditions.

Out-Time Log: The life of a material requiring cold storage depends on the time and temperature of storage. If the temperature is raised (e.g., when the material is out of the freezer for part of a roll to be cut off for use or for a required amount to be removed from a tin of liquid resin, paste adhesive, or compound), then the time out of the freezer also must be recorded because the permitted life under these conditions is much shorter. Most film adhesives and pre-pregs have an out life of one month or, in some cases, much less. If this out life is exceeded before the shelf life in the refrigerator is exceeded, then the material must be retested or scrapped. A log of out time must be maintained to ensure a correct record. Check the data sheet for the limits for each material. Material allowed to remain at room temperature beyond the manufacturer's limitations will not flow or bond properly. An example of an out-time log sheet is shown in Figure 12.15.

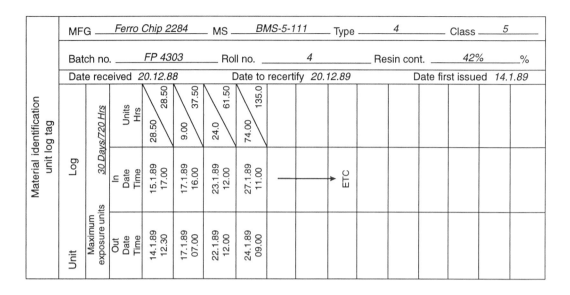

Fig. 12.15 Storage Life Card (Out-Time Log).

Note: When a material reaches either its shelf-life or out-time limit, it must be scrapped or retested in accordance with the manufacturer's instructions and given a further shelf-life and out-time. This new shelf-life and out-time must be carefully recorded, and a new record must be kept similar to the log previously described.

12.4 Component Documentation

Time/Cycle History: This information is recorded on the Component Record Card, AB 104, shown in Figure 12.4. The total flight hours, flight cycles, and calendar time should be recorded on this card. If the component has a fixed life in flight hours, flight cycles, or calendar time, this becomes critical because the item must be scrapped when the first of these limits is reached. Some items may be overhauled and returned to "zero time" if critical parts are replaced. Other parts have a fixed fatigue life and must be scrapped on reaching that time. Cut or otherwise completely destroy such components to ensure that they cannot be returned to service by organizations that supply bogus parts. Many fatalities have been caused by such bogus parts.

Weight and Balance: All aircraft are carefully balanced vehicles that are sensitive to permanent weight changes. Most repairs add insignificant weight to the overall vehicle; however, some components, especially primary flight control surfaces and trim tabs, often have weight and balance limitations. Small amounts of repair material a long way from the hinge line can create a need for large amounts of weight forward of this line to achieve balance. This can result in exceedance of the maximum permitted weight of the whole item if care is not taken at an early stage to minimize the amount of repair material used. These requirements and limits should be obtained from the SRM or other sources, and they must be followed strictly. This should be done before work begins to enable a new item to be ordered if the limits cannot be met. Any work done to restore weight and balance must be recorded on the Component Master Worksheet (Figure 12.1) and the Component Record Card (Figure 12.4). Again, note that excessive paint alone can take a sensitive item out of balance, and this should be considered. On-wing repairs may require careful weight comparison of removed material versus repair material to avoid this expensive and time-consuming off-wing rebalancing, which can lead to additional flight test requirements on some aircraft types.

12.5 Training Records

The quality of composite and bonded metal repair work is highly dependent on the training given to staff members and the skills that they develop. Because of the rapid rate of change in technology, continuing education has become a necessity in the workplace. Formal training in new materials and techniques will be required, in addition to on-the-job-training (OJT) programs. Curricula should be developed and approved by industry experts to ensure that employers have confidence in graduates. Companies and individuals must maintain records of training given and received. This will enable regulatory bodies such as the FAA and JAA to quickly check that work has been performed by suitably trained staff. Knowledge can be taught, but skills and abilities must be developed. Minimum times for skill development should be agreed on and documented in training records. See Figure 12.16 for a sample training record.

UK
PERSONNEL RECORD SHEET

Name	Address
Date of birth	
Home tel	
Age	

Previous employment history

Dates	Company	Position

Relevant training record

Date	Course	Result

Previous company approvals held

Employment commenced	Position	Employment ceased

Form Number AB 109

Fig. 12.16 Personnel Record Sheet. (Courtesy of Aerobond [United Kingdom].)

Shop Equipment and Hand Tools

13.1 Hand Tools and Techniques

The variety of tools used for composite repairs is considerable and ranges from homemade items to sophisticated jigs and tools for producing scarf joints and stepped lap joints. Correct hand-tool selection and skilled usage are essential to good repair quality. Proper usage and maintenance of tools will also reduce health risks and increase the service life of the tools and equipment. Use of electrically powered tools is discouraged because of the conductive nature of carbon fiber dust and the risk of sparks igniting combustible materials and dusts. Pneumatic tools with rear exhausts are recommended and are beneficial for two reasons:

1. Exhaust air is directed away from the work, and dust is not blown around the shop
2. Venturi-type vacuum attachments to the tooling can be used to provide dust extraction

Tools with built-in dust extraction should be specified for health reasons. When performing field repairs, battery-powered tools can be used. These should be designed to be sparkproof and must meet local safety regulations for work on aircraft. See Chapter 6, Safety and Environment.

13.1.1 Drills

Small hand-held pneumatic variable-speed drills with 6-mm (0.25-in.) chucks are the best choice for most hand drilling. Versatility and weight, rather than power and torque, are the important features. Most are made with aluminum bodies, which tend to cool in operation and can cause discomfort to operators suffering from arthritic-type hand problems. Wearing a glove with the fingers removed often is helpful and can improve comfort, while leaving the operator in full control of the drill. Experience has shown that using the second finger on the trigger, with the index finger pointing along the barrel, gives the best balance and minimizes unconscious tipping of the drill motor. See Figure 13.1.

Drill Bit and Cutting Tool Selection: The type of bit required and the angles to which the cutting faces are ground will depend on the material being cut. The SRM should be carefully studied for advice on each type of material. For example, Kevlar requires special drilling and cutting techniques and special drills and scissors because its machining characteristics are poor. Fibers tend to fuzz at the cut edges and do not give a clean cut edge. Special Brad-point drills are required, as well as special carbide-tipped countersinking bits. Always check the SRM for the type of tool, the drill RPM, and the feed rate before starting work. Difficulties in machining can be minimized by the correct selection of tools, feeds, and speeds. Ref. 13.1 from Du Pont contains much useful information. A video also can be supplied on request from Du Pont to illustrate proven practical techniques.

Problems can be reduced to an acceptable level with the correct tools; however, aramids are more difficult to drill, sand, and machine than glass and carbon fibers. See Boeing B.757 SRM 51-70-16 and equivalent information for other types. This SRM chapter contains useful details and should be studied prior to drilling and cutting operations. Reaming is often required to achieve the specified hole accuracy. Boron composites are difficult to drill and machine. They require diamond-coated drills, cutting tools, saws, grinding

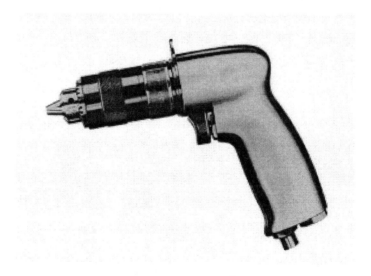

Fig. 13.1 Air-powered drill. (Courtesy of ATA Engineering Processes.)

wheels, and a good coolant supply. Alcohol-based products from the BOELUBE range of cutting fluids from Boeing provide heat reduction at the point of cut through lubrication of the cutting tool without contamination. Wax, paste, or liquid formulations are available for convenience of application.

Speed and Feed Rates: Improper cutting speed and feed rates can cause heat damage and/or delamination of composite parts. Check the SRM 51-70-16 for the correct values. Generally, a high-speed, slow-feed technique is preferred. Small adjustable hydraulic clamping cylinders from KINECHECK are available as accessories for portable tools and handfed machinery to provide feed control and to reduce breakout delamination. Avoid using excessive pressure because this generates heat, and vary the drill speed to maximize material cutting. Do not look for clean, spiral shavings emanating from each flute in the drill as is the case when drilling metal. Too much pressure will result in internal ply delamination and fiber breakout on the far side of the material. Always back up the part being drilled with a piece of wood, plastic, or other softer yet fairly hard material to support the last plies as the drill breaks through the rear face. High speed combined with high pressure will result in high temperatures, and this excessive heat can embrittle the surrounding resin and cause fracture. See the next paragraph on cooling. It follows that sharp tools must be used to ensure cutting rather than rubbing if overheating is to be avoided. Composite fibers, especially carbon, rapidly wear cutting tools; therefore, sharpening must be done more frequently than when drilling metals. This sharpening must be done correctly, preferably by the tool manufacturer.

Lubrication and Cooling: Composite materials react differently to lubricants and coolants than metals react. The temperature should remain below 60°C (140°F) to minimize the risk of delamination. The machining temperature should remain below the glass transition temperature (T_g) of the matrix resin to ensure clean cutting. Drilling above the resin T_g may cause clogging. For many room-temperature curing systems, this will be approximately 60 to 80°C (140 to 175°F), and in some cases lower. These drilling aids may contaminate composite parts, particularly if petrochemical or oil-based products are used. If incorrect lubricants are used on composites, they are certain to affect any subsequent adhesive bonding operations; thus, they should be avoided completely because removal by cleaning is difficult. Coolants may saturate any exposed honeycomb or other core materials. The real advantage of a lubricant is in removal of material from deep holes; however, such holes are not found in most composite parts, which have thin skins. This situation will change as more solid laminate structures are used. Coolants lower the temperature of the drill bit or cutter and also the temperature of the material being cut. Most repairs are made to thin monolithic or sandwich parts, but the number of repairs to thick solid laminates is likely to increase. Composite parts are not good at conducting heat away from drilling; thus, keeping the drill bit cool is the major concern. Air cooling using clean, filtered air may be used. Approved lubricants are alcohol lubricants from the BOELUBE family of cutting agents. These will not ingress the fibers or resins, will not cause outgassing in honeycomb structures, will not contaminate the adhesive or bonding agents, and may be removed easily by alcohol solvents or a mild detergent rinse. BOELUBE materials are not soluble in water. Advantages will be gained in tool life and surface finish, but these will not be improved further by excessive application of these lubricants. Care should be taken to use the minimum at all times. Cooling through lubrication will provide the best results because composite cutting tools must be carbide- or diamond-tipped and may be easily damaged if subjected to thermal shock.

Other Accessories: When holes must be drilled accurately for the attachment of metal fittings or other parts that may be supplied predrilled, special drill jigs must be provided to ensure the precise location of these parts. Drill bushings mounted in ZEPHYR collapsible drill guides attached to the tool will prevent damage to carbide- or diamond-tipped drills and reamers, while providing positive location through a strip template.

13.1.2 High-Speed Grinders

These machines are used for light sanding, feather edging, and cutting. Thus, they must be small and light enough for operators to handle with ease. Most repair preparation when using these tools and accessories requires a high skill level that is more often associated with surgery. High-speed grinders usually are air powered, and they run at 18,000 RPM or faster. They are available in both straight and 90° configurations. See Figure 13.2. With practice and care, they are used to cut scarf joints and step-sanded joints.

Sanding Discs: Always consult the SRM for the correct type of sanding disc, in terms of both the grit type and the grit size. Use the finest grade that will do the job. Paint removal may require a more coarse grade of grit. Aluminum oxide and silicon carbide are the most commonly used grit types. Abrasive papers with grit sizes ranging from 80 to 1200 are available. Generally, 120 grit should be coarse enough to remove paint; 240 to 320 grit is suitable for preparing surfaces for adhesive bonding. Finer grits can be used if a particularly good finish is required. Several practical details must be considered when sanding. Most sanding discs are used with a special mandrel that permits quick attachment and removal. Unfortunately, some detach too easily and fly across the room. The half-twist mandrel/pad system has proven to be the most reliable. Mandrels and pads are available in different diameters. To prevent expensive divets or grooves from being made in the work surface, use a mandrel with a diameter that is smaller than the sanding pad. This will provide flexibility at the extremity of the pad, which will compensate for the operator's unintentional tipping of the grinder. Many technicians prefer nylon pads because nylon pads carry a nonclogging open-coat abrasive. Both types are available with fine, medium, and coarse abrasives. For quick identification, the nylon pads are color coded—gray is the finest, and brown is the most coarse. New diamond-impregnated sanding pads, which are on the market, are more durable but do not appreciably improve workmanship; furthermore, many are rigid and unforgiving if the technician slightly leans the grinder. Regardless of the choice of pad, the pad must be held parallel to the work surface, and the technician

Fig. 13.2 High-speed grinder. (Courtesy of K&C Moldings.)

must avoid any tendency to tilt the pad. Use the finest abrasive that will do the job, but avoid overheating. This will require patience. An abrasive that is too coarse will make it difficult to identify individual composite plies when scarfing or step sanding.

Cutting Wheels: Thin abrasive cutting wheels, such as those used by welders, are useful during initial damage removal and post-repair trimming. These wheels are more durable and are most effective at a high RPM. Special reverse-threaded mandrels are available for using these wheels in small high-speed grinders. A straight grinder is more comfortable to use when cutting materials. A helpful hint when cutting with this combination is to use both hands on the grinder, with one of these hands resting on the work surface to provide stability and control. The tool should be positioned so that rotation of the wheel throws cut material away from the operator. Always wear safety goggles and protective gloves. When shallow cuts must be made without damaging underlying material, a thin stainless steel sheet is sometimes slid beneath the underlying material and the layer being cut. When a spark is first seen, remove pressure from the grinder. Diamond wheel trimmers are recommended for cutting most composites, and several models can be supplied. Most common are 50- to 70-grit diamond wheels, but 40- to 50-grit wheels can be supplied if a slightly faster cut is needed.

Other Accessories: A die grinder kit is supplied with a set of 10 mounted stones of different shapes and sizes to grind edges and radii as required. The speed is 20,000 RPM. See Figure 13.3.

Diamond Wheel Trimmers: Diamond wheel trimmers are used mainly for cutting panel edges to shape, and they use diamond-tipped cutting wheels. They are designed to cut quickly and to give a good edge finish. See Figure 13.4.

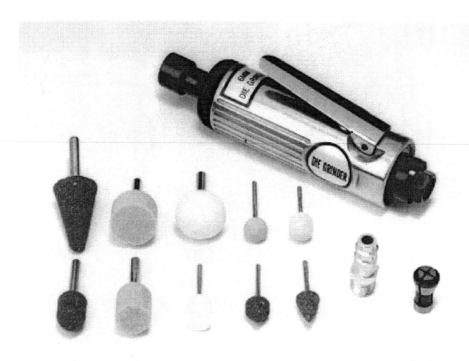

Fig. 13.3 Die grinder kit. (Courtesy of K&C Moldings.)

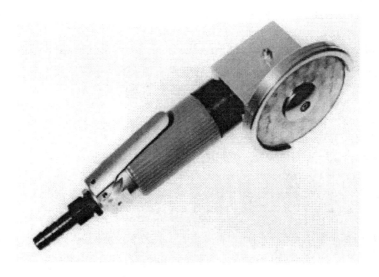

Fig. 13.4 75 mm (3 in.) diamond wheel trimmer, type AC. (Courtesy of K&C Moldings.)

13.1.3 Cutting Utensils

Scissors: Carbon fiber should be cut with sharp, serrated scissors or a scalpel. Do not drag the fibers or weave construction. Special scissors also are required for cutting Kevlar. See notes supplied by Du Pont in the next paragraph on hand shears.

Hand Shears: Hand shears are simply large scissors, but the comments made in the preceding paragraph on scissors also apply to hand shears. The appropriate type must be used for each material, and they must be sharp and the blades must be strong to ensure that they cut the fibers and do not merely bend and drag the fibers or completely fail to cut them. Specially hardened shears are preferred for Kevlar and other aramids. Tungsten carbide or ceramic shears are recommended. These have serrated blades, are specially hardened, and must be kept clean. Clean the blades with acetone after use on pre-pregs. Because aramids need especially sharp blades, limit the use of these special shears only to aramids because glass fiber and carbon fiber will quickly blunt them. Regular shears or scissors can be used, but they will require frequent sharpening. Different weights of cloth may require different shears. If working mainly with one weight of fabric, ensure that the best shears for this weight of cloth are selected. Ceramic shears made for aramids cost more, but they are more durable. Conversely, several manufacturers supply low-cost disposable snips made from hardened pressed steel with solvent-resistant plastic handles. These are excellent on all fiber materials and may be discarded when they become blunt. Power shears are available with carbide inserts and close operating tolerances, and they work well on pre-pregs and heavier fabrics.

Knives: The knives used in many places for cutting through the thin aluminum skins of honeycomb panels, thin doubler plates, or thin composite skins often are locally made from old, large hacksaw blades. They seem to serve their purpose well. Stanley knives sometimes are used. Razor knives, rotary knives, and ultrasonic knives have all been used successfully with aramids. The use of templates to guide the blades is recommended in all cases. Blades must be kept clean and should be replaced as often as necessary.

Skin Peelers (Large Sardine-Can Openers): These homemade tools are used to peel strips of metal skin approximately 25 mm (1 in.) wide from honeycomb cores. They also may be used to peel doublers from sheets of skin. See Chapter 10.

13.1.4 Hand Routers

Similar to drill motors, hand routers usually are air powered, for the same reasons. Models fitted with dust extraction are recommended. See Figure 13.5. These are fitted with diamond-coated burrs of 4.8 mm (0.2 in.) diameter in the case of the model illustrated and may also be fitted with special-purpose cutting tools from 1.5 to 10 mm (1/16 to 3/8 in.) diameter. Spindle speeds should range from 23,000 to 35,000 RPM, and the cutter must be properly supported by a bearing in the routing attachment. BOELUBE will greatly increase the life of routing cutters in all composites.

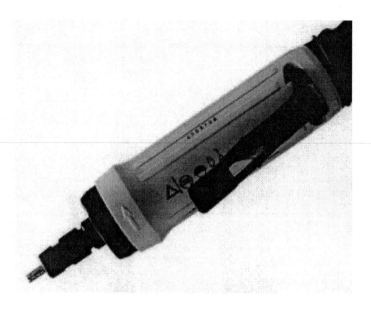

Fig. 13.5 Air-powered hand router. (Courtesy of ATA Engineering Processes.)

Templates: Routing techniques require the use of templates to guide the cutter around the required shape. Templates must be made accurately from suitable materials. The application of BOELUBE to the template will greatly ease the movement of the router guide.

Router Bits: Diamond burrs are typically used with hand routers for work on composite materials and should be changed when they no longer cut effectively. Various grit sizes are available according to the speed of cut and/or finish required. Solid carbide, diamond cut, or special geometry tooling for aramids should be selected according to the material to be cut. However, electroplated diamond burrs should not be used on Kevlar or other aramids, and they tend to run hot in thick conventional composites. Care must always be taken to avoid overheating because this may lead to delamination. Use BOELUBE to prolong tool life and to reduce cutting temperatures where possible. Air cooling also may be employed.

13.1.5 Orbital Sanders

These sanders should be used carefully and with the appropriate grade of abrasive paper. See the SRM for the recommended grade for each job. Fortunately, all abrasive papers become less abrasive and develop a finer grade as they wear. However, this also can lead to overheating in an attempt to maintain the rate of cutting. Always fit a new sheet of abrasive paper when the cutting rate becomes slow. Do not use excessive pressure. Figure 13.6 shows a typical flat bed orbital sander. Random orbit sanders are also available, and they use a circular abrasive pad.

13.1.6 Cast Cutter (Oscillating Saw)

This electrically powered tool, used primarily by the medical profession to remove plaster casts from broken limbs, is safe because the blade does not rotate. Its cutting action is produced by high-speed oscillation. The device cuts on both the forward and backward strokes. In most cases, contact with the operator's skin does no harm. It can be used with Kevlar and other fibers, and the following guidelines apply:

- Use blades #840-40-300 Ti-Ni Coated SST, 2 in., or #840-40-350 Ti-Ni Coated 2.5 in.
- Never use a dull blade; it may cause injury.
- Tighten the blade-retaining nut before each use.
- Clamp the composite to minimize vibrations and to improve the quality of cut.
- Too much pressure may cause the blade to break.
- Clean the blade before each cutting operation.

See Figure 13.7. for examples of cast cutters.

Fig. 13.6 Flat bed orbital sander. (Courtesy of K&C Moldings.)

Fig. 13.7 Examples of cast cutters (oscillating saws). (Courtesy of ATA Engineering Processes.)

13.1.7 Painting Equipment

Painting must be done in properly equipped shops that meet local safety regulations. The required standard of fume extraction must be provided. Masks, goggles, and suitable overalls must be worn. The paint used must be within its shelf life and to the correct specification for the job. Paint shakers normally are used to ensure that the paint is properly mixed and that any solid additives have not settled at the bottom of the container. Where specified, the correct primer also must be used. Particularly with primers, it is important not to use an excessively thick layer. For aircraft purposes, remember that paint is a heavy material and that only the stated amount should be used. The correct weight and balance of some control surfaces can be taken outside specification by the use of too much paint.

13.1.8 Resin Applicators

Some adhesives, resins, and sealants are supplied in packages that allow the material to be mixed in the correct ratio by bursting a diaphragm and then mixing. They must be thoroughly mixed and then can be applied using a nozzle provided. Other types of applicators are special rollers using a spiral-wound wire to control the thickness of the resin layer. These are used to apply adhesives or resins over fairly large areas. In cases where adhesives are used on production lines for the assembly of parts, special purpose applicators are used, often controlled by robots.

13.2 Shop Equipment

13.2.1 Bandsaws

Bandsaws are useful for cutting composites, but a special, high-speed type is useful for cutting honeycomb in the expanded condition. For each material to be cut, use the correct blade and speed. Low-speed bandsaws cannot cut expanded honeycomb successfully. If the high-speed type is unavailable, bond a piece of honeycomb that is too thick into the repair and then sand it carefully to the level of the skin. Diamond-coated blades may be used to cut cured composite parts, but low-speed bandsaws should be used for this purpose. BOELUBE wax or liquid cutting lubricant will prolong blade life without contaminating the material. The best performance with a bandsaw will be obtained only if the blade guides are properly adjusted, the blade tension is correct (to ensure straight cuts and to reduce blade fatigue), the appropriate blade for the job is fitted and in good condition, and the feed rate is correct. Bandsaw machine and blade suppliers offer useful handbooks on bandsaw operation.

Blade Selection and Speed: The choice of blade and its speed will depend on the type of material to be cut. Honeycomb may be cut either in the HOBE (**HO**neycomb **B**efore **E**xpansion) mode as a block of metal or in the expanded condition. In the expanded condition, high speeds are required because only the inertia of the material is available to resist the cutting action of the bandsaw blade. It is usual to cut honeycomb, either aluminum alloy, Nomex, or other types, with a special high-speed bandsaw that has a blade speed of 5000 m (16,000 ft) per minute. The blade also should have a special reduced blade offset of 0.05 mm (0.002 in.) to prevent tearing the honeycomb. Approximately 5.5 teeth per centimeter (14 teeth per inch) is normal for this type of blade. Again, a sharp blade is important. This type of blade should be reserved only for honeycomb cutting because other materials will blunt the blade and make it unsuitable for the next cutting of honeycomb. The special offset used on the teeth will not give enough clearance for swarf removal during the cutting of solid materials. The blade should be changed to the correct type for solid materials and changed back again to fit the special blade when honeycomb must be cut.

Aramid Fiber Considerations: Cutting aramid fiber composites is a problem because the fibers fibrillate (i.e., break down into thinner fibers) during the cutting action. They are tough rather than brittle, and they tend to tear and pull out instead of fracturing cleanly, as do carbon and glass fibers. Sharp tools are required, and Du Pont has provided special instructions for cutting Kevlar. These instructions are equally applicable to aramids made by other suppliers. The cutting of aramids in woven fabric form requires the use of special serrated, sharp shears. See Section 13.1.3. on cutting utensils and hand shears.

13.2.2 Air Compressors

Air compressors require a tank into which compressed air can be pumped. These tanks must meet local safety regulations and may be subject to periodic inspection and pressure testing. Pressure relief valves must be fitted to vent the tanks if the air pressure reaches the design safety limit. Pressure gauges also are required to ensure that line pressure is known. Compressors can be used to drive paint spraying equipment, air-powered tools, or venturi-type vacuum pumps if a shop air line at the required pressure is unavailable.

Venturi-Type Vacuum Pumps: Venturi-type vacuum pumps can be used to produce the vacuum inside a vacuum bag required to apply pressure to a composite repair during cure. The principle is similar to that in an automobile carburetor, in which a flow of air at high speed through a venturi tube, with a reduced cross-sectional area at the center, results in a reduced pressure at this point. If the fuel jet is located at this position, fuel is drawn through the jet to mix with the air flow in the correct proportion for good combustion. In the case of a venturi-type vacuum pump, the same principle is used to produce a reduced air pressure in a vacuum bag. Figure 13.8 illustrates the Airvac 22 model from Airtech International Inc. This type of pump is included in most hot bonders and avoids the need for a separate, electrically powered vacuum pump. However, it relies on the assumption that the shop will have a compressed air line for its power tools and other purposes. Many composite repair shops also have a few electrically powered vacuum pumps for use when a hot bonder is unavailable or is unnecessary for a room-temperature curing repair.

Water/Oil Contamination and Filtration: Most compressed air lines in composite and metal bonding repair shops are fitted with filters to remove oil and water. This benefits the life of the power tools because it helps to prevent corrosion but is very important if compressed air is used to blow swarf or dirt from difficult areas. The compressed air filter should remove from the air line all solid materials, such as pipe scale and dirt, which will accumulate because of corrosion inside the pipe and because of dirt entering from the atmosphere. The filter also should be capable of removing liquids such as water and oil because these may have a detrimental effect on operations such as spray painting. Use vacuum cleaners to remove dust or dirt because blowing it around the shop is undesirable. Compressed air also may be used to dry surface water; however, this is of no benefit if it causes contamination in the process. Generally, do not use compressed air for any purpose

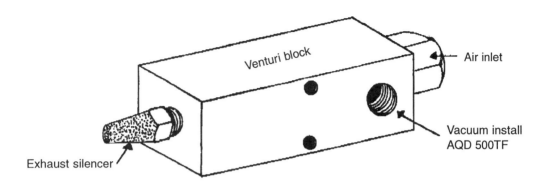

Description: The Airvac 22 is a venturi block and silencer that converts shop air lines to vacuum without the use of a vacuum pump. Airvac 22 will draw 22 inches of mercury. See diagram below.

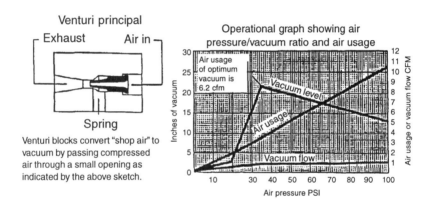

Fig. 13.8 Airvac 22 venturi-type vacuum pump. (Courtesy of Airtech International Inc.)

other than driving air-powered tools, spray guns, or vacuum pumps. The proper design and installation of air lines is important. Air mains and lines should be large enough to avoid excessive pressure loss under conditions of maximum flow. They should be installed with as few restrictions and as few sharp bends as possible. Air mains piping should be pitched downward slightly in the direction of the air flow to ensure that both gravity and air flow will carry the water to traps located at frequent intervals. These traps should be drained regularly and should never be allowed to become full and inoperative. To minimize the possibility of condensed moisture reaching the tools, down pipes or hose connections should never be taken directly from the bottom of air pipes or mains. Connections should be made at the top of the main, and a long-radius return bend should be used. Air-powered tools should have an airline lubricator fitted slightly ahead of them to provide lubrication for the tool. Clearly, any tools fitted in this way must not be used in a clean-room area where lay-up for bonding is being done. For composite and bonded metal repair work, the lubrication should be minimized or, preferably, tools should be used which do not require lubrication by this method. Alternatively, BOELUBE liquid will provide excellent lubrication for all types of air motors without contaminating the workpiece.

13.3 Heating Devices

Ovens: Ovens are commonly used to cure hot-bonded repairs. They must be well designed to give a uniform temperature throughout the oven, or they must be of the air recirculating type to ensure an even temperature distribution. Thermocouples should be used when parts are cured in ovens because it cannot be emphasized too strongly that the glue-line temperature is most important, not the oven temperature. Many have learned this the hard way. See also Chapter 4.

Autoclaves: Autoclaves are the most common method of original manufacture and are covered in Chapter 4, Section 4.9.

Portable Bonding Equipment (Hot Bonders): This type of equipment is covered in Section 10.5.9 of Chapter 10.

Heat Lamps: Heat lamps are mentioned in most SRMs and can be used to speed the cure of wet lay-up adhesives or resins. Care must be taken not to overheat the surface; otherwise, the repair area will be blown apart by gas bubbles resulting from the excessive heat. The resin also may be embrittled in this way. Thermocouples should be used to check the maximum temperature and the uniformity of the temperature across the repair area. Thermocouple-operated temperature controllers can be used to maintain a constant temperature by switching the lamps on and off. Lamps can be used in conjunction with heater blankets to warm spots that are cold. They also can be carefully used to warm a resin or adhesive that must have its viscosity reduced to improve flow. This must be done carefully with the minimum of heat; otherwise, the resin or adhesive may start to cure too soon and its viscosity will rise to the point where fabric impregnation or spreading is no longer possible. Lamps can also be used, with great care, to reticulate a film adhesive onto the cell ends of a sheet of honeycomb. The minimum temperature that will cause reticulation should be used because the resin must be heated again to bond the honeycomb to the skin, and it must flow and wet the surface properly during this operation.

Heat Guns: Heat guns, also called hot air blowers, are available either with or without heat control. Clearly, those without heat control must be used more carefully. Generally, heat guns can be used in the same ways as described in the preceding paragraph on heat lamps.

Heater Blankets: Heater blankets typically are made with suitable wires embedded in silicone rubber and can be used at temperatures up to approximately 200°C (400°F). They are used in conjunction with hot bonders or other temperature controllers. See Chapters 6 and 10.

Other Heating Devices: Although not commonly used, some chemical heat packs are available and will provide heat for a limited time while an exothermic reaction occurs. They are similar to domestic hot-water bottles, except that the heat is supplied by a chemical reaction.

Moens high-velocity heating and cooling systems use temperature-controlled hot air, and piping can be tailored to suit the part. This method is used most frequently in production for thermoplastic and thermoset resin filament winding. The method also may be useful in repetitive repairs. Heat Transfer Technologies Inc. supplies this equipment.

13.4 Measuring Devices

Micrometers/Calipers: These standard engineering instruments are often used in composite and bonded metal repair work to check the thickness of cloths or cured laminates and sheet metals and the diameter of fasteners. Today, they may be used in either their metric or imperial versions, although the move toward complete metrication is proceeding steadily. Even in 1997, American-designed civil aircraft were being built in imperial units (inches).

Chart Recorders: Chart recorders are discussed in Chapter 12, Section 12.2.

Paint Film Thickness Checker: Suitable meters are available to measure paint thickness, and they are worth using. In the case of radomes, excessive paint thickness can reduce signal transmission below acceptable levels. However, no meter exists that can work on fiberglass, aramids, or other nonconductive composites. On critical control surfaces, excess paint can take a component outside its weight and balance limits. For metal parts and carbon fiber composites, the Elcometer 300 (an eddy current type of instrument) can be used and will assist in keeping paint thickness within the required limits. See Chapter 15 and Figure 15.9.

Color Charting: This method is often used for primers to avoid using excessive thicknesses. See Chapters 2 and 15.

Transmissivity Tester: This complex tester is used to measure the transmissivity of radomes before and after repair. Only the large airlines have their own transmissivity testers. Two radar scanner dishes are used—one as transmitter, and the other as receiver. See Figure 13.9. A meter is used to measure the signal strength with only air between the dishes. The radome for test then is placed between the two dishes, and the decibel loss is measured. The radome is held in a stand designed to allow the radome to be moved in both azimuth and elevation. This allows the transmission to be checked and recorded in marked areas of the radome. Any areas where transmission is below specification are carefully marked and checked for moisture content, excessive sandwich thickness, excessive paint thickness, and the use of the wrong type of conductive paint for static charge removal.

Meg-Ohm Meter: This standard electrical meter is useful for checking the resistance of lightning diverter strips on radomes, the resistance of any connections on the radome, and especially the resistance of the complete system from the nose of the radome to the aircraft fuselage. It also is useful for checking any troublesome electrical equipment.

Weighing Scale: These items are critical for the accurate weighing of two-part mix adhesives and matrix resins. The modern types are usually electronic and weigh to an accuracy of 0.1 g (0.0035 oz). They typically have a tare weight button. This button allows the weight of the mixing cup to be zeroed out to enable only the resin and hardener to be measured, which is helpful in achieving correct mix ratios. Scales should be covered with paper or plastic sheeting during weighing to ensure that they are not contaminated with resin all over the scale pan. These items should be recalibrated annually or at any time that readings may be suspect. They are critical to obtaining best performance with wet lay-up resin and adhesive systems. Any spilled adhesive should be wiped immediately with a suitable solvent for the material being weighed.

Radar "wave test" equipment

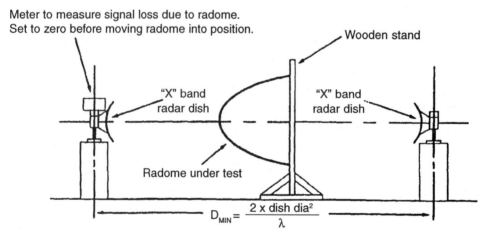

Fig. 13.9 Transmissivity tester for radomes.

13.5 Health and Safety Equipment

Respiratory Protection—Dust and Vapor Masks: Appropriate masks must be worn, depending on the hazard presented by the materials in use. When sanding, dust masks should be worn. When masks are worn for protection against gases or chemical vapors, advice should be sought concerning the correct type for the hazard concerned. Full face (negative pressure) masks are available and can be fitted with a wide range of filter canisters to suit the gases, solvents, or particulates against which protection is required. See Figure 13.10. The Material Safety Data Sheet (MSDS) should be consulted, and the recommended protection should be used. See also Chapter 6.

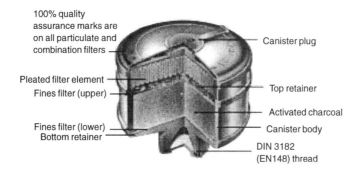

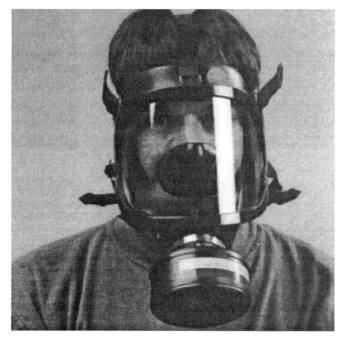

Fig. 13.10 Crusader full face (negative pressure) mask. (Courtesy of Seton Ltd.)

Gloves and Coveralls: Chapter 6 covers the most suitable kind of glove for each chemical type. Gloves should also be worn when working with boron and carbon fibers to prevent them from puncturing the skin. Rubber gloves should be worn to provide adequate puncture resistance. If a serious risk of cutting fingers exists, then special Kevlar gloves are available with a high resistance to sharp cutting tools. These gloves are also resistant to heat. See Figure 13.11.

Overalls giving complete body protection are known as coveralls and should be worn when the situation demands that level of care. If the materials being used are toxic by absorption through the skin, then coveralls providing complete protection should be worn. Chemical-resistant types are available.

Eye and Ear Protection: As previously mentioned, goggles giving all-around eye protection should be worn when easily splashed chemicals are being used. Resin and adhesive curing agents are sometimes of low viscosity (e.g., water thin), and splashing is always a possibility with low-viscosity liquids. Solvents also are in this category. Likewise, face shields and visors are produced. Helmets covering the ears may have to be worn for protection when paint spraying or for protection against easily splashed materials. Ear defenders should be worn for noisy processes such as grinding or riveting.

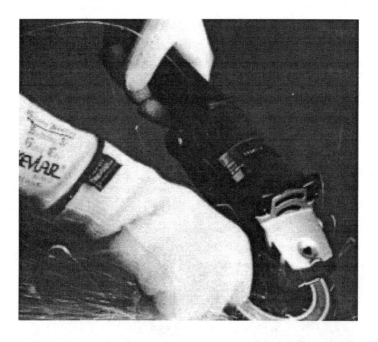

*Fig. 13.11 Kevlar gloves for protection against cutting, abrasion, and heat.
(Courtesy of Seton Ltd.)*

Fume Cupboards: These cupboards, which are fitted with fume extractors, should always be used when mixing adhesives or resins and other chemicals that emit toxic or unpleasant fumes. They are helpful to health, make the mixing process more pleasant, and protect other people in the same area who may also be affected by fumes. By mixing materials in the fume cupboard, which draws air from the room, the fumes are carried away from the person doing the mixing and from anyone else in the vicinity.

Barrier Creams: These special creams should be rubbed into the hands before commencing work with resins and adhesives. They provide protection and facilitate washing later. The hands should be washed and dried before applying barrier creams and again after the work has been done. Check the material data sheet to ensure that the appropriate type of cream is available for the work being done. Two types of barrier creams are available: water-repellent creams and solvent-repellent creams. See Chapter 6, Section 6.5.

13.6 Reference

13.1   *A Guide to Working with Kevlar Aramid Fiber*, Du Pont External Affairs, 1111 Tatnall St., Wilmington, DE 19898.

Tooling and Mold Making

14.1 Introduction

Tooling and mold making is a necessary part of the procedure for many major repairs and several small ones. Large repairs to flat panels often can be made with little tooling or sometimes with none. However, composite parts frequently are used because they lend themselves to the manufacture of smoothly curved shapes. This may be necessary for reduction of drag to improve performance and economy or to give a pleasing shape to the part. If the part in need of repair is a damaged solid laminate or a honeycomb sandwich panel that has suffered damage penetrating both skins, then tooling will be required to support the repair during cure to ensure that the repaired part is maintained to the correct profile. The types of tooling described below are limited to those involved in repairs. Production tooling will be designed to take account of the number of parts expected to be made. Repair tooling is often used only once and must be accurate, although it does not have to be used several hundreds or thousands of times. Except at approved repair stations, the number of repairs made using one tool is seldom large. If major rebuilds are considered, then tooling will have to approach production quality. In this case, it will be necessary to know the number of parts likely to be rebuilt in order to decide whether to perform the repair in-house or to send the items to a repair station that already has the tooling required.

Factors involved in choosing the type of tooling for each application are as follows:

1. The process cure schedule for the repair and, in particular, the maximum temperature required must be considered. For thermoplastic parts, the temperature capability of the tool must be carefully considered because suitable materials are expensive and the choice is limited.

2. The shape, complexity, and required accuracy are important factors.

3. Consider the coefficient of thermal expansion (CTE) of the part to be repaired. The material used for tooling should have a similar value of CTE to that of the part.

4. The choice of tooling dictates the quality of the finished part or repair.

5. The thermal properties of the tool are important:

    •   The tool must have low thermal mass (density x specific heat) to minimize the heat energy required to achieve cure and to obtain a sufficiently rapid heat-up rate. With film adhesives and pre-pregs, the heat-up rate must fall within specified limits to enable melting, flow, and wetting to occur before gelation begins. If the heat-up rate is extremely slow, then gelation may occur before melting, and the proper flow and wetting of the bonding faces and full compaction of composite layers will not occur.

    •   For repairs, the tool must have similar thermal conductivity to the part under repair to ensure that the heat from the heater blanket or other source can provide a uniform temperature across the repair area.

6. The tool should be of minimum weight to aid in ease of handling and to minimize heat energy requirements.

For production purposes, some thermal requirements for tooling materials vary according to whether the part is made using hot press tooling or autoclave tooling. See Ref. 14.1. The requirements are summarized here:

•   Low thermal mass (density x specific heat)
•   Low thermal conductivity (autoclave tools)
•   High thermal conductivity (press tools)
•   Minimum weight (low density)
•   Low CTE

14.2 Caul Plate and Dam Fabrication

Caul Plates: Caul plates are defined as, "Smooth metal plates, free of surface defects, the same size and shape as a composite lay-up, used immediately in contact with the lay-up during the curing process to transmit normal pressure and temperature and to provide a smooth surface to the finished laminate." Thin aluminum alloy sheets of approximately 0.4 mm (0.016 in.) thickness are commonly used because they have a high thermal conductivity. Composite and bonded metal repairs are frequently performed without caul plates; however, when they are used, caul plates improve the finish and assist in the even

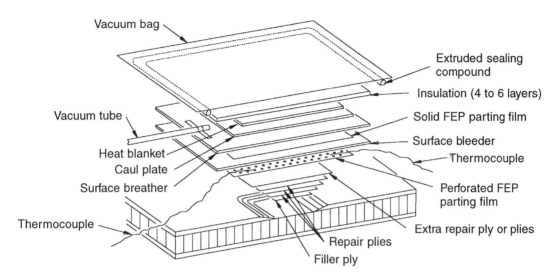

Honeycomb panel shown
Solid laminate similar

Bagging sequence for skin ply repair

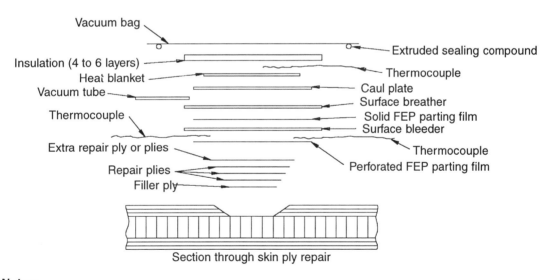

Section through skin ply repair

Notes:

[A] One ply of adhesive film (for full depth core replacement, where damage does not extend through both skins). For partial core replacement, use two plies of adhesive film with one ply of glass fabric between them.

Fig. 14.1 Typical bagging lay-up and caul plate location. (From Boeing B.737 SRM 51-70-08 Sheet 1, Fig. 5. Courtesy of Boeing Commercial Airplane Group.)

distribution of heat. They are not normally used on single- or double-curvature surfaces because of the difficulty of producing plates to a sufficiently accurate contour. See Figure 14.1 for a typical bagging lay-up and caul plate location.

The Boeing SRM states, "When using a heat blanket larger than 12 inches on one side, an aluminum caul plate (0.016 in. thick) can be used under the heat blanket to minimize localized heating. Make the caul plate slightly smaller than the surface breather."

Dam Fabrication: This procedure is used during bonded metal repairs to prevent the spread of chemical paint strippers or surface treatment fluids or gels beyond the area they are intended to cover. Make a dam approximately 6 mm (0.25 in.) thick from vacuum bag sealing tape, plasticene, or other suitable removable compound. This dam should be located around the repair area to contain the paint stripper or surface treatment chemical fluid, thereby avoiding contamination of other areas. Replacement honeycomb should be covered with Speedtape to prevent chemicals from entering the honeycomb. Residual chemicals are likely to cause corrosion if not completely removed; therefore, keep them out of the honeycomb area and any other areas where they could become trapped. If any doubt exists, rinse carefully with distilled water several times and dry thoroughly before applying primer or adhesive.

14.3 Splash Mold Making

A splash is an intermediate tool, made with a fiber-filled synthetic plaster material. One such material is Ludor 500L, a chopped glass fiber-filled synthetic "plaster" material. The procedure for making a splash mold is as follows:

- An undamaged area of the airframe or other component of the same profile, or the same area as the damage on another identical component, is covered locally with an adhesive-backed Teflon-coated glass fabric to act as a release agent. A suitable material such as Tooltech A-005 may be used. This gives good release and masks out any weave pattern on the component.

- A gel coat is applied to the release sheet mentioned above in the local area on the component, and a synthetic plaster splash of Ludor 500L or similar material is built up on top of the gel coat and allowed to cure.

- The intermediate splash is then removed from the component. A tooling pre-preg lay-up is made and cured in the gel-coat-faced intermediate splash. This forms a more durable tool, resistant to high temperatures, from which the repair patch can be produced.

- The tool made from tooling-grade pre-preg can then be post-cured in steps, according to the manufacturer's instructions, to approximately 20°C (36°F) or more above the temperature at which the repair patch must be cured, and the tool is ready for use.

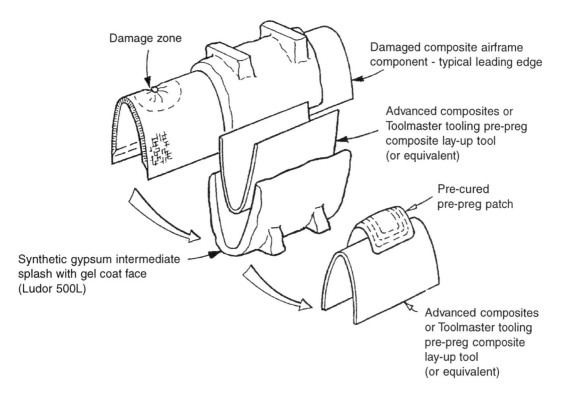

Damage zone

Damaged composite airframe
component - typical leading edge

Advanced composites or
Toolmaster tooling pre-preg
composite lay-up tool
(or equivalent)

Pre-cured
pre-preg patch

Synthetic gypsum intermediate
splash with gel coat face
(Ludor 500L)

Advanced composites
or Toolmaster tooling
pre-preg composite
lay-up tool
(or equivalent)

Fig. 14.2 Use of splash molds. (Courtesy of Aero Consultants [United Kingdom] Ltd.)

- This tool can be coated with Tooltech A-005 from Aerotech or equivalent material, and the repair patch can be laid up to the SRM requirements. The patch can then be cured at the required temperature and pressure for the specified time, as directed in the material data sheet.

- After the bonding surface has been abraded with 320 grade silicon carbide paper or similar and a water break check has been found satisfactory, the patch can be given a drying cycle before being bonded to the damaged component. See Figure 14.2.

14.4 Release Agents/Films

These are supplied by several companies, and some details are provided here.

Note that no silicone-containing compounds may be used because these materials are good release agents and can easily contaminate surfaces prepared for bonding, thereby preventing adhesives and resins from adhering to them. The problem with choosing release agents is that they are necessary to the task, but they must not be so good that the whole objective of making the correct parts stick together is defeated. Liquid release agents must be sprayed onto tools outside the repair area and allowed to cure before they are brought into the lay-up area. Surfaces that have been in contact with a release agent must be lightly abraded and given a water break test before bonding to ensure that they are not contaminated.

14.5 Simple Room-Temperature Tooling

This tooling can be made in several ways, but two methods are most common:

1. Solid laminate tools can be made from a suitable number of fiberglass layers and epoxy or other room-temperature curing resin systems. These are similar to the pre-preg tools made from an intermediate splash, as previously described. The difference is that no splash is used. The tool is made simply by laying a release sheet, or a spray coat of release agent, on an area of the part with the same profile as the repair area or on the same area of an undamaged part and then applying a sufficient number of layers of fabric or chopped strand mat onto the part after impregnating those layers with a suitable resin. After curing, this solid laminate forms the tool against which repairs can be made. If necessary, a gel coat can be applied to improve the finish. Ribs or other reinforcing may be applied to the back face of the tool if required.

2. Composite skinned sandwich-type tooling can be made using a honeycomb or other core material. These tools are lighter and stiffer for their weight and are made in a way that is similar to solid laminate tools. After coating the chosen area of the component with a suitable release agent, two or three layers of resin-impregnated fabric are applied to the selected area of the part, which has the required profile. While this is wet, a layer of honeycomb or other core material is coated with resin and laid on the skins. Two or three more resin-impregnated skins are then laid onto the core material to form a sandwich panel, and the tool is held under vacuum pressure until it is cured. The tool face may be given a gel coat to improve the finish if necessary. When cured, the tool is ready for use with repairs made using room-temperature laminating resins. Ribs or stiffeners may be added to the back face if required. In both of the above cases, mild heat may be used to speed the cure, but the temperature should remain at least 20°C (36°F) and preferably 30°C (54°F) below the glass transition temperature (T_g) of the laminating resin used to make the tool.

14.6 Room-Temperature Curing Pre-Preg Tooling

Around 1984, special tooling pre-pregs were introduced, which overcame all of the disadvantages previously associated with wet lay-up tooling. These pre-pregs cure at room temperature over several days and have the further advantage that they can be post-cured in steps up to high temperatures so that they finally have a high T_g. The post-cures allow repairs to be made in these tools at the original cure temperatures of the parts concerned, without problems of creep or distortion if these tools are properly designed and supported. Tooling pre-pregs now are supplied by several companies. Check the data sheets to ensure that correct procedures are followed because every company procedure may not be the same. The advantages of these systems are as follows:

- Composite tooling pre-pregs allow a clean, controlled tool fabrication process to be achieved.

- Fiber/resin variability in the finished laminate is eliminated.

- Resin/hardener mixing problems are eliminated.

- Pre-pregs are available using all common types of fabric; thus, another advantage is that tooling can be made with a CTE that is identical to the CTE of the original part.

- High-fiber-content laminates are achieved in the finished tool, with low void content and improved thermal and mechanical properties. This can be achieved with pre-pregs having high-viscosity resin systems because they suffer almost zero bleed of the resin from the laminate, giving a fiber content exceeding 55% by volume.

- These tools will cure to a good strength at room temperature and can then be post-cured. This facilitates the manufacture of tools directly from the parts in need of repair.

So many advantages also carry some penalties. Material cost is higher than with wet lay-up, but labor cost is reduced. These pre-pregs must be transported in refrigerated vehicles. After removal from freezer storage, they must be laid up fairly quickly. (See time limits in Section 14.6.5 for storage life and usage after removal from the freezer.) Some of these materials will achieve initial cure after several days at room temperature, but other versions require a minimum of 30°C (86°F) or 40°C (104°F) for initial cure. They all require post-curing in steps. When correctly cured, they may be used at temperatures up to 200°C (392°F). Many tools have been made by this method, and many remain in use several years after being manufactured. See Ref. 14.2 for useful details of tool making with pre-pregs.

14.6.1 Lay-Up

The following details are from information supplied by The Advanced Composites Group Ltd. for its materials known as LTM 10, LTM 12, and LTM 16. Procedures for pre-pregs from other suppliers may differ in some aspects. Consult information from the pre-preg supplier before commencing lay-up. Work strictly within the procedures given for the material in use. The first thing to consider before commencing lay-up is the material against which the lay-up will be made. LTM pre-pregs can be cured against most non-porous materials (e.g., aluminum and syntactic foams) and some porous materials (e.g., wood and plaster) if a suitable sealant is applied. The use of the correct sealing and release treatment for a given model is essential to ensure a satisfactory result. Refer to The Advanced Composites Group Ltd. for advice if necessary, or conduct a trial.

The curing reaction of LTM pre-pregs can be inhibited when the material is molded in contact with certain master model materials, lacquers, resins, and sealants. Cure inhibition results in a sticky, uncured layer at the laminate surface, which generally is extremely thin but is sufficient to impair the cosmetic quality of the surface finish. Materials known to give such a reaction include:

- Acid-catalyzed resins
- Urethane-based tooling blocks
- Polyurethane lacquers and sealants
- Some phenolic resins
- Some polyester resins
- Some acrylic paints

Where possible, the reaction should be prevented by the use of CS727 model sealer (PDS 1013), which forms a physical barrier against the migration of potential reactants. Where the use of CS727 is either inappropriate or not possible, a model pre-release treatment may be used to neutralize the potentially reactant material prior to tool lay-up. The use of such pre-release treatment has been shown to either eliminate or minimize the effects of inhibition reactions, depending on the severity of the problem. When these factors have been considered and suitable action has been taken, the lay-up of the tool may begin.

- The LTM 10 pre-preg may be prepared by cutting preforms as appropriate; the material may then be refrozen in airtight packages. When removed from refrigerated storage, allow the material to thaw until it is frost free, or for a minimum of a half hour, before it is used. Fresh pre-preg removed from refrigerated storage should be sufficient to complete only that day's laminating. Do not remove more than can be used.

- The tool lay-up will consist of outer surfacing plies with several heavier core plies. Recommended lay-up configurations are given for both carbon and glass fiber laminates.

- The pre-preg material should be cut and jointed to allow consolidation into female corners without bridging. Under no circumstances should any single piece of pre-preg be laid up around both a male and a female radius.

- The first layer of pre-preg should be butt jointed. Joints in subsequent layers may be overlapped. The maximum desirable overlap is 3 mm (0.12 in.). Joint positions should be staggered in adjacent layers by at least 15 mm (0.6 in.).

- Individual squares or rectangles of pre-pregs may be used, if desired, to facilitate handling.

- Layer orientations are critical and should produce a mid-plane symmetric laminate.

- Warp directions should be as detailed. Alternative lay-up configurations are permissible. Consult The Advanced Composites Group Ltd. for details.

- The lay-up must be completed and the autoclave cure must be completed within the maximum work life of the LTM material.

- For vacuum bag cured tools, it is not necessary to complete the lay-up within the maximum work life of the LTM material, if the partially completed lay-up is vacuum debulked at the end of each day and fresh material is removed from the freezer and used for each day's laminating. In this way, large laminates can be produced in stages over several days.

14.6.2 Debulk Procedure

This procedure is performed after the first ply of carbon pre-preg has been applied to the model. With glass laminates, two layers of pre-preg should be applied before the initial debulk is done. Additionally, a debulk is performed after the fifth ply on carbon and the sixth ply on glass lay-ups. Vacuum debulks may be done more frequently if required to assist consolidation in the case of difficult model geometries, and they should be conducted daily in the case of tool lay-ups with laminating times in excess of two days. To perform the debulking operation, proceed as follows:

- Apply one layer of pin-pricked release film (hole spacing 10 mm [0.4 in.], minimum hole size 0.4 mm [0.016 in.]). Cut and splice materials as necessary to avoid bridging.

- Apply a nonwoven bleeder/breather of suitable conformability. Use two layers of 130 to 160 gsm or one layer of 340 gsm. Cut and splice material as necessary to avoid bridging.

- Vacuum bag the lay-up, ensuring sufficient slackness is provided in the bagging material to avoid bridging.

- Apply a full vacuum (no less than 25 in. Hg), and debulk the laminate for at least 15 minutes (12 hours or overnight in the case of an intermediate debulk in multistage lay-ups of more than two days duration).

Although the lowest-temperature curing of these three systems (LTM 10) will cure to a good strength at room temperature, it is necessary to initially cure LTM 12 and LTM 16 at slightly higher temperatures. The same applies to LTM 10 if a curing time less than seven days is required. If the end usage temperature must be higher, then post-curing is described after the bagging procedure and initial cure.

14.6.3 Final Bagging Procedure

This procedure is performed after the lay-up has been completed in preparation for curing. Position one or more thermocouples between the first two plies of the tool lay-up.

- Apply a layer of solid release film. This should overlap the edge of the lay-up by at least 15 mm (0.6 in.). Cut and splice the material as necessary to avoid bridging. Note: If the material is to be cured only under vacuum bag consolidation, use pin-pricked release film at this stage.

- Apply a nonwoven bleeder/breather of suitable conformability. Use two layers of 340 gsm. Cut and splice the material as necessary to avoid bridging.

- Vacuum bag the lay-up, ensuring sufficient slackness is provided in the bagging material to avoid bridging.

- Apply a full vacuum (no less than 25 in. Hg), and debulk the laminate for at least 20 minutes at room temperature.

14.6.4 Autoclave Cure

With full vacuum applied to the bag, increase the autoclave pressure to 21 psi. Vent the bag to atmosphere. Raise the autoclave shell pressure to 90 psi. Care should be taken on pressurization not to exceed the 0.5°C (0.9°F) per minute heat-up rate.

Slowly increase the air temperature (not the lay-up temperature), at a rate not greater than 0.5°C (0.9°F) per minute to the desired initial cure temperature. Note: LTM pre-pregs contain highly reactive resin systems, which can undergo severe exothermic heating during the initial curing process if the recommended curing procedures are not followed. Care must be taken to ensure that the recommended heating rates, dwell temperatures, and lay-up/bagging procedures are followed. Also, recognize that the model material, laminate thickness, and insulating effect of the breather/bagging materials are factors that affect the cure behavior. Consult The Advanced Composites Group Ltd. for further information on exotherm behavior.

Initial cure time for these materials, after a slow rise to the chosen temperature, is given at various temperatures for different times. For example:

LTM 10	4 to 5 hours dwell at 55°C (131°F) or
	7 hours dwell at 50°C (122°F) or
	12 hours dwell at 40°C (104°F) or
	19 hours dwell at 35°C (95°F) or
	35 hours dwell at 30°C (86°F) or
	7 days at 21°C (70°F)
LTM 12	3 hours dwell at 70°C (158°F) or
	5.5 hours dwell at 60°C (140°F) or
	12 hours dwell at 50°C (122°F) or
	15 hours dwell at 45°C (113°F)
LTM 16	12 hours dwell at 60°C (140°F) or
	18 hours dwell at 55°C (131°F) or
	26 hours dwell at 50°C (122°F)

14.6.5 Time Limits

Time limits should be checked for each type of material and should be carefully observed. For some of the LTM materials from The Advanced Composites Group Ltd., the following details are tabulated:

Resin System	Work Life at 22°C (72°F)	Shelf Life at -18°C (0°F)
LTM 10	2 to 3 days	>1 year
LTM 12	5 days	>1 year
LTM 16	10 to 14 days	>1 year

Note: The cure times given above are absolute minima. Actual cure times should always exceed those quoted, particularly where temperature deviations, however small, from the intended conditions may have occurred.

On completion of the required dwell time, remove the laminate from the autoclave and allow it to cool slowly to ambient temperature in air before demolding.

14.6.6 Step Post-Curing Procedure After Room-Temperature Initial Cure

If the mold is large and requires a rigid support structure to prevent elastic distortion of the mold surface under its own weight, this should normally be attached before removing the tool from the model. Alternatively, the tool may first be post-cured and then reclamped to the original model for the attachment of the support structure. Care should be taken during demolding because the partially cured tool will be brittle. Full resin strength is attained only after the tool has been post-cured. The correct procedure will vary, depending on the geometry and the size of the tool. If in doubt, seek advice from the pre-preg supplier.

Post-Cure Procedure "A":

- Heat to 60°C (140°F) at <1°C per minute—dwell for 2 hours, then
- Heat to 80°C (176°F) at <1°C per minute—dwell for 1 hour, then
- Heat to 120°C (248°F) at <1°C per minute—dwell for 1 hour, then
- Heat to 150°C (302°F) at <1°C per minute—dwell for 1 hour, then
- Heat to 180°C (356°F) at <1°C per minute—dwell for 1 hour, then
- Heat to 200°C (392°F) at <1°C per minute—dwell for 8 hours, then
- Cool to 60°C (140°F) at 3°C per minute maximum

Post-Cure Procedure "B":

- Heat to 200°C (392°F) at a maximum rate of 20°C per hour—dwell for 15 minutes, then

- Cool to 190°C (374°F) at a maximum rate of 20°C per minute—dwell for 4 hours, then

- Cool to 60°C (140°F) at a maximum rate of 3°C per minute

The following table should be used to determine the post-cure temperature and time required.

Maximum End-Use Temperature	Recommended Maximum Post-Cure Temperature
80°C (176°F)	120°C (248°F) for 10 hours
120°C (248°F)	140°C (284°F) for 10 hours
140°C (284°F)	160°C (320°F) for 10 hours
160°C (320°F)	180°C (356°F) for 8 hours
175ºC (347°F)	190°C (374°F) for 8 hours

Note: All intermediate dwells up to the recommended maximum post-cure temperature must be included. If a support structure is to be attached after completion of the post-cure (as previously detailed), the mold should be supported in the oven to merely prevent gross elastic distortion under self weight. However, LTM pre-preg laminates do not undergo significant softening during post-cure, providing the recommended procedures are followed; hence, the temporary support is not required to hold the molding exactly to profile.

14.6.7 Support Structures

The support structures may be manufactured from solid laminate or honeycomb sandwich panels. If the support structure is too rigidly attached to the tool (e.g., using wet-laminated cleats), the support structure should be manufactured from the same type of reinforcement as the tool itself. When attaching the support structure to a tool, the tool should be clamped to the original model to avoid the possibility of mechanical distortion. If an egg-crate support structure is to be used, a nominal gap of 2 mm (0.08 in.) or more should remain between the support boards and the tool back face, in order to prevent mark-off at the support positions. See Figure 14.3. The support structure should be manufactured from the same type of fiber reinforcement as the tool.

A high-temperature laminating resin should be used for manufacture of the cleats. A full range of premolded sections offered by The Advanced Composites Group Ltd. includes solid and honeycomb panels, round and square tubes, and a range of angles, "I" beams, gussets, tees, and capping strips. These are made from the same resin systems and fiber volume fractions as used in the tooling pre-pregs to ensure that the CTE and T_g are matched.

14.6.8 Initial Release Priming of New Composite Tools

To achieve optimum service life, the following procedure is strongly recommended for all new LTM tools:

1. Remove all traces of release agent and other contaminants from the tool surface. Ideally, the tool should be vapor degreased. Chlorinated solvents may be used.

2. Apply either one coat of release primer or three coats of release agent in accordance with the manufacturer's instructions. Bake at 80°C (176°F) for 15 minutes.

3. Apply a second coat of release primer or three further coats of release agent. Bake at the component cure temperature for 15 minutes.

4. Apply the standard component release procedure in addition to the above procedures.

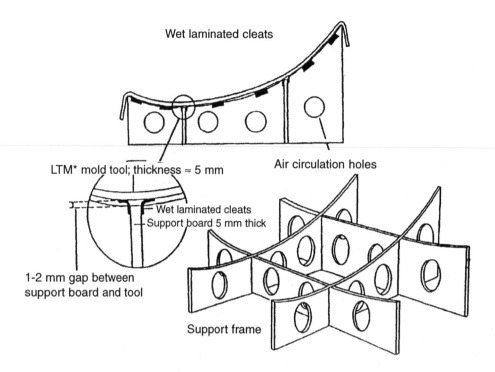

Fig. 14.3 Egg-crate tooling, with a nominal gap between the support boards and the tool back face. (Courtesy of The Advanced Composites Group Ltd.)

5. In the event of a buildup of release agent, resin, or other surface contamination developing on the tool surface, remove such buildup by using a fine rubbing compound. The first four steps must be repeated before any further component cures are performed on the tool. In no other circumstances should abrasive materials be used on LTM composite tools.

The release priming procedure should be repeated at intervals that will depend on the specific tool and process parameters used. The above information has been extracted from Ref. 14.3, courtesy of The Advanced Composites Group Ltd. Although extensive, these extracts are only a part of the information supplied, and they have been included at length here to show that tool making is a serious business requiring considerable knowledge and experience to obtain good results. As with any complex activity, a learning curve exists, and experimentation with small tools is strongly recommended before attempting the manufacture of large items. Similar pre-pregs are available from Cytec, Aerotech (Toolmaster), and several other companies. Ref. 14.4 is another useful book on this topic.

14.7 References

14.1 Bishop, W., "Tooling for Advanced Composite Materials," ETH Zurich Composites Seminar, Aero Consultants (United Kingdom) Ltd., Stonehill, Stukeley Meadows Industrial Estate, Huntingdon, Cambridgeshire PE18 6ED, UK, 1989.

14.2 Vacarri, John A., "Pre-Preg for Tooling Composite Parts," *American Machinist,* October 1991.

14.3 "Manufacturing Procedure for LTM 10, LTM 12, and LTM 16 Mould Tools," The Advanced Composites Group Ltd. Composites House, Adams Close, Heanor Gate Industrial Estate, Heanor, Derbyshire DE7 7SW, UK.

14.4 Murphy, John, *The Reinforced Plastics Handbook,* Elsevier Advanced Technology, ISBN 1-85617-217-1, 1994.

Chapter 15

Metal Bonding

15.1 Introduction

Adhesive bonding has been used longer than most of us realize. Casein glues were used in ancient Egypt, although mainly for items that carried no load and that were stored in a dry environment (e.g., items from the tomb of Tutankhamen). In the real world of wind, weather, moisture absorption, and large temperature variations, adhesives are not as durable. More accurately, their ability to "stay stuck" is not good. Animal glues used in furniture do not last long if the furniture is allowed to remain outdoors in the rain. The durability of adhesive bonds seems to be a greater problem than the durability of the adhesives. This aspect will be covered later in this chapter.

15.1.1 History and Requirements

Some historical dates in adhesives usage and development are as follows:

1500 BC   Casein glue is used by the ancient Egyptians.

1910 AD   Phenol-formaldehyde glue is introduced.

1930   Casein is rediscovered and developed.

1942   Urea-formaldehyde is an improvement on casein.

1942   Polyester is used with fiberglass on early radomes.

1943   Phenol-polyvinyl formal Redux 775 is developed for metal bonding.

1948   Epoxy resins are first used, and their production expands rapidly.

1972   Polyimides and bismaleimides are developed to meet the need for high-temperature adhesives and resins.

1978   Cyanate ester resins marketed for printed circuit boards

According to Hartshorn in Ref. 15.1, the production of adhesives on an industrial scale began approximately 300 years ago. The birth of modern structural adhesives can be dated from around 1910, with the introduction of phenol-formaldehyde resins. The animal hide glues used in the construction of wooden aircraft in World War I were known to have problems with moisture. Thus, attempts were made to protect joints by binding them with varnished tape, but these were not completely successful. Most aircraft were shot down or otherwise damaged before this became a problem; however, it would be interesting to know how many were lost as a result of structural failure caused by disbonded joints. Casein glues were a significant improvement and came into use in aircraft and yachts around 1930. In the second year of World War II, the first De Havilland Mosquito flew on November 25, 1940. See Figure 15.1.

The Mosquito flown at the Biggin Hill Airshow (United Kingdom) in 1993 was in very good condition, more than 50 years after it was made. This high-performance aircraft was known as "The Wooden Wonder." It was made from plywood and balsa core sandwich construction for the fuselage and a double plywood sandwich with spanwise stringers of spruce (later Douglas fir) for the wings and tail. The wings had two spars with tip-to-tip top and bottom booms of laminated spruce boxed with plywood webs. The grain of the balsa core was along the fuselage length and parallel to the skins, rather than at 90° to the skins as it is today in sandwich panels. The earliest Mosquitoes were made with casein

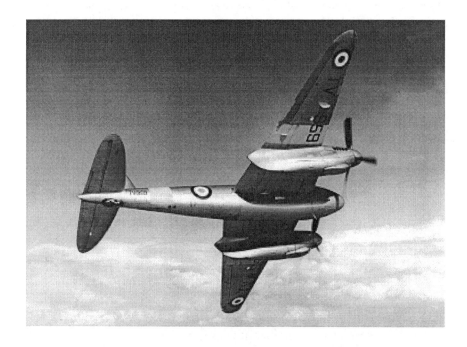

Fig. 15.1 Photo of the De Havilland Mosquito, taken in the late 1940s.

glue. This casein glue suffered from fungal growth and was quickly replaced with a new urea-formaldehyde glue with the trade name of Beetle resin. Beetle resin was developed by Dr. Norman de Bruyne at Aero Research Ltd. at Duxford, near Cambridge, United Kingdom. It was introduced in 1942 and was more resistant to humidity and temperature cycling than casein. Similar materials are produced today by Ciba under the trade name of Aerolite. Even with the greatly improved urea-formaldehyde glue, the lifetime was not too long when the Mosquito was used in Malaya under the hot/wet conditions prevailing there. However, under good storage conditions, the life of a wooden airplane may be longer than that of a metal one. The oldest flying Mosquito is now more than 50 years old.

The expected service life of a military or civil aircraft is always a vital factor to be considered in design and may differ according to the circumstances of the time. During World War II, Lancaster and other bombers had a maximum expected life of 600 hours. Some were lost on their first mission, with a total flight time of even less than 50 hours. For this reason, the inside of the structure was not painted. After World War II, similar aircraft, which were operated for many years, suffered corrosion problems because they were unpainted and needed complete replacement wings. Military aircraft in peacetime commonly have a service life of 40 years; civil aircraft are expected to exceed approximately 30 years or 100,000 hours of flying. Many Boeing 747s and other types remain in service after 27 years of continuous use. A large number of Douglas DC-3s (Dakotas) also are used regularly and are more than 50 years old. These aircraft are too old to have used structural adhesive bonding. Because durability is the essential factor, surface preparation treatments have become critical in achieving the lifetime required.

15.1.1.1 Metal Bonding

Metal bonding began in 1943 when it was used for the first time on the De Havilland Hornet, the successor to the famous Mosquito. On this aircraft, metal bonding was used to bond the aluminum alloy spar boom to wooden structure. The adhesive used was a phenolic resin, toughened with polyvinyl formal, and was given the number Redux 775. This was also developed by Dr. Norman de Bruyne at Duxford, near Cambridge, United Kingdom; hence, the name "Redux" (**Re**search at **Dux**ford) was derived. Today, Redux is the generic term for all Ciba epoxy structural adhesives, in addition to Redux 775. (This trade name has now passed to Hexcel.) Fortunately, Redux 775 has proven to be durable and remains more durable than many epoxies that have been tried to replace it. On the other hand, Redux 775 requires high pressures during bonding because of the water vapor emitted during cure—approximately 100 psi (6.8 bar). This makes production more expensive than for epoxies. De Havilland continued to use metal bonding on the Dove and Heron and later on the Comet, the world's first jet-powered commercial transport. Because of its durability advantages, Redux 775 continues to be used on the Fokker F-50 and F-100 and the BAe 146, now called the RJ Series. Redux 775 was used on the Comet, Nimrod, Trident, Fokker F-27, and VC 10. In 1995, one Comet continued to fly but has since been retired. Many

Nimrods, VC 10s, and F-27s continue in service. Many of these aircraft have been in service for 30 years or longer. The reason for the durability of Redux 775 bonding has not been firmly established, but one possible explanation is that Redux 775 is of approximately neutral pH but slightly acidic. Epoxies contain chlorine because of their chemistry and processes of manufacture. Many epoxy curing agents are fairly alkaline, especially those used with room-temperature curing systems, which can have pH values as high as 11. To avoid corrosion in printed circuit boards and microchips, the epoxies used for these purposes must have a low chlorine content and are specially purified to achieve the low chlorine and other ion contents required. Specifications for aerospace structural use are not as tightly controlled. See Refs. 15.2 through 15.4.

Tests on Redux 775 have shown that its water content at saturation is much higher than the worst epoxy tested; however, its durability is good. This suggests that the chemical reactions at the interface and the importance of chlorine and other chemicals require further study. Aluminum alloys are amphoteric; that is, they react with both acids and alkalis. Experiments have shown that general corrosion proceeds as soon as pH 7(neutral) is exceeded. There is evidence that pitting corrosion also will occur below pH 5.5 (i.e., in the mildly acid range). This means that:

• A pH between 5.5 and 7.0 must be maintained at the bond line, or

• Corrosion inhibitors must be incorporated into the adhesive and/or primer, or

• The phenolic resin makes chemical bonds with the oxide produced on the aluminum alloy by anodizing, which increases the resistance of the oxide layer to hydration

Brockmann (Ref. 15.3) believes this results from the production of surface chelate complexes with the metal oxide. Pocius *et al.* (Ref. 15.5) state that phenolics are known to bond strongly to aluminum oxide and that their slightly acidic pH helps this. Therefore, epoxies require carefully controlled chemistry, pH adjustment, and the incorporation of corrosion inhibitor to help them to compete with the original Redux 775. The electrical conductivity of an adhesive, the same adhesive when saturated with water or other fluids, and the conductivity of any materials leached from an adhesive may also be important to corrosion.

Part of the aluminum alloy boundary ring of a radome was recently found corroded in many places. All of these places were located where the ring had been in contact with a conductive paint that should have been removed at the points of contact. This tends to confirm the above comment. One of the values of chromic acid anodizing may be that the oxide layer is not electrically conductive.

Analysis of distilled water, in which a cured epoxy pre-preg (Ciba Fiberdux 914) had been immersed for 400 days, gave the following results:

pH	7.3
Chlorine Content	2.63 ppm
NO_3	0.1 ppm
PO_4	< 0.1 ppm
SO_4	0.55 ppm

Although these figures are low, some other trials showed that a sample of 7075-T6 corroded rapidly in distilled water alone but suffered only slight pitting in the same water in the presence of an acrylic adhesive, which reduced the pH to 5.5. Compared with other epoxies, these results are low in terms of ppm of ions; however, the pH is high enough to be a potential problem. Bolger *et al.* (Ref. 15.2) quote chlorine contents as high as 943 ppm for some epoxies tested. Such materials certainly would aid corrosion. They also note that some curing agents emit acid vapors, and others emit alkaline ones; thus, the choice of curing agent is important when the possibility of corrosion exists.

15.1.1.2 Epoxy Adhesives

Epoxies are the most common adhesives used for metal-to-metal bonding in aircraft and other applications, if we exclude the vast amount of polyester resin used in composite boat construction. Epoxies began to be used in 1948, and their production and development has increased at a colossal rate. Development work on epoxies continues to improve toughness, water resistance, and long-term durability. Several versions are produced according to the end-use performance required. Epoxies normally are made to cure at room temperature, 120°C (248°F), and 180°C (356°F). Materials in each curing temperature range are produced by many manufacturers. The room-temperature versions are two-part pastes; the hot-curing versions are film adhesives or pre-preg matrix resins. Hot-curing one-part pastes with latent curing agents also are used. Some repair materials have been developed to cure at 95°C (203°F) and have adequate high-temperature properties to repair parts originally made with 120°C (248°F) and 180°C (356°F) curing film adhesives and pre-pregs. These have been made as films and as matrix resins. Two-part paste adhesive and matrix resin systems curing between room temperature and 80°C (176°F) and having good hot/wet properties also have been produced. See Chapter 2 for more details of adhesives.

15.1.1.3 Requirements for Adhesives

The following could be called a "wish list" for adhesives and varies greatly, depending on the end use. Nonetheless, the following are desirable attributes for adhesives:

- Low cost
- Easy mixing (two-part types)
- Easy application
- No health hazards

- High strength
- High toughness
- Good high-temperature performance, including hot/wet performance
- Unlimited durability
- Short cure time
- Simple cure cycle and low cure temperature
- Long storage life at room temperature
- Electrically nonconductive

15.1.2 Principles of Adhesion

Cleanliness is critical at all stages of adhesive bonding. Unfortunately, adhesive bonding demands understanding and application of the appropriate conditions. Cleanliness alone is insufficient for good bonding.

15.1.2.1 Adhesion Theory

It is important to consider why any material adheres to any other material. Why does the adhesive stick to the wood, metal, plastic, leather, or other material? Why use an adhesive? If two pieces of cured adhesive are put together, they will not adhere to each other, nor will two pieces of wood or metal. A clue can be found by considering that very finely ground and polished plates show some degree of adhesion when wrung together. In spacecraft, doors and hatches will stick shut unless designed not to stick shut. This suggests that part of the answer is in placing materials very close to each other. In space, the vacuum is advantageous for adhesion.

Recognize also that only some of the following theories apply to each type of adhesive bonding. We shall consider carefully those theories applicable to metals.

- **Electrostatic Theory:** This theory considers the two parts of an adhesive joint to be similar to the two plates of a condenser. The electrostatic charge arises only between two dissimilar materials (e.g., adhesive and adherend) and is electronic in nature. The theory is not well proven but applies, at least in part, with some materials. For example, if an adhesive tape is attached to an insulating material when both are dry and then the tape is stripped rapidly in the dark, sparks can be seen. Bond strength of silver on PMMA (Perspex) increases with elapsed time, can be reduced to zero by removing the charge with a dose of radiation, and will build up again when radiation ceases. See Ref. 15.6. For polymers, there is an electrostatic component of adhesion.

- **Diffusion Theory:** This theory assumes that if long-chain polymeric materials such as adhesives are brought into contact, the surface molecules diffuse from one polymer to another. When two lumps of masticated rubber join together in this way, it is called autohesion. The ease of diffusion depends on factors such as temperature, molecular weight, and degree of branching. Evostik and other contact adhesives work in this way.

- **Mechanical Anchoring Theory:** An instinctive belief is that a rough surface is good for adhesion. This is not true in the case of wood because planed surfaces provide stronger joints than sandpapered ones. It is true of titanium, where chemically produced rough surfaces are better than smooth ones. Two descriptions of rough surfaces would be either pits or inkwells. Pits will increase the surface area to be bonded and thus can explain some of the benefits of roughening. Inkwells, which are the shape that a dentist produces before filling teeth, obviously rely on mechanical interlocking rather than any form of adhesion. Inkwells increase bond strength considerably but are less likely to occur. See Figure 15.2.

- **Physical (Adsorption Theory):** This theory, based on interactions between molecules, depends on van der Waals forces (named after a Dutch physicist). These forces are explained in terms of quantum mechanics and are the statistical average attractions of electrons in a molecule at a given instant. The most important factor is that these are extremely close-range attractive forces, and anything that increases the distance over which they must work is detrimental to bond strength.

For metal bonding, only two of the above theories apply, namely, the adsorption theory and the mechanical anchoring theory. The electrostatic theory cannot apply because any charge would be conducted away by the metal. The diffusion theory cannot apply because the polymeric material of the adhesive cannot diffuse into the metal, although the theory could apply slightly to diffusion into fine pores in unsealed anodized surfaces. Except for a little mechanical keying, we rely on adsorption of the adhesive on clean, reactive, high-energy metal surfaces that may, in some cases, be rough and have an increased surface area. Two points then become important. First, any contamination will reduce the surface energy and the possibility of any chemical bonds, while increasing the distance over which the van der Waals forces must operate. Second, van der Waals forces are close-range attractive forces. They operate over a limited distance, which explains why two rough surfaces placed together will not adhere to each other. For this reason, low-viscosity primers, which flow and wet a surface, also provide a good base for an adhesive. The reason for using a liquid

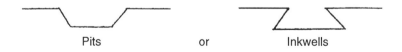

Fig. 15.2 Pits and inkwells.

adhesive is that it can get into this necessary close-contact range in which van der Waals forces operate. The adhesive also must have a lower surface energy than the surface to be joined.

Table 15.1 shows forces involved in adhesive bonding. The first three listed are included in van der Waals forces. The last three are different types of chemical bond.

Table 15.1
Physical and Chemical Forces Involved in Adhesive Bonding

Dispersion forces (Fritz London)		10 kcal/mole
Polar forces	Dipole/dipole forces	5 kcal/mole
	Dipole induced forces	0.5 kcal/mole
Chemical bonds	Hydrogen bonds	12 kcal/mole
	Covalent bonds	15–170 kcal/mole
	Ionic bonds	140–250 kcal/mole

Polar forces are not always present but are helpful when they are present.
Chemical bonds are even more short range than van der Waals forces. See Ref. 15.7. When they can be achieved, they are also stronger and more durable.

Figure 15.3 shows results obtained by Tabor and Winterton, who measured van der Waals forces directly, using crossed cylinders of mica. Their results show how close range these attractive forces are. Bearing in mind that the normal or nonretarded van der Waals forces are the ones most important to adhesion, good bonding will occur only if the adhesive is within 0 to 10 nanometers of the surface. One nanometer is one millionth of a millimeter (10^{-9} meter), or approximately one thousandth of an inch divided by 25,000. This is approximately the same as the thickness of a human hair divided by 75,000. Obviously, these are short-range forces.

Tabor and Winterton in Ref. 15.8 had to allow for an adsorbed monolayer of water (a layer of moisture one molecule thick, or half a nanometer thick) on the surface of the mica cylinders. The attractive forces of interest operate only to approximately 10 nanometers, or the thickness of 20 water molecules. Because the attractive forces weaken with distance, the adhesive must be closer to 0 than 10 nanometers from the surface. Even the monolayer of water makes a difference; it occupies that most important first half nanometer. On the

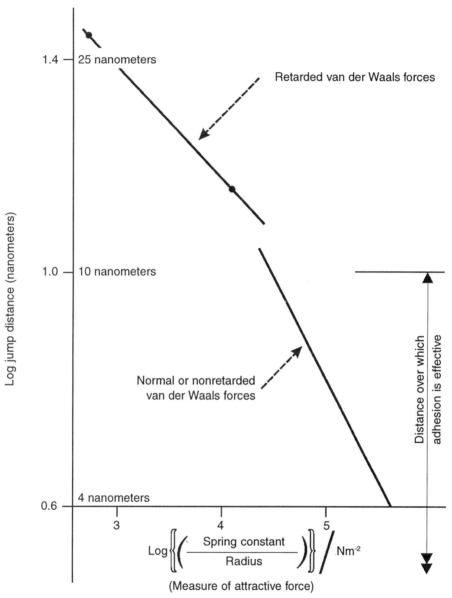

Distance effect on van der Waals forces
One nanometer = thickness of human hair ÷ 75,000

Fig. 15.3 van der Waals forces. (Courtesy of The Royal Society [United Kingdom].)

surface of the earth, adhesion is difficult. However, in the hard vacuum of outer space, metals such as aluminum and copper must be treated to prevent them from sticking. Clearly, that last half nanometer is important.

15.1.2.2 Environmental Durability

Having learned how to make materials adhere to each other, we want them to remain that way, sometimes for 30 years or longer in a hot/wet environment. That is a real challenge. Obtaining a strong initial bond can be difficult, but it is easy compared with obtaining a durable bond, especially if sustained stress is involved with high temperature and humidity. Only the best chemical treatments have any hope under these conditions. Figure 15.4 from the work of Minford (Ref. 15.9) compares different surface preparations. These tests used a freeze/thaw cycle on bonds made with a room-temperature curing epoxy, bonding test pieces of 6061-T6 aluminum alloy. The test cycle used was relevant to aircraft and is detailed as follows:

Wet Soak	24 hours at 73°C (165°F)
Freeze	24 hours at -34°C (-30°F)
Thaw	24 hours at 77°C (170°F)

Although the paper cited in Ref. 15.9 was published in 1974, the results confirm the importance of good chemical surface preparation. Since then, both chromic acid anodizing and phosphoric acid anodizing have been found to be the most durable surface preparations and superior to the original FPL chrome/sulfuric etch if they are used unsealed. This means rinsing in clean water at a temperature less than 60°C (140°F). Clean water is defined here as purified water having a conductivity below 10 mS/m (milli-Siemens/meter), sometimes expressed as 100 micro-Siemens/cm and/or when the silica content exceeds 5 ppm w/w (as SiO_2). See Ref. 15.10. Fluorine content must be below 1 ppm. Rinse water used after any treatment process must meet the requirements of that particular process specification. (See Ref 15.12, pages 149 to 151.) Any distilled or deionized waters used for rinsing must meet these requirements. Spray rinsing is recommended in preference to using a rinse tank in which contamination slowly accumulates. Most people believe that unsealed anodizing provides a better mechanical component of adhesion than sealed anodizing. Minford's results as discussed in Ref. 15.9 show the sealed anodizing to be less effective than an etch. With all these treatments, an optimum exists for chemical content of the bath solution, immersion time, and current density when anodizing is used. The best results are obtained only under optimum conditions. Because these points are important, it is possible to produce bad surface treatments, in spite of all the expense, if the optimum specification is not maintained.

Sealed anodizing is not permitted for adhesive bonding and should be used only when corrosion protection and nothing more is required. Clad 7075 doublers or repair parts should not be used in bonded metal repairs. Corrosion may occur at the bond interface.

15.1.2.3 Temperature Effects

The effect of temperature on the durability of bonded joints can occur in several ways:

- Moisture is absorbed into the adhesive faster at high temperatures and reduces T_g.
- Reduced T_g means that more creep will occur at low temperatures in joints under load.
- High temperatures mean that any chemical reactions in the joint will occur more rapidly. These could be degradation of the adhesive itself or corrosion at the bond line.

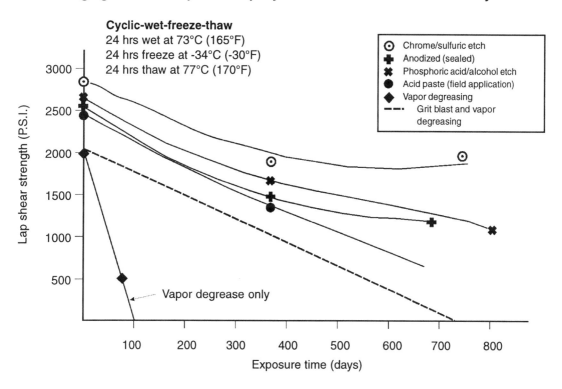

Fig. 15.4 Effect of surface preparation on durability. (From Ref. 15.9. Courtesy of Adhesives Age.*)*

15.1.2.4 Humidity Effects

Humidity means moisture uptake by the adhesive, which reduces T_g. It also may cause corrosion in the bond line and, in the case of poor surface treatments, will eventually displace the adhesive from the metal surface. If mechanical abrasion is used as the surface preparation, disbonding may require only a day under total immersion in water if high stress is applied (e.g., in a wedge test).

15.1.3 Advantages and Disadvantages of Metal Bonding

As with all other processes in industry, advantages and disadvantages exist for metal bonding. Good design means choosing the best process for each job. This means that cost, performance, and durability must be considered with many other factors for each application. Adhesive bonding will be used if it is perceived to have an overall advantage when all the various factors have been considered and, in some cases, where no other choice is possible (e.g., the bonding of honeycomb and other core materials to thin skins for the manufacture of sandwich panels).

Advantages of Adhesive Bonding of Metals:

- Adhesive bonding provides a more uniform distribution of stress than with bolts and rivets. However, the stress distribution in a bonded joint also is not uniform.

- Adhesive bonding has no refinishing cost as, for example, occurs when spot welding or riveting must be filled and refinished.

- Adhesive bonding improves fatigue resistance.

- Good vibration damping occurs as a result of adhesive bonding.

- Stiff and strong structures can be made with adhesive bonding.

- Adhesive bonding makes it easy to join different materials, and a wide range of materials can be bonded.

- Adhesive bonding can easily be combined with other fastening methods (e.g., used in combination with spot welding or riveting).

- Large and small areas can be bonded.

- Adhesive bonding seals and joins in one operation.

- Adhesive bonding is easily automated.

Disadvantages of Adhesive Bonding of Metals:

• Bonded joints cannot easily be separated and thus should not be chosen if regular dis-assembly is required.

• Surface preparation is critical to durable, strong bonding.

• Care is required with joint design.

• The maximum and sometimes the minimum service temperature of the product may be limited by the mechanical properties of the adhesive selected. For metal bonding, the upper service temperature is more likely to be limited by the adhesive than the metal.

• The curing mechanism required for the adhesive may not be practical for the intended application.

• With two-part and other room-temperature curing adhesives, time is required before handling strength is reached. Critical parts cannot be used until full bond strength has been reached; this may require twenty-four hours to seven days for room-temperature curing systems. Hot-curing adhesives must be given the correct cure cycle of temperature and pressure for the required time, and the part cannot be used until this has been completed.

• Bonded joints cannot be used in the presence of solvents that damage the adhesive being used.

• Bond lines should be sealed from water and other fluids as much as possible.

• Health and safety regulations must be observed when using adhesives and surface preparation methods.

• Adhesive materials require careful storage and the observance of shelf-life limitations. Some adhesives must be stored at low temperature, whereas others must be stored at room temperature if that is specified.

15.2 Surface Preparation Methods

Although many factors affect the performance of adhesive bonding, one key factor will decide the success or failure of adhesive bonds, especially under hot and/or wet conditions. This factor is surface preparation. Both the type of treatment selected and the quality to which it is performed will have more effect on durability than any other factor and possibly more effect than all other factors combined. However, surface preparation cannot be taken too seriously. The best available treatment should always be used for structural parts and nothing less than the minimum specified in the SRM. The following methods are listed in an approximate order of merit. The first method listed is the worst, and the methods improve progressively throughout the list. This order is: abrasive cleaning, Pasa-Jell, hydrofluoric acid etching, grit blast/silane, Alodine or Alochrom 1200, FPL etch, chromic acid anodizing, phosphoric acid anodizing, and others as detailed in the following pages. This order may not be the same for all material surfaces and should be treated as a rough guide. What is certain is that the chemical methods generally are better than the mechanical methods and that anodizing of all types and methods is the best of the chemical treatments for those materials that can be anodized. To achieve good preparation, the right equipment must be provided and properly maintained, sound training must be given, and the surfaces must remain free of all contamination until bonding occurs. Note that Boeing has reduced the recommended time between anodizing and priming from four to two hours maximum after its phosphoric acid non-tank anodizing process (PANTA). Boeing B.747-400 SRM 51-70-09, page 8, states:

> Caution: Do not touch the dried anodized surface. Protect surfaces from contamination after drying the anodized surface. Primer must be applied within two hours maximum after anodizing. Do not touch the finished surfaces with bare hands. Wear clean, white cotton gloves when handling parts. Reduction in quality of anodized surfaces will occur if these conditions are not met.

Experience has shown that skin oils and acids will drastically reduce adhesion; thus, clean, white cotton gloves must be worn. Do not touch the surfaces even with these gloves. At the risk of seeming pedantic, it is necessary to define the meaning of "handling only by the edges." This is shown in Figure 15.5.

Frequent reference to cleaning prior to bonding is made in the data sheets supplied by adhesive manufacturers. However, cleaning alone is inadequate, although many data sheets inadvertently imply that it is enough. Allen, Chan, and Armstrong in Ref. 15.11 tested several adhesives on a range of surface preparations. All of these preparations would have been said to provide clean surfaces, but the bonds produced were better in some cases than in others. Dirt on the surface is always bad for adhesion. However, in addition to

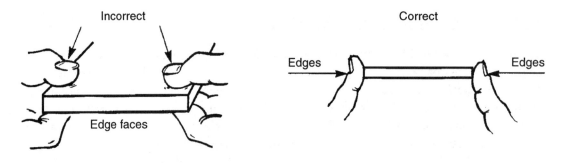

Fig. 15.5 Definition of "handling by the edges."

cleanliness, a surface treatment often is required to raise the surface energy of the part so that it is higher than the surface energy of the adhesive. The greater the difference between the surface energy of the bonding surface and the adhesive, the better the adhesion. This can be achieved by raising the surface energy of the part, lowering the surface energy of the adhesive (e.g., by heating), or both. If the surface free energy of the part is low, the adhesive will not spread or wet the surface properly. This need to increase the surface energy of the part to be bonded applies to all surfaces and especially to plastics, which naturally have a low surface energy. The use of a thin liquid primer may slightly improve adhesion when the surface treatment is poor. If the surface treatment is good, the primer will not increase the bond strength; however, the durability should improve, especially if the primer contains a corrosion inhibitor appropriate to the metal being bonded. Good surface preparation will also greatly improve durability, compared with poor surface treatments.

15.2.1 Abrasive Cleaning

This process is one stage better than solvent wipe, but it should not be used for metal parts of any importance or when moisture is present. It can be done with abrasive paper, Scotch-Brite abrasive pads, or abrasive powders in the grit size range of 180 to 320. This process can remove dirt and some surface oxides to expose a fresh surface. It can be used for some lightly loaded parts if they are in a dry environment.

15.2.2 Pasa-Jell

Several types of Pasa-Jell are supplied for treating different metals. They are an improvement on abrasion, and most (if not all of them) contain hydrofluoric acid (HF). This dissolves human tissue and therefore must be used only with the utmost care and all specified safety precautions. Snogren in Ref. 15.12 says that they are proprietary blends of mineral acids, activators, surfactants, inhibitors, and thickening agents for treating metal surfaces. Pasa-Jell 107 and 107M are formulated specifically for prebond treatment of titanium alloys (not pure titanium) and are reported to deposit a conversion coating that provides a receptive surface for adhesion. Type 107 is a thixotropic Jell for application to surfaces by brushing. Type 107M is a liquid for immersion applications. Pasa-Jell 105 is a paste for aluminum alloy; Pasa-Jell 105M is the liquid version for aluminum alloys. Pasa-Jell 101 is a thixotropic Jell for stainless steels. All these materials should be used carefully and in accordance with their data sheets and the local safety regulations for the materials involved.

15.2.3 Hydrofluoric Acid Etching

This process uses a dilute solution of HF in water and is more dangerous than Pasa-Jell, which is a mixture of several materials. Its use should be restricted to applications having no acceptable alternative, and it should be used only by well-trained and properly equipped staff, in accordance with the SRM.

Warning: Goggles giving full protection, masks, and protective clothing must be worn when working with HF. Supplies of HF antidote gel, a calcium gluconate gel, must be available before starting work.

Method of Use: For external use only. Store above 8°C (46°F). Treat acid splash immediately. First, wash with clean water for one minute. Apply the antidote gel freely, and massage into affected areas. Reapply with massage until pain is entirely relieved. Continue applications and massage for another 15 minutes to prevent reversion.

Warning: Use as many tubes as required by directions, but discard all tubes after opening once.

This gel contains 2.5% calcium gluconate and is available from IPS (Industrial Pharmaceutical Service Ltd.) Healthcare, Manchester WA14 1NA, UK.

In the Boeing 757 SRM, the use of HF is detailed in conjunction with a chemical conversion coating (Alodine) as follows:

- Solvent wipe the surfaces surrounding the repair location. Remove all visible oil, dirt, and grease.

- Mask off the repair area with aluminum foil tape for approximately 76 mm (3 in.) beyond the area to be bonded. Do not allow paint stripper to enter existing bond lines because damage to bonds will occur.

- Remove the organic finish from the repair area per 51-20 of the 757 Maintenance Manual.

- Abrade the faying surface until a water-break-free surface results. (For a definition of the water break test, see Chapter 10, Section 10.1.6.) Use Scotch-Brite aluminum oxide pads or Tycro Type 3A aluminum oxide wheels.

- Verify the water-break-free surface for 30 seconds.

- If the film does not remain continuous for 30 seconds, repeat the preceding two steps (i.e., abrade the faying surface and verify the water-break-free surface for 30 seconds).

- Dry wipe with clean gauze to remove dust or debris.

- Proceed with HF etching to Section 2D, paragraph 3 of Boeing 757 SRM 51-20-9, immediately following the cleaning procedure.

Warning: Do not allow acid to contact the skin or eyes. Wear rubber gloves, eye and face protection, and protective clothing. If contact occurs, immediately flush with large quantities of water and seek medical attention. Acid will burn the skin and eyes if contact is made.

Caution: Do not allow HF solution to contact component surfaces other than the repair area for which it is intended, because damage to paint and resin systems will occur. Do not allow HF solution to contact steel parts, because structural damage will occur.

Section 3. Hydrofluoric Acid Etching Procedure from Boeing B.757 SRM 51-70-09, Page 8, Section 2D, Paragraph 3:

* Moisten (do not saturate) clean gauze with 2% by volume HF solution. Wipe the abraded area briskly with the pad. Allow the solution to remain on the surface for 15 to 30 seconds.

* Remove the solution by wiping with a gauze pad dampened with clean water. Apply the chemical conversion coating surface treatment before the surface dries.

Apply the Alodine chemical conversion coating surface treatment per SRM 51-20-01 and the following:

* Check crevices with blue litmus paper for possible acid contamination. If the litmus paper turns red, acid is present, and the rinsing and blotting steps must be repeated.

* Check the surface for the presence of a powdery coating by wiping with a clean cotton cloth immediately after processing. If a powdery coating is found, the area must be stripped and chemical conversion coated again.

Follow all subsequent procedures for priming and bonding, which follow in the SRM.

Snogren in Ref. 15.12 gives HF as a surface treatment for magnesium. In this case, a 25% solution by weight in water is specified. The procedure given is as follows:

* Immerse the parts for 10 minutes at a temperature of 25°C, ±3°C (77°F, ±5°F).
* Rinse in deionized water and dry at 65°C (150°F).

15.2.4 Grit Blast/Silane

This method has been used extensively in Australia and is particularly useful for repair work when a patch must cover aluminum skin and existing steel fasteners. A good grit blast using 180 to 320 aluminum oxide grit and a distilled water rinse can be followed by the application of one of the few silane solutions found to be suitable. A range of silane treatments are made by Dow Chemical, and others are made by Union Carbide. An epoxy/silane typically is used and acts as a coupling agent; one end of the molecule bonds to the metal oxide, and the other end bonds to the epoxy resin. In Australia, it is common to make a solvent solution of silane using a variety of solvents. Methanol, ethanol, MEK, and 1,1,1-trichloroethane all were successful, as were water-based solutions. See Ref. 15.13. In the United Kingdom, a water-based silane solution is more commonly used. In their paper cited here as Ref. 15.13, Pearce and Tolan studied eight different organofunctional silane coupling agents as adhesion promoters. Both the solvent solutions and the water solutions of silane have been shown to be successful. As with most of these surface preparations, the solution must be mixed correctly. In addition, it must be used within the specified time scale and dried properly before applying the adhesive. In service, grit/blast and silane has proven to be effective. Pearce and Tolan also found that the pH of the silane solution affected durability. This is particularly interesting in view of the remarks previously made concerning the durability of phenolic adhesives compared to epoxies. See Section 15.1.1 of this chapter.

15.2.5 Alodine or Alochrom 1200

Alodine is the name used in the United States; Alochrom is the name used in the United Kingdom. Alodine is the registered trade name of a proprietary procedure marketed by the American Chemical Paint Company. A range of products is made, and they are chemically identical, provided that the number following the name is identical (i.e., Alodine 1200 and Alochrom 1200 are the same product).

The Alochrom 1200 process has long been used to restore local areas on aluminum alloys where the anodizing has been damaged or lost through corrosion. Its primary purpose is to restore corrosion protection. This process converts the aluminum surface into a complex aluminum chromate, which provides good corrosion resistance and is an excellent key for paint and adhesive bonding. This process may be applied only to aluminum and aluminum alloys if there is no risk of entrapping the Alochrom solution, and it should always be supplemented with an organic coating.

Many similar products are available, and some are specified in SRMs. Two such products are Iridite 14-2, marketed by the Witco Chemical Company, and Turco Alumigold, marketed by Turco Products Inc. Many others are listed in Ref. 15.12 and are qualified to MIL-C-5541. When called up in an SRM, the detailed procedure for each type is given. This includes pH adjustment using concentrated nitric acid. This procedure must be performed strictly to the instructions given, especially when this treatment is used as a preparation for adhesive bonding.

Parts to be treated off the aircraft shall be thoroughly degreased to SRM standard procedures using high flash cleaning solvent, followed by a wipe with prepaint cleaning gauze wetted with prepaint cleaning solvent.

For aircraft exteriors or parts of the exterior, the area to be treated shall be washed and rinsed to SRM standard procedures using a fine abrasive pad for the cleaner application. If a water break appears on the wetted surface within 20 seconds or less, the area shall be wiped with prepaint cleaning gauze wetted with prepaint cleaning solvent and then retested for water breaks. (See Section 10.1.6 for a discussion of the water break test.) This process shall be repeated until a water-break-free surface is achieved, after which the Alochrom treatment shall immediately follow.

The Alodine or Alochrom solution must then be made strictly in accordance with the SRM, using all specified safety precautions, and should be allowed to stand for at least one hour before use. It is usual to adjust the pH of the solution by adding concentrated nitric acid until the pH is within the required range. This requires strict observance of all the safety precautions in the SRM. The Alodine/Alochrom process may be applied by any of the following methods:

- **Brush or Swab:** Alochrom may be applied by either brush or swab to the surface, which shall be kept wetted with the solution until a satisfactory golden film is obtained. This usually occurs within approximately two to five minutes, depending on the temperature. The surface shall then be rinsed thoroughly.

- **The Paper Technique:** The area to be treated is covered with a strong absorbent paper soaked in Alochrom solution, which is pressed down firmly to achieve good contact with the metal surface. The paper shall be kept moist by brushing more solution on it as required. After approximately 1.5 minutes, the corner of the paper should be carefully lifted to observe the progress of the treatment. When a pale golden film is visible, the paper is removed carefully and may be used on another area. The treated area shall then be thoroughly rinsed.

- **Spray Application:** Alochrom solution may be sprayed using approved equipment. The minimum air pressure required to produce a droplet spray must be used. Care shall be taken to ensure that the air pressure and nozzle size are not such as to produce atomization of the Alochrom solution. Any sign of mist formation indicates either too high an air pressure or too small a nozzle. Operators shall wear full respiratory and skin protection. The surface shall be kept wet with the Alochrom solution until a light golden film has formed. The treated area shall then be thoroughly rinsed.

Rinsing: All rinsing shall be done with clean running water until there are no signs of contamination of either the water or the treated surface.

Drying: The wet film is normally air dried, but this may be hastened with clean warm air. Wet Alochrom film is fragile, and no rubbing or abrasion should be used.

Painting or Adhesive Bonding: The treated surface shall be primed as soon as possible and, in any case, within four hours of the Alochrom treatment. During this period, the surface shall be kept under clean, dry conditions.

The above is a summary of the requirements for using this process. Check with the SRM and the manufacturer's instructions, and follow the correct application and safety procedures. Wedge testing is necessary to confirm the performance of this process as a preparation for adhesive bonding and any fine details that may have to be observed. Discussion with a technician, who has used this process, revealed that the strength of the bond was not good if bonded directly to the Alochrom surface. The technician found that a light abrade with a fine grade of Scotch-Brite after the coating had dried gave a stronger bond. This could remove any loose material. Further testing would seem to be required to refine this process. The instructions for Alochrom given above state that if a powdery surface is found, the surface should be stripped and the process repeated.

15.2.6 FPL Etch

This is the oldest treatment for aluminum alloys and was developed by the Forest Products Laboratory (hence, the name FPL) because, in the early days of adhesive bonding, those accustomed to gluing wood were the most expert at bonding and thus were given the job of developing a suitable system for treating aluminum alloys. The FPL etch is a sulfuric acid–sodium dichromate mixture. It has served well and has been bettered only by anodizing processes. FPL etch must be performed to an approved specification using the correct solution mixture, temperature, immersion time, and rinsing procedures. A modified FPL etch is now used; this requires the addition of a small amount of copper to the solution before use. Check the specification to ensure that the solution content is correct in every way. See Ref. 15.14.

15.2.7 Chromic Acid Anodizing

This process also must be performed to an approved specification such as the UK Specification DEF STAN 03-24/Issue 2 (Ref. 15.10) after precleaning to DEF STAN 03-2/Issue 2 (Ref. 15.15). Again, the correct solution mixture, immersion time, temperature, and rinse procedures are vital, and the voltage/current density must be maintained to specification. This is an excellent treatment for aluminum alloys prior to bonding if the cold rinse is used, leaving the pores in the oxide unsealed. The standard chromic acid anodizing process is unsuitable for use on aluminum alloys containing a specified maximum total of 7% copper + nickel + iron, or more than 6% copper. A silicon content of greater than 7% in castings may adversely affect the properties of the film. See Ref. 15.10 for alternative procedures for these materials. See Ref. 15.16 for a comparison of durability with different surface treatments. See also Ref. 15.14.

15.2.8 Phosphoric Acid Anodizing

This is the most favored method for aluminum alloys at the time of this writing and may be performed to Boeing Specification BAC 5555 or equivalent specifications. It produces a larger pore size than does chromic acid anodizing in the unsealed condition. Debate has continued for many years about its performance compared to chromic acid anodizing. The results are sufficiently close that if a chromic acid anodizing plant already exists in a company, it would be difficult to justify changing it to phosphoric. However, the toxicity of chromic acid is greater; thus, it presents more serious waste disposal problems. Environmental concerns are causing a move toward phosphoric acid anodizing. For long-term durability, an unsealed anodizing treatment obviously is the best available surface preparation for aluminum alloys. See Ref. 15.14. The tank method is used on the production line for new parts and is to be preferred when possible in service. However, this is impossible on the flight line, in the repair workshop, or on assembled parts. Thus, portable methods have been developed to take the surface preparation to the job in those situations where the job cannot be taken to the anodizing tank. Some methods of doing this are discussed here.

Three methods of non-tank anodize are described. The first two methods have been developed and tested by Boeing. The third method has been tested by the author in conjunction with Selectrons (now Sifko) and also was found to be durable, although with a performance slightly below a tank anodize. This slightly reduced performance compared to a tank anodize was also found for the first two methods. Similar equipment is available from the Metadalic Company.

Note that all chemical prebonding treatments for aircraft structures and components must not be used where fasteners exist, for two reasons:

1. The chemicals may not be compatible with the fasteners

2. The chemicals could penetrate the structure through the bolt or rivet holes and cause corrosion

Use grit blast and silane at these positions. See Section 15.2.4.

15.2.8.1 Phosphoric Acid Containment System (PACS)

This system is somewhat similar to the PANTA process described in the next section. However, in this case, the phosphoric acid is in liquid form, without any thickening agent, and is drawn across the surface to be anodized by vacuum during the anodizing process. The equipment is supplied by ATACS. The new Model 0810 Phosphoric Acid Anodizing Containment System (PACS) is a self-contained, portable system used for the preparation of aluminum alloy surfaces prior to bonding. It is designed to be used with a 12% phosphoric acid solution, but the power source and materials also may be used for the PANTA process. The system may be used on-aircraft as well as in the workshop. A double vacuum bag contains any leaks or ruptures of the inner bag, preventing exposure or dripping of the 12% phosphoric acid solution. The outer vacuum bag also holds the inner bag in place, especially if the work is being done on the underside of a wing or on a vertical surface. After application, the acid solution can be reused a few times, if several jobs must be done within a short time, or it can be neutralized with a powder supplied by ATACS and then safely poured down an ordinary drain. See Figure 15.6 .

Model 0810 Data Sheet

ATACS new Model 0810 (PACS) is a self-contained, portable system used for preparation of metal surfaces prior to metal or composite bonding. It is designed to be used with a 12% phosphoric-acid solution, but the power source and materials may also be used for the PANTA process. Easy set-up and selectable voltage or current regulation make this unit ideal for on-aircraft and shop application.

The lightweight, compact size of this unit allows more flexibility in applications where time and space are a problem (i.e., "AOG" repairs).

Designed, manufactured, and patented in the United States, the Model 0810 is readily available for shipment.

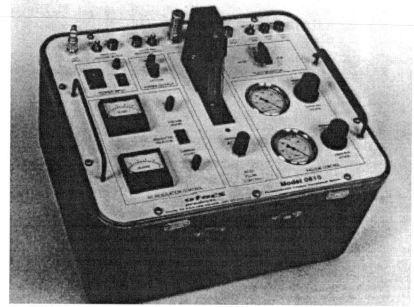

The ATACS Model 0810 - patent number 4,882,016

Features:
- Compact, lightweight, rugged
- Easy to read instrumentation
- Simple, clean operation on-aircraft or at depot level
- Pre-measured process materials

The 0810 design incorporates high reliability with corrosive-resistant gauges, fittings, and instruments. Initial unit purchase includes ATACS 0813 Bulk Materials Kit and 0814-6PK Individual-Jobs Pack. The combination of kits provides sufficient material to complete six applications. Additional pre-measured kits are also in stock.

Benefits:
- Easy to use and store
- Less hazardous application process
- More environmentally safe disposal
- Aircraft downtime and expenses reduced
- Process savings mean fast payoff

Fig. 15.6 Phosphoric Acid Containment System (PACS). (Courtesy of ATACS Products Inc.)

This method is easier to use under a wing or on a vertical surface than the PANTA process because it can be completely set up before acid is drawn through the system. It cannot be used where acid could leak through a crack or penetrate honeycomb. The treated area should always be slightly larger than the area to be bonded because the flow of acid around the edge of the area is not as good as it is at the center, and the anodizing may be of lower quality around the edge. A suitable allowance of extra area will help to ensure that the area to be bonded is of the best quality for a good structural bond. It will also help to ensure that the extra treated area around the repair will give good adhesion for paint. See Ref. 15.17.

15.2.8.2 Phosphoric Acid Non-Tank Anodizing (PANTA)

This was the first non-tank anodizing method developed by Boeing. The 757 SRM 51-70-09, page 4, states, "Use the PANTA method when making an elevated-temperature cure repair with BMS 5-101 adhesive." The PANTA method is used to prepare aluminum surfaces for metal-to-metal and metal-to-honeycomb bonding, to obtain a more durable bond than the other non-tank methods of surface preparation. This process is not to be used as a substitute for or an alternative to chemical conversion coating to produce a protective finish for aluminum skins. When use of the PANTA method is impossible, treat metal-to-metal bonding surfaces with HF/Alodine. The ideal room temperature for anodizing ranges from 21 to 30°C (70 to 85°F); however, 21 to 38°C (70 to 100°F) is an acceptable range.

The Boeing 757 SRM 51-70-09 gives the following procedure for the PANTA process. See also Figure 15.7.

- Mask off undamaged areas, crevices, and fasteners with aluminum foil tape and plastic film to prevent acid contamination.

- Protect repair parts and work areas by placing plastic film (Mylar) over bench tops and other working areas.

- Apply a uniform coating of gelled 10 to 12% phosphoric acid to the aluminum surfaces. Gelled phosphoric acid compound can be made by adding Cab-O-Sil, Grade M-5, or Cab-O-Sil, Grade PTG, to a 10 to 12% phosphoric acid until thickened.

- Place a layer of gauze over the acid coating. Alternate layers of acid and gauze until two or three layers of gauze are built up. The gauze must be completely saturated. The gauze can be saturated prior to application by squeezing the acid through the gauze by hand, using rubber gloves for protection.

- Cut a piece of stainless steel screen to the size of the repair area.

- Place the stainless steel screen over the coating, and apply another coating of the gelled phosphoric acid over the screen. See Figure 15.7. Note: Make sure that the stainless steel screen does not contact any part of the aluminum surface being anodized.

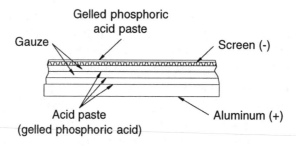

Section through anodizing lay-up

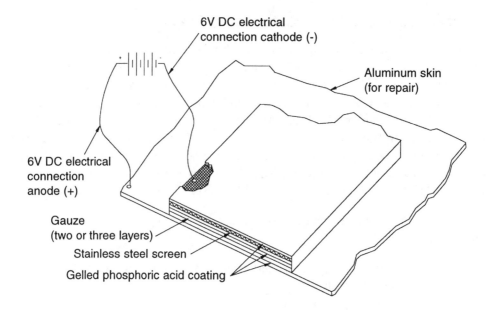

*Fig. 15.7 Boeing Phosphoric Acid Non-Tank Anodize (PANTA) process.
(Courtesy of Boeing Commercial Airplane Group.)*

In the case of patch doublers or skin details in which both surfaces are being anodized, it is preferable to anodize and prime one side at a time. When one side has been anodized and the primer has cured, mask off that side and repeat the procedure for the other side. Parts should be oriented so that the surface being anodized faces upward. This allows gases to escape which, if trapped against the surface, could cause smutting. Check setup before proceeding: screen as cathode (-); aluminum surface as anode (+).

- Apply a DC potential of 4 to 6 volts for 10 to 12 minutes. Current should range from 1 to 7 amps/ft^2. Note: A dry cell battery may be used to supply the voltage during anodizing.

- At the end of the anodizing time, open the circuit and remove the screen and gauze.

- After immediate removal of the screen and gauze, flush the area with a spray of clean water to remove the phosphoric acid from the anodized surface. Begin to remove the phosphoric acid in 2.5 minutes or less from the time the anodizing current is switched off, preferably within one minute. Flush the area for at least five minutes. If you cannot flush the area, lightly rub the area with gauze that is soaked in clean water.

- Air dry for at least 30 minutes at room temperature, or force air-dry at 60 to 71°C (140 to 160°F). Do not touch the dried anodized surface. Do not apply tape to the surface. Protect surfaces from contamination after drying anodized surface. Primer must be applied within two hours after anodizing. Do not touch the finished surface with bare hands. Wear clean, white cotton gloves when handling parts. Reduction in the quality of the anodized surface will occur if the above actions are not observed.

- Check the quality of the anodized surface. A properly anodized surface will show an interference color when viewed through a polarizing filter, rotated at 90° at a low angle of incidence to fluorescent light or daylight. An interference color is when the original color changes to a complementary color when the polarizing filter is rotated 90°. See Figure 15.8. Machined or abraded surfaces sometimes are difficult to inspect for color. Rotation of the polarizing filter is required because some pale shades of yellow or green are so close to white that, without a color change inspection, they might be considered no-color, which would falsely indicate no anodic coating.

- If no color change is observed, repeat the steps described above.

- Prime the surfaces as required by the SRM.

15.2.8.3 Metadalic/Sifco Selective Plating (UK) Ltd.

While at British Airways, I tried the Selectrons process (now Sifco) using chromic acid anodize. Wedge tests have shown it to be successful, and the process remains in use. In this method, a cotton gauze pad is soaked in chromic acid solution, and the handle to which the pad is attached is wired as the cathode to a power pack. The work surface is the anode. A time is calculated, depending on the area to be treated, and set on the power pack. The pad is dipped frequently in chromic acid to keep it and the surface wet, and it is moved gently and evenly around the surface to give all parts of the area an equal treatment time. At the end of the calculated time, the area is treated with distilled or deionized water until all traces of acid have been removed. Then it is dried. Selectrons is now part of Sifco, and the Metadalic Company produces similar equipment. It is reported that phosphoric acid anodizing by this method is unsuccessful because the oxide produced is weaker and is damaged by the movement of the pad. Further testing may be necessary to confirm this.

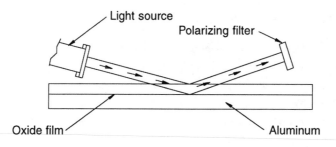

Polarized light test - Verification of anodic oxide film

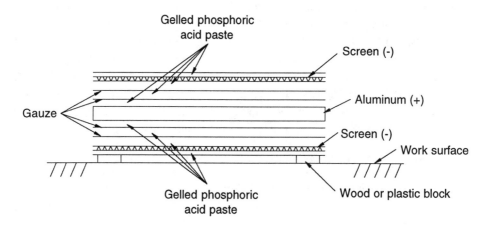

Section through anodizing lay-up for both sides of aluminum detail

Fig. 15.8 Polarized light test for verification of anodic coating.
(Courtesy of Boeing Commercial Airplane Group.)

15.2.9 Other Surface Preparations

Other surface preparations that have been tried are Deoxidine 202, a phosphoric paste etch, and Deoxidine 624, a liquid phosphoric acid etch. These are a considerable improvement on solvent wipe and mechanical abrasion but are not as good as grit blast and silane or Pasa-Jell processes. Performance is far below FPL etch and either chromic or phosphoric acid anodizing, but the processes are cheaper and easier to perform. Many specialist treatments exist for steels, stainless steels, and titanium and its alloys. Magnesium requires special treatment, as do copper and zinc surfaces. For aircraft applications, the SRM will provide details if these materials require treatment. In other industries, consult the adhesive manufacturers for the appropriate treatment for each material and end use. For a good outline of available treatments, see Refs. 15.12,15.15, 15.18, and 15.19.

15.3 Primers

Primers are commonly used in metal-to-metal bonding, for the reasons given below. Although primers from one manufacturer often can be used with adhesives from another, it is common to use primers, honeycomb potting compounds, and adhesives from one manufacturer to ensure compatibility.

15.3.1 Reasons for Using Primers

There are several reasons for using primers, and each makes its contribution to good bonding. See Refs. 15.7, 15.20, and 15.21.

Treated Surface Protection: Always prime an anodized or otherwise treated surface within two hours of treatment and drying. This minimizes the risk of contamination by dust or vapors and ensures a good bond between the primer and the surface. The use of a primer reduces the surface free energy from a high value to a lower one, which reduces the attraction of contaminants to the surface and also provides a surface that is highly compatible with the adhesive. Primed parts can be stored for some time before bonding if they are kept in a clean, dry area and are covered. An exception to this is when the primer data sheet recommends bonding within 24 hours. Protection of the surface should continue to be provided between priming and bonding, even when the period of time is less than one day.

Wetting: Primers are always of a lower viscosity than adhesives; therefore, they wet the surface more readily. This helps to ensure a good bond. Primers are most commonly used with film adhesives. These adhesives must melt and flow before they gel; thus, the heat-up rate is important, and primers can help to achieve a good bond. Primers sometimes are used with two-part paste adhesives.

Bond Performance: The use of a primer does not always significantly increase the initial bond strength. The main value of a primer is bond durability, as discussed in the next section.

Bond Durability: Durability of bonds, especially under hot/wet conditions, is the main reason for using primers. Phenolic primers have served well on aluminum alloys. Corrosion-inhibited primers have been shown to improve bond durability. Silane-containing primers also are useful and improve the stability of the substrate in moist environments.

15.3.2 Primer Types

Primers are usually developed to suit particular adhesives and therefore will be of the same chemical type as the adhesive or of a compatible chemical type.

Corrosion-Inhibiting Adhesive Primers (CIAP): If the reason for using a primer is to increase long-term durability, a corrosion-inhibiting adhesive primer (CIAP) will be used. These primers previously contained strontium chromate or similar materials that today are considered toxic, and efforts are being made to produce organic corrosion inhibitors or other types that are safe but equally effective. Ensure that all safety procedures are used when applying primer coatings.

Non-CIAP Types: Some primers are made without inhibitors. These are used for surface protection prior to bonding and may improve durability if they contain organosilanes.

15.3.3 Application of Primers

Primers must be applied strictly in accordance with their data sheets if good results are to be obtained. This means that the correct air temperature and humidity range must be observed.

Primers also must be used within any shelf life limitations recommended by their manufacturers. Most primers are hygroscopic and absorb water. Therefore, tins of primer requiring refrigerated storage should be allowed to warm to room temperature before use. This will ensure that condensation does not occur when they are opened. For the same reason, lids should be replaced on the tins as soon as the required amount has been taken. Tins of primer should be returned to refrigerated storage as soon as possible. The type of primer specified may also depend on the metallic or other surface to be bonded. Check the SRM to ensure that the correct primer and adhesive are used for each repair task.

Agitation: Primers usually are solvent solutions and therefore should be put into a paint shaker or other agitation device to ensure that they are thoroughly mixed before use. Some data sheets emphasize the importance of thorough mixing before use.

Method of Application: Primers are most commonly applied by spraying, especially if large areas must be primed, because spraying is the most efficient and economical method of applying the solution. Apply a light and continuous coating, enough to ensure complete coverage of the area to be protected but not enough to give it a pronounced color. Aim to achieve the lightest coating that will cover the surface. Primers also may be applied by brush or roller if small repair areas must be primed.

Thickness: The thickness of a primer coating is critical and must be controlled carefully. The greatest and most common error when priming is to use too much. The feeling that a little more is helpful is definitely not true. The easiest method is to obtain a color chart from the primer manufacturer, which will contain three sections: too thin, appropriate thickness, and too thick. If any error is made, it should be on the side of making the primer layer too thin. A thick coating will cause a reduction in the strength of the bond by preventing proper evaporation of solvents in the solution nearest to the metal surface. If the coating is applied by brush or roller, extra care will be needed to avoid an excessively thick primer coating. The permitted thickness for primers normally ranges from 0.004 to 0.01 mm (0.00015 to 0.0004 in.), with a maximum in a specified percentage of the area of 0.02 mm (0.0008 in.). Primer tends to be thicker at edges and around bolt holes because of surface tension effects; therefore, it should not be allowed to exceed 0.01 mm (0.0004 in.) in these areas. Primer and paint thickness may be measured using an eddy current type thickness gauge made by Elcometer Instruments Ltd. See Ref. 15.21. Choose an instrument that can measure in the thickness range required and on the type of material required. These instruments are made to operate on ferrous metals, nonferrous metals, or both. Thus, the correct model must be specified. The correct probe also must be selected to suit the surface shape involved (i.e., there are limits to the minimum concave and convex radii that can be measured). They can be powered by dry or rechargeable batteries and can be fitted with computerized printout if required. Primers are seldom used on composite materials, but use a suitable instrument if this must be done.

Paint thickness also may have to be measured. If so, note the following points for composite materials. This instrument will work on carbon-fiber but will not work on fiberglass, aramids, or other nonconductive composites for which no nondestructive method of paint thickness measurement is known at present. Also, note that conductive coating thicknesses cannot be measured. Eddy current modules can measure nonconductive coatings on nonferromagnetic metals including paint, anodizing, vitreous enamel, plastic, epoxy, varnish, powder coatings, glass fiber, or rubber on aluminum, copper, brass, austenitic (nonmagnetic) stainless steel, and other similar materials. Fortunately, they also can measure primer and paint thickness on carbon fiber composites. See Figure 15.9.

Fig. 15.9 Elcometer 300 primer and paint thickness gauge.
(Courtesy of Elcometer Instruments Ltd.)

15.3.4 Curing/Drying

Some primers only require drying by solvent evaporation and are noncuring. Others may specify a curing cycle either separately or as a co-cure with the adhesive. Some noncuring primers require that the adhesive shall be applied within 24 hours. Check the data sheet to ensure that each primer used is treated as specified. Drying may be done at room temperature or in a well-ventilated flameproof oven or drying tunnel at the specified temperature. When a curing cycle is specified, a suitable oven may be used.

15.3.5 Thickness Verification

The thickness of a primer coating is important and can be verified by either of the following methods. In the case of primers, ensure that the primer is not too thick.

Isoscope: This instrument is an eddy current type that effectively measures primer coating thickness. It is made by Fischer in Germany and is specific to nonferrous materials. The name Isoscope has tended to become generic for this type of instrument, but it is actually the trade name of Fischer for this product. Fischer calls its instrument for ferrous materials the Deltascope and its instrument for both ferrous and nonferrous materials the Dualscope.

Similar instruments made in the United Kingdom are the Elcometer 300 Series, made by Elcometer Instruments Ltd. in Manchester. The instruments are supplied with probes for use with ferrous or nonferrous materials, and they indicate which type of probe is connected. They can print maximum, minimum, and average thicknesses from a range of readings taken over the coated area. Readings can easily be switched from imperial to metric units. See Ref. 15.21. These instruments must be calibrated for the range of thickness to be measured, using standard samples supplied by the manufacturer. The thickness of adhesive primer coatings falls at the bottom of the range of measurement; therefore, calibration should be done carefully to ensure accuracy. These instruments can be used to develop a color chart by measuring the thickness of one, two, three, or more spray coat passes. They also can be used to establish the number of passes needed to achieve a given thickness. In this case, the passes would require automatic control with a specified primer viscosity, nozzle size, and spray pressure. Each primer has a different viscosity range; therefore, the trials would have to be repeated with every new primer introduced. See Figure 15.9.

Color Samples: A color chart from the manufacturer is the simplest method of checking primer thickness, but it does not perform any measurement.

15.4 Handling Primers

See the data sheet and MSDS for safety precautions. See also Chapter 6, Safety and Environment. Most primers are flammable because they are solvent solutions and thus should be stored and handled accordingly. Some modern primers are water based, but they require careful handling because of their epoxy content. Primed surfaces should not be handled between priming and bonding because they are bonding surfaces. If possible, avoid touching them, even with clean gloves. If contact is inadvertently made, a light solvent wipe should be used immediately before bonding. Some primers require refrigerated storage. Check each data sheet, and store as required. When refrigerated storage is specified, the primer must be warmed to room temperature and thoroughly mixed before use.

Safety Precautions: See Chapter 6 for safety precautions associated with solvents and resins because primers are usually resins in a solvent. See also precautions for static discharge because primers are flammable. When working with primers, use eye protection and the correct type of gloves. Ensure good ventilation, or wear a breathing apparatus if necessary.

Protection of Prepared Surfaces: To avoid contamination, primed surfaces need protection until bonding occurs. Primed parts should be stored in a clean, dry area and should be covered with clean kraft paper or polyethylene sheeting, or they should be wrapped in a clean plastic bag. In some cases, the primed surface can be covered with a protective tape that must be removed before bonding. The 3M Company makes suitable tapes. Only the correct tapes, designed to not leave residues on the bonding surface, are suitable for this task. 3M-855 tape is an example, but similar tapes may be available from other suppliers. Check that the type intended for use is suitable for the purpose and is compatible with the primer in use.

15.5 References

15.1   Hartshorn, S.R. [ed.], *Structural Adhesives—Chemistry and Technology,* Plenum Press, New York, ISBN 0-306-42121-6, 1986.

15.2   Bolger, J.C. and Mooney, C.T., "New Epoxy Adhesives for Compliance with MIL-A-87122," 17th National SAMPE Technical Conference, Kiamesha Lake, New York, October 22–24, 1985, published by the Society for the Advancement of Material and Process Engineering (SAMPE), Covina, CA.

15.3   Brockmann, W., Hennemann, O.D., Kollek, H., and Matz, C., "Adhesion in Bonded Aluminum Joints for Aircraft Construction," *International Journal of Adhesion and Adhesives*, Vol. 6, No. 3, July 1986, pp. 115–143.

15.4   Sedlacek, B. and Kahovec, J. [eds.], "Cross-Linked Epoxies," Proceedings of the Ninth Discussion Conference, Prague, Czechoslovakia, July 14–17, 1986, Walter de Gruyter, Berlin, ISBN 3-1-010824-0 (Europe), ISBN 0-89925-401-2 (USA), 1987.

15.5   Pocius, A.V., Wangsness, D.A., Almer, C.J., and McKown, A.G., "Chemistry, Physical Properties, and Durability of Structural Adhesive Bonds," *Adhesive Chemistry*, Lee, Lieng Huang [ed.], Plenum Publishing, New York, 1984.

15.6   Wake, William C., *Adhesion and the Formulation of Adhesives,* Applied Science Publishers, London, ISBN 0-85334-660-7, 1976.

15.7   Packham, D.E. [ed.], *Handbook of Adhesion*, Longman Scientific and Technical, ISBN 0-470-21870-3 (USA), ISBN 0-582-04423-5 (UK), 1992.

15.8   Tabor, D. and Winterton, R.H.S., "The Direct Measurement of Normal and Retarded van der Waals Forces," Proceedings of The Royal Society, A-312 Figure 5, 1969, pp. 435–450.

15.9   Minford, J.D., "Effect of Surface Preparation on Adhesive Bonding of Aluminum," *Adhesives Age*, July 1974.

15.10   "Chromic Acid Anodizing of Aluminum and Aluminum Alloys," Defence Standard DEF 03-24/Issue 2, September 14, 1988, Ministry of Defence, UK.

15.11   Allen, K.W., Chan, S.Y.T., and Armstrong, K.B., "Cold-Setting Adhesives for Repair Purposes Using Various Surface Preparation Methods," *International Journal of Adhesion and Adhesives*, October 1982, pp. 239–247.

15.12   Snogren, R.C., *Handbook of Surface Preparation*, Palmerton Publishing Co. Inc., New York, 1974.

15.13   Pearce, P.J. and Tolan, F.C., "Epoxy Silane—A Preferred Surface Treatment for Bonded Aluminum Repairs," National Conference Publication No. 91/17, International Conference on Aircraft Damage Assessment and Repair, Melbourne, Australia, August 26–28, 1991, Australia, The Institution of Engineers.

15.14   Wegman, Raymond F., *Surface Preparation Techniques for Adhesive Bonding*, Noyes Publications, Park Ridge, NJ, ISBN 0-8155-1198-1, 1989.

15.15   "Cleaning and Preparation of Metal Surfaces," Defence Standard DEF 03-2/Issue 2, 1st March 1991," Ministry of Defence, UK.

15.16   Armstrong, K.B., "The Long-Term Durability in Water of Aluminum Alloy Joints Bonded with Epoxy Adhesive," *International Journal of Adhesion and Adhesives,* Vol. 17, 1997.

15.17   Wegman, R.F. and Tullos, R.T., *Handbook of Adhesive Bonded Structural Repair,* Noyes Publications, Park Ridge, NJ, ISBN 0-8155-1293-7, 1992.

15.18   "Common Bonding Requirements for Structural Adhesives," Boeing Specification BAC 5514.

15.19   "Phosphoric Acid Anodizing of Aluminum for Structural Bonding," Boeing Specification BAC 5555.

15.20   Ellis, B. [ed.], *Chemistry and Technology of Epoxy Resins*, Blackie Academic and Professional, Chapman and Hall, ISBN 0-7514-0095-5, 1993.

15.21   Elcometer 300 (leaflet on a coating thickness gauge), Elcometer Instruments Ltd., Edge Lane, Droylsden, Manchester M43 6BU, UK.

Design Guide for Composite Parts

16.1 Introduction

There is a need to design and produce balanced laminates without distortion after cure that can meet all the in-service loads and environmental conditions expected. In addition, many details should be considered when all the laminate analysis has been done, and preferably before it is done. I hope that the following study will prove helpful to those involved in the design of composite parts and structures.

A survey by the Design Task Group of the Commercial Aircraft Composite Repair Committee (CACRC) received 55 responses detailing problems with composite materials in service. A wide range of problems was raised, and the results are of concern and interest in their variety and in those which occurred most frequently. The sample is not statistically sound because there were too many different problems to ensure that all problems occurred often enough to indicate their true importance. For example, the problem of moisture entering honeycomb structures at inserts through the honeycomb is more important than the number of occurrences reported here would suggest. The findings are listed below in order of frequency of reporting and then are discussed individually. Some reports mentioned more than one problem; thus, the total number of comments exceeds 55. After comments on the 55 reports are made, other factors are included from previous experience and study. Similar problems have been grouped together.

16.2 Analysis of Reports Received

The following list summarizes the 55 reports received. These reports cover 24 different topics, although items (j) and (k) are similar. The reports indicate the many different factors that must be considered when designing composite parts. Following the list, each item is covered in detail.

Item	Topic	Number of Reports
(a)	Impact damage to honeycomb panels	8 reports
(b)	Honeycomb punctures due to badly located pipe clips	2 reports
(c)	Leading edge of panels eroded	7 reports
(d)	Damage at latch positions	3 reports
(e)	Damage at fastener holes	3 reports
(f)	Damage when drilling out blind fasteners	3 reports
(g)	Water wicking into composite from rivet holes or impact damage in aramid components	4 reports
(h)	Moisture/fluid ingress in radomes and other composite parts	3 reports
(i)	Water ingress due to bad edge sealing of honeycomb section	2 reports
(j)	Holes through honeycomb, water ingress	1 report
(k)	Tooling holes through honeycomb allow moisture ingress— See also item (j)	1 report
(l)	Sealing honeycomb panel inside skins against Skydrol— See also item (h)	1 report
(m)	Heat shield inadequate	3 reports
(n)	Cracking in surface filler	2 reports
(o)	Damage due to cowlings being slammed shut	2 reports
(p)	Fuselage ice protection plates too small	2 reports
(q)	Parts too large or access panels too small	2 reports
(r)	Lightning damage	2 reports

Item	Topic	Number of Reports
(s)	Damage caused by hot exhaust air	1 report
(t)	Pre-preg tape used for an outer ply	1 report
(u)	Filament-wound parts difficult to repair	1 report
(v)	Panel interference with structure due to flexure under load	1 report
(w)	Acoustic panel, inadequate water drainage	1 report
(x)	Metal fittings, thermal expansion problem	1 report

(a) Impact Damage to Honeycomb Panels: Eight reports were received, and this is a known problem requiring design attention. If weight savings are to be retained, training in careful handling also is necessary.

Several factors must be considered when composite parts are made using honeycomb sandwich structures:

- In the case of panels of sandwich construction, the best bending performance is obtained if core weight is equal to skin weight. If further improvement in indentation resistance is required (e.g., for floor panels), this is best achieved by increasing the honeycomb core density rather than the skin thickness. See Refs. 16.1 and 16.2. See also Figure 16.1.

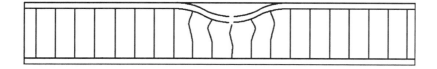

Fig. 16.1 Honeycomb panel.

To improve indentation resistance with a minimum weight penalty, increase the core density because it gives a greater improvement than adding the same amount of weight to the top skin. This principle also should apply if greater impact damage resistance is required of any type of honeycomb panel. Impact damage to current designs of honeycomb structures is a common complaint. Such lightweight structures cannot be made resistant to vehicle or other heavy impacts. If impact damage is expected from runway stones or other unavoidable sources and the occurrence of damage is likely to be frequent, this should be reflected in the design. Impact damage from dropped tools during maintenance also should be considered, and suitable padding should be provided and called up in maintenance instructions. Damage by vehicle impact during turnaround servicing at airports can be handled only by good training and staff discipline. Damage found on honeycomb panels has one advantage: a skin puncture is visible and can be easily repaired. Impact damage to solid laminates is not always visible. Consequently, a level of barely visible impact damage (BVID) requires careful inspection if it is to be found, and a level of unseen damage must be accepted in design.

• Nomex aramid honeycomb has a greater deflection before failure than metal honeycomb; however, if impacted too severely, it fractures. If pressure is applied to the damaged area, the sound is similar to the crushing of a biscuit. The core fails completely. Aluminum alloy honeycomb buckles under impact and retains approximately half its compression strength. For this reason, freight floor panels that carry the freight directly, as opposed to floor panels in freight holds using pallets that serve only as walkways, are best made from aluminum alloy skinned panels with aluminum alloy honeycomb. The metal skin is better able to withstand abrasion from metal parts, and the metal honeycomb is better under impact conditions.

Composite skinned Nomex honeycomb panels are easier to repair because only light abrasion is required to prepare the surface after suitable cleaning and drying. For floor panel repairs, offcuts from the boards used to make the original panels can replace damaged areas and joints made by potting the honeycomb together, followed by overlays of glass fabric to join the skins. Good room-temperature curing resin systems can be used for these repairs.

(b) Honeycomb Punctures Due to Incorrectly Located Pipe Clips: Two reports were received and are also mentioned in (g). Although the problem could be resolved by staff training, design should eliminate it. According to the Murphy's Law principle, which states, "If a fault can possibly occur, then at sometime somewhere it will," it is always better to eliminate a problem by design. That principle is especially important in this case because unseen damage will neither be sealed nor repaired until the problem becomes large. At this stage, the problem will have to be repaired and will probably cause operational delays. Figure 16.2 illustrates an air-conditioning bay door, in which primary damage caused by the clip screw on the air-conditioning duct resulted in secondary problems that increased both the size of the repair and the time required to complete the repair. After the skin was punctured, the aramid layer in the graphite/aramid face sheets wicked water into the core, and large areas of honeycomb had to be removed to clear contaminants. The electromagnetic interference (EMI) shielding layer of aluminum on the inside skin was required to be hot bonded. An 0.05 mm (0.002 in.) foil was found to be difficult to apply or repair.

(a)

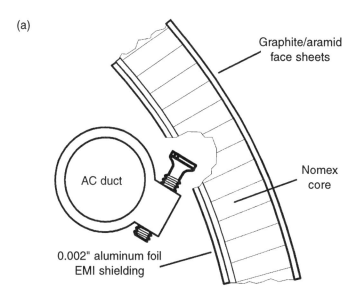

Graphite/aramid
face sheets

Nomex
core

AC duct

0.002" aluminum foil
EMI shielding

Air-conditioning bay door

(b)

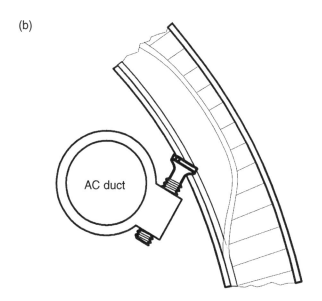

AC duct

Recess in panel to clear screw

Fig. 16.2 (a) Interference problem with a pipe clip;
(b) Proposed solution to the interference problem.

(c) Leading Edge of Panels Eroded: With seven reports, this was the second most frequently reported problem and has been seen in service on many other occasions. Thus, the problem must certainly be given more consideration in future design. The problem was reported for many types of panels, including access doors, engine cowlings, nose landing gear doors, and any panels with a forward edge in the airstream. Correct rigging was mentioned as being important, as well as the possibility that a panel that was flush when closed might be pulled partly into the airstream by aerodynamic loads. Any projection of the panel edge into the airstream will cause more rapid deterioration. Cowlings and any panels that are held shut by latches, rather than screws all around their edges, seem to suffer most. The damage begins as erosion but, if not repaired at that stage, soon progresses to delamination and pieces being broken off. This requires considerable time to repair. The problem sometimes is aggravated by new panels being supplied oversize and then being trimmed to fit. If this is done badly, some delamination may be caused in the process. This has to propagate only for a while before layers are broken off. Design must include one of the solutions illustrated in the diagrams below, or something similar. If a panel must be trimmed to fit, then instructions for fitting leading edge protection, together with part numbers, materials, and techniques, should be included in the relevant Aircraft Maintenance Manual (AMM). This edge protection may be required on other edges in addition to the leading edges, which suffer the most. See Figure 16.3. Alternatively, a metal edging as shown in Figure 16.5(d) may be used.

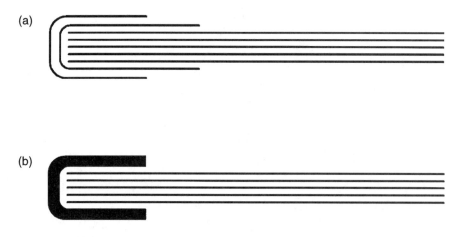

*Fig. 16.3 (a) Wraparound layers of fiberglass at the leading edge;
(b) A metal or plastic edge channel.*

Engine cowlings often are made with several layers of fabric running to the edge, with no wraparound layers to prevent splitting. These items frequently are dragged across the floor on these edges, causing considerable splitting and delamination. Alternatively, plastic blocks or wheels could be built into the design to lift the edges above the floor. This has been done on the Rolls-Royce Trent engine cowling made by BP (now GKN/Westland Aerospace), and the plastic wheels should prevent significant damage. These large and heavy items should be designed with handling in mind. The fitment of comfortable handholds at suitable positions and slinging points for crane lifting would be appreciated.

(d) Damage at Latch Positions: This problem was reported three times and is a known problem that should have received attention long ago because similar problems occurred with metal panels, although to a lesser degree. Latch areas on composite parts often suffer delamination and other damage resulting from screwdrivers being used to open them. Latches for cowlings and access panels should have a metal surround for at least 50 mm (2 in.) beyond the latch all around. This is necessary because, under cold winter conditions, small and stiff latches frequently are opened by stabbing a push button or levering with large screwdrivers (i.e., the 300-mm [12-in.] variety or larger). The second method is shown in Figure 16.4. One alternative would be a larger latch that can be operated easily with a gloved hand. Open latches by hand, rather than with screwdrivers. Design latches that can be opened with a gloved hand because large gloves may be worn in cold weather. Bare fingers cannot be used at subzero winter temperatures. If this cannot be done, ensure that adequate metal plating protection is provided all around the latch so that composite areas are unlikely to be damaged. This may include protecting the next panel or surrounding structure as shown in the proposed alternative design in Figure 16.4.

(e) Damage at Fastener Holes: This problem was reported three times and also has been found on numerous previous occasions. Therefore, the problem requires more consideration at the design stage. More attention must be paid to the specification of correct values and the control of torque loading when tightening fasteners in composite parts, especially countersunk fasteners, and to the specification of lower torque loading values than those used with metals. This is needed to avoid hole wear and edge splitting resulting from over-tightening. Training in this area is also required. Torque values for fasteners in composite components are given in the SRM for each aircraft type and should be followed. For panels requiring regular removal for inspection or maintenance, there may be a need to fit metal edging strips or to use some type of quick release fastener. Where large panels are involved, cut them into small sections at the design stage to ensure that only a few fasteners must be removed frequently. The most frequently removed panels then could be fitted with metal inserts and/or quick release fasteners. Large washers should be used at the back of rivets, special fasteners, or bolts. Percussion riveting should never be used with composites because it causes delamination. Tinnerman washers are recommended under countersunk bolt heads to avoid abrasion of the countersinks during bolt rotation if inserts are not already

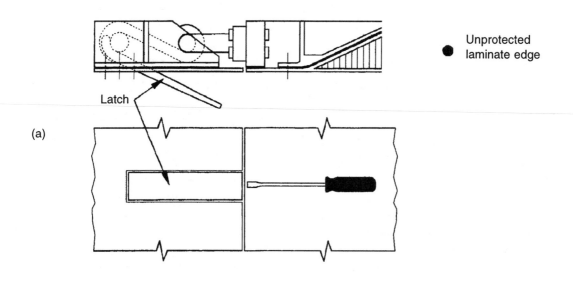

(a)

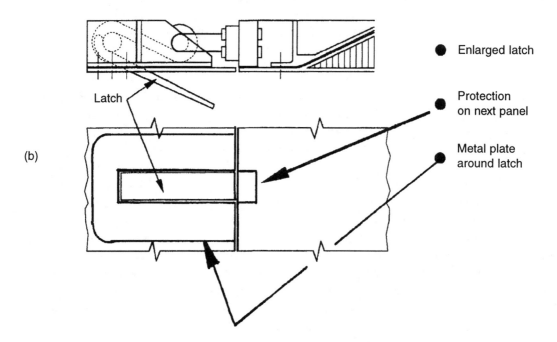

(b)

Fig. 16.4 (a) Original latch design; (b) Proposed alternative latch design.

fitted. Fasteners for composites (especially carbon composites) should be made from corrosion-resistant steel or titanium. Always consult the SRM, and use the specified type. Figure 16.5(a) shows typical fastener hole damage.

Possible Solutions: Inserts at all bolt positions, as shown in Figure 16.5(b), are preferred to Tinnerman washers, which are shown in Figure 16.5(c). Figure 16.5(d) shows two types of metal edges to accommodate countersunk holes.

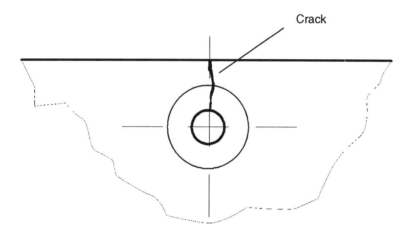

Crack

Overnightening edge - cracks

(a)

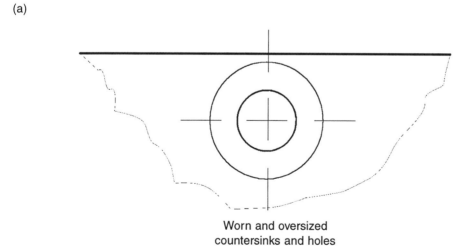

Worn and oversized
countersinks and holes

Fig. 16.5 (a) Typical fastener hole damage.

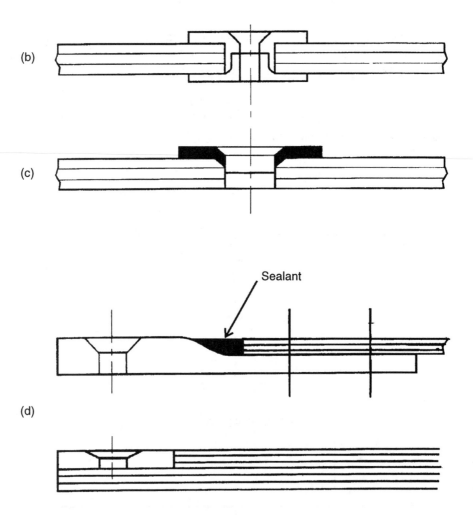

Fig. 16.5 (b) Inserts at all bolt positions; (c) Tinnerman washers; (d) Two types of metal edges to accommodate countersunk holes.

(f) Damage When Drilling Out Blind Fasteners: This was reported three times and is another well-known problem.

Another source of trouble when making repairs is the use of blind fasteners to attach panels to structure. This is common for all control surfaces and engine cowlings and has significant costs in manhours and downtime if panels must be removed for repair. Drilling out titanium fasteners from a relatively soft carbon fiber panel is risky and often causes additional damage and repairs. Blind fasteners should be avoided for any items requiring disassembly for repair.

On rudders and elevators on some aircraft, composite skin panels are both adhesively bonded and riveted where they meet at the trailing edge, in addition to the use of blind rivets at rib and spar positions. This design often causes more damage during disassembly

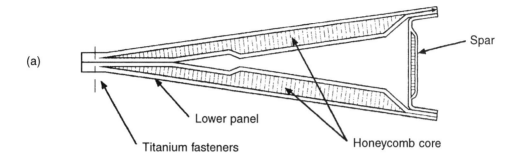

(a)

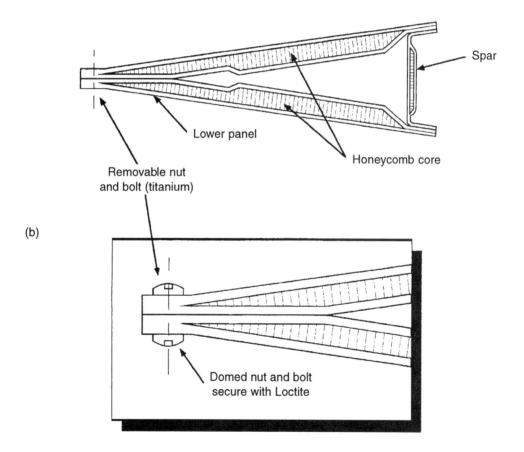

(b)

Fig. 16.6 (a) Avoid this design of trailing edge; (b) Use this design of trailing edge to allow disassembly when necessary.

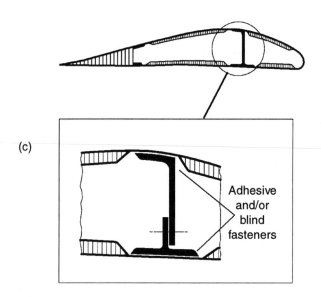

(c)

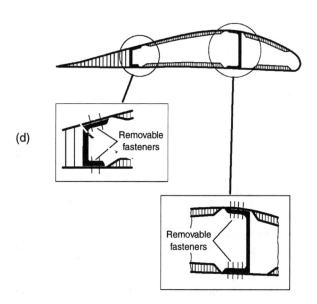

(d)

Fig. 16.6 (c) The existing design of flap structure. This repair requires the breaking of bond lines and/or drilling out fasteners to allow removal of parts. (d) This proposed solution uses bolts and anchor nuts. In this case, all parts of the flap structure that require removal are attached with bolts and anchor nuts to simplify removal for repairs.

and leads to more work than the original damage requiring repair. See Fig. 16.6(a). Use removable nuts and bolts. If one side is in accessible, use anchor nuts. See Fig. 16.6(b). Figures 16.6(c) and 16.6(d) illustrate other examples of this problem with blind fasteners in flap structures.

(g) Water Wicking into Composite from Rivet Holes or Impact Damage in Aramid Components: This problem was raised in four reports. In one case, the primary damage was to a honeycomb-cored pylon fairing caused by pipe clips that were left in a position where fouling occurred when the panel was closed. A secondary observation stated that, "rivets that attach the fairing also provide an avenue for moisture ingress that is greatly aggravated by the wicking action of Kevlar. Fairings became saturated with water and Skydrol." In a report on an air-conditioning bay door, the primary damage was also a result of clamps being incorrectly positioned. However, the report further stated that, "The Kevlar isolation layer (between the outer carbon fiber layers and the honeycomb) wicks fluids into the core of the door. Large chunks of core have to be removed to clear contaminants. The SRM requires 110°C (230°F) cure, and any contaminants will cause more damage at this temperature. Wet lay-up repair is not always approved on graphite. It could be reconsidered, especially for lightly loaded components. Aluminum foil EMI shield also must be hot bonded."

A report on a flap track fairing stated, "Aramid face sheets over aluminum alloy Flex-Core receive much abuse from [Foreign Object Damage] F.O.D. Little speckles develop, and dents and dings occur which cause cracking of the Kevlar face sheets. Kevlar acts as a wick, drawing moisture and Skydrol into the honeycomb. A case occurred of an entire inner ply being separated from the honeycomb." Another report, on an overwing escape slide door, made a similar statement. "Aramid skinned honeycomb door. Doors must be trimmed on installation to fit the existing opening. See also notes on leading edge erosion in paragraph (c). Anything made with Kevlar must be sealed well to prevent fluids from being wicked into the parts. Sealing with resin or an overnight repair is a problem. The wicking problem clearly needs attention because it has been reported in many completely different parts." See Figure 16.7.

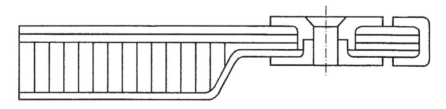

Fig. 16.7 Edge sealing and bonded two-part inserts to minimize the wicking problem.

This wicking problem seems to require the use of adhesively bonded inserts at each bolt position to seal the ends of fiber bundles. Metal or plastic edging strips, adhesively bonded in place with tough flexible adhesives or sealants, also would seem essential with aramid parts, especially if they are trimmed to size on site. These should be called up in the SRM and/or the AMM.

Unfortunately, the problem is not limited to aramids, although they appear to suffer more. Any cut edge of a composite is susceptible to moisture wicking along the fibers. Design with all fiber types should consider this. For example, when making blocker doors for the Boeing 747 and 767 and others, BP Chemicals' Advanced Materials Division (now GKN/ Westland Aerospace) produced holes in the acoustic attenuation face by running the fibers around plugs, during lay-up of the dry pre-form, to avoid drilling holes later. The reasons for this were as follows:

- Reduce the amount of moisture wicking along the fiber bundles that would have been caused by cutting the fibers in a drilling operation.

- Avoid loss in strength of the panel because of cutting fibers.

- Reduce the chance of edge delamination.

Proposed Solutions:

- Change to fiberglass; however, this incurs a weight penalty.

- Use wraparound layers or edge sealing channels. In the case of panels requiring trimming to size before being fitted by an airline or maintenance base, the edge sealing requirements must be included in the SRM and/or the AMM.

- Use a low-viscosity resin to encapsulate Kevlar fibers more effectively.

- The fouling problems, resulting in clips causing damage, could be overcome by either better staff training or designing recesses into the parts to avoid them.

- Rivet and bolt holes could have metal or fiber-filled plastic inserts adhesively bonded in place to seal rivet and bolt hole edges and to minimize damage at these positions. This seems essential with aramid parts and is highly desirable with all fibrous composite parts. These inserts should have thermal expansion coefficients (CTE) as close as possible to those of the composite and should be bonded with tough, flexible resins if they are not to disbond due to thermal stresses. The parts and processes required also should be specified in the SRM and the AMM.

- The case of disbonding from the honeycomb could have been aggravated by the use of aluminum alloy honeycomb without a good surface treatment such as corrosion treatment and resin dip or anodizing. A film adhesive layer is recommended rather than complete reliance on excess resin in the pre-preg when bonding pre-preg skins to honeycomb. This film adhesive should have the lowest possible water uptake. It is usually significantly tougher than the matrix resin in the pre-preg and therefore is more resistant to impact.

- In these cases, the problem has occurred mainly in areas of parts that are not visible, or where the damage may be small and receive no immediate attention. Fortunately, all of these problems can be eliminated by improved design or production processes. The impact damage could be reduced by using a high-density honeycomb core, which would add little weight for the benefit gained. In cases of moisture ingress at points of wear or damage, the best solution (after all redesign has been done and especially before that point has been reached) is to seal allowable damage as soon as possible after the event if repair must be delayed.

(h) Moisture/Fluid Ingress in Radomes and Other Composite Parts:

1. **Radomes:** This was reported three times. One report concerned an aramid radome, but this problem will occur on all radomes if small lightning punctures are allowed to continue in service without repair. Lightning protection strips are seldom carried around the nose because of the effects on radar transmission. Small strikes occur at the nose, where rain impact is hardest. Polyurethane overshoes and painted protection schemes can help to resist rain impact. However, after a small pinhole has occurred, rain can penetrate. Ensure that radomes have an adequate volume fraction of resin and at least three layers of skin so that the skins are not porous because they are resin starved. Also, use a high honeycomb density (i.e., 6 lb/ft^3 is normal) to give the radome some hail and rain impact resistance. However, a radome must transmit the radar signal with minimal losses in order to give the pilot a good picture. Therefore, it becomes a difficult problem in design optimization to produce a radome that will give a good service life while giving a good radar picture. Regular inspection and the correct repair of any damage as soon as possible is the best way to ensure a good life and a good picture.

2. **Other Composite Parts:** To prevent water or other fluid ingress into the honeycomb through woven composite skins, experience has shown that it is necessary to use at least three layers of skin with a 50% resin content. One layer of skin and a shortage of resin leads to considerable moisture ingress and a serious weight gain, which defeats the purpose of using composite materials to save weight. To achieve sealing without significant weight increase, use at least three layers of thin fabric instead of one thick layer. This principle applies to all types of fibers and all fabric weaves. Although this will likely increase the cost, water ingress must be prevented. Similar problems have occurred with cross-plied unidirectional skins when only one layer in each direction

has been used. In such cases, a medium-power magnifying glass is sufficient to show pinholes between the fibers, and removal of the skin has shown large quantities of engine oil in the honeycomb.

Alternative methods for inner skins, which may be cheaper, are to seal the skin by:

- Applying a layer of Tedlar to the surface
- Painting a brush coat of Thiokol or polythioether sealant onto the surface
- Applying an aluminum foil coating to the surface

For outer skins, a good primer and paint finish should be used; for radomes and leading edges, erosion-resistant paints or thin polyurethane rubber anti-erosion coatings often are helpful. These items must cope with rain and hail impact and usually have an outer skin consisting of three layers of a fairly heavy fabric. When thinner skins are used (i.e., fewer than three layers), ensure a good resin content and preferably an external resin coating before painting. If a surface filler is used, it should have a good elongation to failure. In addition, the CTE and hygro-expansion of the surface filler ideally should be similar to those of the skin being filled. Brittle fillers soon suffer cracking, spoiling the paint finish and its appearance.

(i) Water Ingress Due to Bad Edge Sealing of Honeycomb Section: This problem was mentioned in two reports. The first of these was a main landing gear door, which has a square-edged honeycomb design with Nomex honeycomb core at the leading and trailing edges of the door. In this case, the core was recessed approximately 3 mm (1/8 in.) and filled with a potting compound. The edge was then wrapped with two layers of fiberglass (wet lay-up) to seal the edge. The report states, "The two plies are porous and are inadequate to seal out moisture, and potting compound is brittle and cracks readily. Fluids and moisture ingress through the wrapped fiberglass and potting cracks are severe. Moisture and Skydrol cause deterioration. Damage may extend 100 mm (4 in.) into the core." The report unfortunately does not state whether the leading edge suffers more than the trailing edge, but other reports suggest that it should. Landing gear doors suffer considerable aerodynamic buffeting during takeoff and landing; thus, redesign of the leading and trailing edges seems necessary.

Another report deals with a similar problem; however, in this case, the component was an aileron. "The designed edge closeout of a honeycomb core aileron was accomplished by locally potting the core at the edge and bonding a fiberglass end cap over the skins and core. The core is built up of several spliced core pieces, which create a septum at the splice. One septum is located at the edge closeout. The forward edge can be as deep as 125 mm (5 in.) and taper to nothing at the aft edge. Potting of the core at the forward edge can be difficult. If an insufficient seal is created at the bonding of the end cap, moisture will travel to the core edge. The potting is intended to prevent further ingress of moisture, but a moisture path was found through the edge cap and into some cells that were not completely potted. However, the potting will crack over a period of time. The core septums may provide another path for moisture." See Figure 16.8.

Proposed Solutions:

- Use a formed closeout, at least three fabric layers thick and pre-molded with a good resin content.

- Use more layers of thin fiberglass (at least three), a tough resin, and a high resin content.

- Use a tough potting compound that can withstand some flexure after recessing the honeycomb to a greater depth, say 12.5 mm (0.5 in.).

- Paint the close-out with polysulfide or other sealant after application of the fiberglass layers or the pre-molded part.

Honeycomb edge sealing demands careful attention. At the ends of control surfaces of the full-depth honeycomb type, the use of potting compounds to seal the honeycomb followed by an overlay of fiberglass has not always been successful. Even in such an apparently simple case, the compound must have both the fluidity to seal effectively in the uncured state and the flexibility to accept some movement in the cured state. It also may be necessary to match the CTE and the hygro-expansion coefficient of the compound and the skins.

(j) Holes Through Honeycomb, Water Ingress: Although only one report was received, this is a common problem with both composite and metal parts with honeycomb cores. Likewise, it is a well-known problem with spoilers on at least one aircraft type. The report says, "The fastener holes provide a moisture path into the core. Even if the attachment fittings are installed with generous amounts of sealant, moisture finds its way into the core."

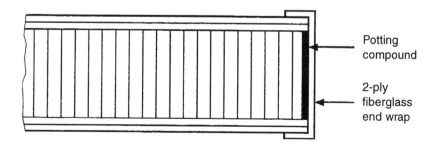

Fig. 16.8 Diagram of the edge sealing problem with honeycomb panels.

Experience has shown that if bolt inserts are required through honeycomb sandwich structures, they should be of the double-sided interference fit type with large flanges bonded to the skins. In addition, they should be made from aluminum alloy and should be given a chromic or phosphoric acid anodizing treatment (unsealed), dried, and coated with a phenolic primer before being stored in sealed plastic bags. The inserts should also be potted in place with a suitable tough epoxy compound having a low water uptake. Ideally, bolt holes should not go through the honeycomb area, and this should be avoided in design. Experience has shown that after time, water always penetrates the honeycomb through these inserts. In the cases of aluminum alloy skin and aluminum alloy honeycomb, the ingress of water invariably leads to corrosion around the insert of both the skin and the honeycomb, and then to large disbonded areas and component failure. In the cases of composite panels used for wing trailing edges, the ones on the lower surface usually take in water or hydraulic fluid at bolt insert positions. Even spoilers with bolts that have been well sealed eventually take up many pounds of water. This seems to occur in the region of the bolts. Therefore, it may be concluded that bolts or fasteners should not be fitted through honeycomb areas, and components should be designed to avoid this completely. See Fig. 16.9 for the best design of insert if one must be used.

Most spoilers on one aircraft type have a moisture ingress problem around the actuator fitting. Efforts to establish the route of entry for the moisture used dyes and vacuum pumps. Thermal scans and x-ray photographs defined the area in which liquid water was present. The majority of the water was located around the actuator fitting and also had spread along the spar to the outer hinge fittings. A vacuum test indicated that water had not entered at the hinge attachment bolts. However, when the same test was tried on the core to spar bond line, the dye flowed easily. The spoiler contains a network of interconnecting splice lines of a foaming film adhesive, which extends along the rear face of the spar and around the high-density core in the actuator attachment area. Water was believed to be able to spread through this network of core splice adhesive because of its porosity. The manufacturers of this foaming core splice material insist that is designed to have closed cells to ensure that this will not be a problem. A point that must be covered in training is that these foaming films have an expansion ratio of 1.5 to 3.5. One of these films is known to have

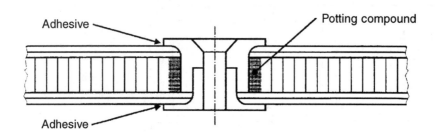

Fig. 16.9 Adhesively bonded, two-part, interference fit type bolt inserts for honeycomb panels.

an expansion ratio of 2.5. Too much foaming film means that honeycomb may be displaced sideways by the foaming pressure. Too little could mean that the foam expands far enough to become porous or does not expand enough to completely fill the gap. Again, the end result is highly dependent on the knowledge, experience, and care of the operator who is assembling these parts. Also, the spar must not be touched or otherwise contaminated on the bonding faces during assembly. If it is, then the bond will be poor and will be easily disbonded in the presence of moisture. Two-part room-temperature curing syntactic foam potting compounds that can accept high-temperature post cure may do a better job. However, one of these in common use was found by test to have a saturation water uptake of 35% after room-temperature cure and 34.4% after cure at 80°C (176°F). This suggests that the weight saving achieved by using syntactic foams with glass or other hollow spheres may be highly unprofitable. Less water might enter if solid room-temperature curing two-part paste adhesives or hot-curing one-part paste adhesives were used for honeycomb potting. This is another component in which bolt holes go through the honeycomb area. Another possibility is that water could get through the bond line between the actuator fitting and the skin for one of several reasons:

- If the bond between the fitting and the composite skin was not good (i.e., contaminated fitting or skin prior to bonding)

- If the bond between the fitting and the skin was damaged during riveting

- If the skins over the fitting area were delaminated by riveting or any other reason, allowing water between the plies and then into the spoiler through the bolt holes

The true cause must be established because this is a well-made part. The bolts are covered with plenty of sealant, and an obvious cause for the problem cannot be determined. It is also true that kilograms of water can be extracted from many of these parts after a long time in service. Even at this stage, after thousands of hours of service, these parts appear good and well sealed; however, the water is undoubtedly there. Again, bolts should not be fitted through honeycomb areas. No percussion riveting should be permitted on composite parts.

The cause of moisture ingress into honeycomb-cored parts is believed to be that the honeycomb cells act as a one-way pump. If any small leakage paths exist, the air in the honeycomb cells will leak outward as an aircraft climbs and the outside pressure reduces. As the aircraft descends, air will leak into the honeycomb again. If the descent passes through clouds, moist air will enter the honeycomb. On the next climb, most of this water vapor will condense to liquid as the air temperature falls, and only relatively dry air will leak outward again. After many repetitions of this cycle, the amount of water in the honeycomb is considerable. In aluminum alloy honeycomb structures, corrosion, disbond, and failure result. In composite skinned Nomex honeycomb structures, opportunities for water ingress increase if water can enter through the skin, in addition to leak paths at edges and inserts. The honeycomb also is slightly porous, and water can move slowly from cell to cell. This reduces bond strength between the skin and honeycomb and considerably increases the weight of the part. In the case of radomes and composite parts at the extremities of an aircraft (e.g., wing, tailplane, and rudder tips), pinholes in the skin can result from small lightning strikes rather than any design faults. These must be found by inspection, and repairs must be made as necessary.

(k) Tooling Holes Through Honeycomb Allow Moisture Ingress: One report states, "Tooling holes in honeycomb structure require sealing. Over a period of time, sealing materials may degrade or crack. This may allow a moisture path into the core."

Proposed Solutions: Provide excess material for location of tooling holes that can be removed during final installation. Alternatively, locate tooling holes in areas without honeycomb. See also item (j) above.

(l) Sealing Honeycomb Panel Inside Skins Against Skydrol: One report was received, but this is a known problem requiring attention. See item (h) for possible solutions.

(m) Heat Shield Inadequate: This was reported three times. In each case, heat shields were made from aramid or glass/epoxy construction and were used at temperatures too high for the resin system. Clearly, this is a design error.

(n) Cracking in Surface Filler: This was reported twice and has been seen on several other occasions.

Proposed Solutions:

- As previously stated, if a surface filler is used, it should have a good elongation to failure over the service temperature range and ideally a similar CTE and hygro-expansion coefficient as the skin it is filling. Brittle fillers soon suffer cracking, spoiling the paint finish and its appearance.

- A thin layer of glass fabric on the outside could be a good solution.

(o) Damage Due to Cowlings Being Slammed Shut: Two reports were received.

Proposed Solution: Both reports suggested the use of gas struts in place of the existing support struts, with sufficient pressure/force to prevent cowlings from being slammed. This is the best solution, but training also could help. See Fig. 16.10.

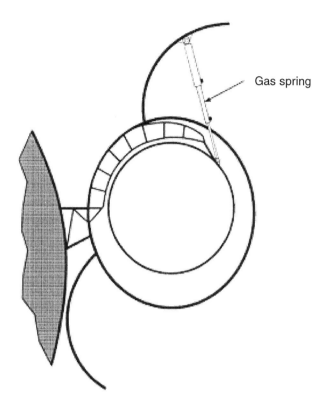

Gas spring

Fig. 16.10 Use of a gas strut to reduce slamming damage to cowlings.

(p) Fuselage Ice Protection Plates Too Small: Two reports of this problem could be relevant to future aircraft if composite fuselages are chosen for new aircraft because another report tells of difficulty in repairing filament-wound parts. The comments are relevant because existing plates have been found to be too small. On the present generation of aircraft, ice protection plates are fitted to the fuselage opposite the propellers. Generally, these are not large enough because ice coming off in flight may be blown back and may strike the fuselage behind the plates. See Fig. 16.11.

Proposed Solutions:

* The true area that requires plating should be established by experience. The faster the aircraft, the larger the plating should be.

* If filament-wound composite fuselages are eventually made, the difficulty of repairing filament-wound components will have to be addressed. Therefore, any protection plates must be made adequately large. See item (u).

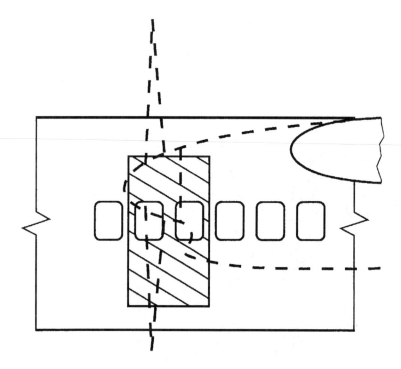

Fig. 16.11 Fuselage ice protection plates.

(q) Parts Too Large or Access Panels Too Small: Two reports were received concerning this problem, which has been mentioned previously, in Ref. 16.3, and in the International Air Transport Association IATA Document DOC.GEN 3043. Report 36 states, "Because the flap is damaged very frequently, all compartments must be easily accessible for repair. Current design has too few access panels." Report 39 states, "In panel assemblies generally, access is provided only for inspection. Additional access is required to fit vacuum bags inside for repair. When designing access panels, maintenance and repair must be considered, in addition to inspection."

Proposed Solutions:

- Reduce the panel size.
- Use more access panels.
- Use larger access panels.

(r) Lightning Damage: This is an important issue having serious safety implications and repair needs. Two reports mention this problem. One of these reports is unusual and merits special consideration because it raises the possibility of a composite part acting as an electrical capacitor. The report states, "Sandwich design with carbon and Kevlar or glass skins and aluminum honeycomb core using the carbon as the outer skin and Kevlar or glass next to the honeycomb acts as a capacitor and attracts lightning strikes. This possibility should be investigated for safety reasons because aluminum honeycomb is being used due to its lower cost. It cannot be used in direct contact with carbon in damp or wet environments because of the corrosion risk."

Proposed Solutions:

- Do not use aluminum core.
- Improve electrical grounding.
- Improve lightning protection.

Note that the Culham Laboratory Lightning Studies Unit (UK) found that metal honeycomb panels with carbon fiber skins are subject to damage in lightning situations. Another designers guide on lightning protection (origin not given) states, "Unprotected [carbon-fiber composite] CFC structures that utilize honeycomb cores are particularly subject to damage. The aluminum honeycomb is much lower in resistance than the CFC skin; hence, lightning currents will puncture the skin at the attachment point and then flow through the honeycomb. The honeycomb, which is constructed from thin aluminum foil and fastened at the nodes with dielectric adhesive, is too thin to carry the lightning current without vaporizing. Explosive forces due to this vaporization of the foil, and heating of the gas in the voids and the adhesive, or by sparking and heating of the air in the honeycomb cells, can cause serious delamination of the CFC skins."

The extent of the hazard will depend on the following factors:

- Thickness of the CFC skins
- Density, thickness, and ribbon direction of the aluminum honeycomb
- Nature of the CFC interface bond
- Position of the component in the aircraft

Insulated Honeycomb Panels: Nomex honeycomb or panels with CFC skins present less of a problem because the core is of insulating or high-resistance material. The current is generally confined to the CFC skins.

Nature of Lightning Damage: From tests on several aluminum alloy and Nomex cored honeycomb panels, the following effects were reported:

- In contrast to solid CFC plates, the application of erosion-resistant polyurethane paint drastically increased damage area and depth.

- In some cases, only minimal damage was found at the attachment point; however, a larger area of damage was found on the opposite side of the sandwich. This effect has been observed on radomes in which a small burn on the outside of 6 mm (1/4 in.) diameter may indicate a 450 mm (18 in.) diameter area of disbond of skin from the honeycomb on the inside.

- In cases where no exterior damage was experienced, interior interlaminar shear failure was discovered on cutting through the test specimen at the attachment point.

- Comparing the skin damage area with that of the core in the case of Nomex sandwich plates, the zone was found to be similar. In the case of aluminum honeycomb, the area of damage was at least twice the extent of the outside skin damage. In regard to a problem with a radome, one report states, "Insufficient grounding. Lightning can weld antenna bearings and/or render radar inoperative. Lightning striking radomes causes damage ranging from pinholes to skin damage." This seems to be a special problem needing attention on the aircraft concerned. Continuity checks are specified in most SRMs, and the whole lightning protection system should be maintained as the SRM requires. Most radomes have effective lightning protection schemes that have greatly reduced damage to these parts. They must be maintained in good condition. Lightning protection is a highly specialized subject, and the advice of specialists should be sought at the design stage. Adequate lightning protection must be provided for composite parts where it is likely to be necessary. Designs should be subjected to testing and confirmed as suitable for the intended function or modified as necessary. Protection schemes should be easily repairable because they must be maintained in good condition. All aircraft extremities are lightning prone, including engine cowlings. See Figure 16.12.

A bank of experience in this subject should now be used. Some past occurrences have shown the following:

1. Induced currents have resulted in the release of bombs and drop tanks by military aircraft. Drop tanks have been blown from aircraft.

2. Radomes have been blown forward and completely lost from aircraft when no lightning protection was fitted. The pressure generated inside the radome, as a result of the temperature generated as the lightning struck the radar antenna, has been sufficient to fracture all the latches. In other cases, pieces of skin 1 ft^2 or larger have been blown away.

3. On light aircraft, control bearings have been welded solid, resulting in the loss of the aircraft and crew.

4. On a medium-sized jet airliner with carbon fiber elevators, several cases of spar damage occurred that could have been more serious. In these cases, several points of interest were noted:

   - The elevator hinge bearings and the main pivot arm bearing felt gritty when operated by hand. The current involved had tried to weld each ball to its race.

   - The jumper lead that was supposed to carry the current had been fractured, presumably by inductive forces. The current had flowed along the carbon fiber spar, through a bolt, and out over the aluminum flame-sprayed coating on the outside.

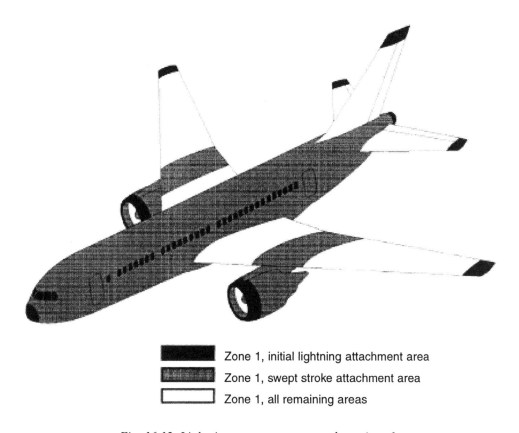

Zone 1, initial lightning attachment area
Zone 1, swept stroke attachment area
Zone 1, all remaining areas

Fig. 16.12 Lightning prone areas on modern aircraft.

The spar was locally delaminated and required a hot-bonded repair. The aluminum flame-sprayed coating was burned, and parts of the skin required repair. The bolt had received an instantaneous spark-eroded undercut of approximately 0.5 mm (0.02 in.). The bolt hole was approximately 10 mm (0.4 in.) in diameter, and the carbon fabric around the bolt hole had no epoxy resin. The temperature generated was considered likely to have exceeded 1000°C (1800°F) and to have instantly gasified the epoxy resin (it begins to char only at approximately 400°C [750°F]). It may be concluded that lightning protection should be on the outside of parts. Lightning prefers to travel in straight lines and will find the easiest path, even if that path ignores the path provided. Composite parts must have adequate protection; otherwise, the consequences could be fatal. Lightning protection schemes should be thoroughly tested. Composite wings, tailplanes, and fins used as integral fuel tanks must be provided with good protection.

5. On a fairly large jet airliner, numerous cases of lightning damage occurred to rudders, ailerons, and elevators until improved protection systems were fitted.

6. Composite fuselages do not naturally provide as much electromagnetic interference (EMI) shielding as aluminum fuselages. Sufficient shielding must be provided, by other means if necessary, to avoid catastrophic interference with automatic flight controls, radio, radar, and navigation equipment.

(s) Damage Caused by Hot Exhaust Air: One report was received on this issue. The item in question was a fan cowl made with a carbon fiber/epoxy skin bonded to an aluminum alloy honeycomb core. The anti-ice exhaust air from the inlet cowl impacted the fan cowl, causing skin delamination. This problem requires redesign, and the following solutions were suggested:

- Reduce the temperature of the exhaust air.
- Redirect the exhaust to another location.
- Fit a heat shield resistant to the temperature of the exhaust air.

(t) Pre-preg Tape Used for an Outer Ply: One report of this problem was received, but the problem is so clearly fundamental that action on all future design is required. The report states, "Outer plies made with tape cannot be drilled or mechanically worked because they split and peel away. Any type of repair or working around where plies truncate at edges and corners causes splitting and peeling of the outer plies."

Proposed Solution: Discontinue the use of tape for outer plies.

(u) Filament-Wound Parts Difficult to Repair: One report raised this problem, which occurred on an engine cowl inlet outer barrel. It states, "Skin penetrated and splintered due to impact damage. Filament-wound structure causes difficulty in removal of damage and determination of repair lay-up orientation. Flush repairs cannot be done because of the construction of the lay-up."

More detail must be known before commenting on this case. However, it requires careful study because of the far-reaching implications for future design. If filament-wound fuselages should be produced in large numbers at a future date, this problem requires thorough investigation. As mentioned elsewhere, fuselages become damaged from propeller ice and also from fork lift trucks and impacts of other vehicles. Small, propeller-driven aircraft are the most likely candidates for experiments with filament-wound fuselages. The repair of filament-wound parts merits further study.

Proposed Solution: A study of the repair problems of filament-wound structures is needed.

(v) Panel Interference with Structure Due to Flexure Under Load: One report was received, which states, "The damage is an abrasion that runs chordwise from the leading edge to the trailing edge, caused by interference with the leading edge slat extending and retracting. The abrasion occurs in the same place on every airplane on both the right and left wings."

Clearly, a design fault fails to provide sufficient clearance between the slat and the fixed panel to allow for deflections under load.

Proposed Solutions:

* Stiffen the deflecting component.
* Increase the clearance if practical to do so.
* Fit a rubbing strip.

(w) Acoustic Panel, Inadequate Water Drainage: One report was received on this subject, but the problem has been found by others and thus merits more action than only one report would suggest. The report concerned an engine inlet acoustic panel having a carbon/glass/epoxy skin construction with aluminum honeycomb core and a perforated aluminum facing skin, as is commonly the case with acoustic panels. The problem was delamination of carbon/glass/epoxy skin from the honeycomb. With the skin perforations required for acoustic attenuation, water easily penetrates the honeycomb in large quantities and must be removed.

Proposed Solutions:

* Provide drainage slots in the honeycomb. This has been used on some designs.

* Check that the aluminum honeycomb specified is of the corrosion-treated and resin-dipped or anodized types. If not, disbond in the presence of moisture in a short time is certain.

(x) Metal Fittings, Thermal Expansion Problem: One report states, "Aluminum fittings are hot bonded on carbon composite components. Because of the difference in thermal expansion rates of aluminum and carbon composite, residual stress remains in the bond line. Disbonding and bond line cracks are frequent. Repair of damaged fittings is complicated and expensive."

Proposed Solutions:

• Attach metal fittings to composite components only with fasteners.

• Bond fittings with toughened two-part room-temperature curing adhesives. A durable bond depends more on an unsealed anodized finish on the aluminum than it does on the adhesive, but the choice of adhesive is important. Toughness and flexibility are necessary.

• Bond fittings with a low-temperature curing film adhesive. Some film adhesives will cure at 80°C (175°F) if a long cure cycle is acceptable.

16.3 Other Required Design Features Not Mentioned in the Analyzed Reports

Several other design features were not mentioned in the 55 reports that were analyzed. These are discussed in the following paragraphs.

1. The edges of panels should have layers of fabric at several angles (e.g., 0 to 90° and ±45°) to improve shear out loads. The edge margin around bolt or rivet holes should be larger than for metal panels.

2. Toughened resin matrices are preferred to brittle ones if the tensile and compression moduli can be maintained. More work is needed to confirm the findings of Palmer of the Douglas Aircraft Company (Ref. 16.4) to allow the relationship between matrix properties and composite properties to be better understood. Standard test methods, preferably International Standards Organization (ISO) methods, for testing neat matrix resins and adhesives are required.

3. Engine cowlings are made to withstand the highest temperature expected. This may occur only in a small area near a hot component; however, the repair requirements often make no allowance for this and specify that all repairs should be made at the maximum temperature. This is highly uneconomical and must be covered in the SRM, preferably with a temperature map for the cowling, indicating which areas require hot repairs and which ones can be repaired at lower temperatures.

4. It may be desirable to manufacture composite-skinned honeycomb-cored parts at 180°C (350°F) to be able to repair them at 120°C (250°F) without suffering disbonding. When only vacuum pressure can be used, many organizations have learned the hard way that the area around the repair often is disbonded by the air pressure inside the honeycomb cells. Vacuum pressure above 10 psi (20 in. of mercury) is seldom achieved. The pressure inside the cells is usually 15 psi and increases with the temperature. If moisture is present in the cells, the vapor pressure of water at the temperature

concerned must be added to this. Above 100°C (212°F), the vapor pressure of steam must be added. At some point as the temperature rises, the pressure inside the cells increases and the strength of the adhesive falls. When the crossover point is reached, the bond fails and the repair must be repeated. At this stage, the repair is larger than it had been. The only ways to avoid this are to increase the external pressure by a variety of means, such as shot bags, pressure bags, presses, clamps, or autoclaves. If additional pressure is unavailable, the only solutions are to use room-temperature curing resins/adhesives, low-temperature curing resins/adhesives, or pre-pregs/film adhesives curing at least 25°C (45°F) and preferably 50°C (90°F) below the original cure temperature of the part. If these requirements could be accepted by OEMs, then composite skinned honeycomb panels would be easy to repair with only vacuum pressure and heater blankets controlled by hot-bonder units. These units are available from several suppliers, and their design is improving continuously.

5. In the design and repair of radomes, the honeycomb or other core material thickness must be one-quarter of the radar signal wavelength for best transmission. Deviation from this causes greater signal losses. As new wind-shear detecting radars now are coming into use, requiring radomes with a higher transmission efficiency than previously needed, greater attention must be paid to achieving the transmission requirements. Most civil aircraft now use x-band radar systems. The precise thickness of the sandwich for best transmission depends on the dielectric constants of the skin and core and on the skin thickness. For this reason, unless approved data is available to permit variation, repair radomes with materials to the original specification because the optimum skin spacing (core thickness) with materials other than the original could be different. The outer skins should be designed with at least three layers of fabric and a good resin content because moisture penetration into radomes is a common complaint.

6. Above a laminate thickness of 2 to 3 mm (0.08 to 0.12 in.), bolted joints are lighter than bonded joints. Below 2 mm (0.08 in.), bonded joints are lighter. These require tapered scarf joints with a 1 in 50 taper; 1 in 20 is acceptable at panel edges where the thickness has been increased to suit countersunk fastener heads. For thin skins, lap joints with an overlap of 50 times the skin thickness often are used. It is common to use 50 times the skin thickness as a parallel portion with a step for each layer beyond that.

7. The fatigue strength of scarf joints is not as good as the parent laminate. However, work at Imperial College on tension fatigue has shown that, after long times, peeling at the ends of the scarf occurs and progresses slowly. Thus, more than adequate warning is given, and the repair can be done again. See Ref. 16.5.

8. When bonding Kevlar to Nomex honeycomb and probably also when bonding to aluminum alloy or any other type of honeycomb, a thin layer of fiberglass and a layer of film adhesive between the Kevlar and honeycomb improves the peel strength in the

climbing drum peel test and the subsequent durability of the bond in service. The use of a layer of film adhesive between the honeycomb and the skin is recommended, regardless of the type of fiber pre-preg used.

9. When using aluminum alloy honeycomb, only the corrosion-treated and resin-dipped variety (etched and dipped), or preferably chromic or phosphoric acid anodized honeycomb, should be used. Lesser treatments give poor bond durability and result in disbonds between the skin and honeycomb after a period of time.

10. Aluminum alloy parts bonded to composites should be anodized (unsealed) by either the phosphoric acid or chromic acid process. If bonded to carbon fiber parts, a layer of glass fiber should be placed between the carbon fiber and the aluminum part as an insulator to reduce the risk of galvanic corrosion.

11. Testing of composite parts and test pieces should seek to relate the performance of the part or piece to the neat resin mechanical properties, the fiber properties, and the surface treatment of the fibers. A data bank should be built using ISO test methods to allow easier comparison of one resin with another when selecting alternative materials for either manufacture or repair. See Refs. 16.4 and 16.6.

12. A data bank of resin physical properties is also required, as follows:

    *   Diffusion coefficient of water and other fluids likely to contact composites is needed. Note that moisture or other fluids already in a composite may result in some solvents or fluids being absorbed faster than they would be absorbed by a dry composite. All work should be done on thoroughly dried specimens or specimens with a known moisture or other fluid content.

    *   Solubility coefficients (i.e., saturation water uptake values) for water are required for all resins, adhesives, and fibers (where appropriate). The saturation uptake value is important because some resins have a high diffusion coefficient and low saturation uptake; others have the reverse. The 24- or 48-hour water uptakes sometimes quoted could be misleading and could give a reverse order of merit to the true one. Uptake values for any other fluids in regular contact with composites would be useful.

    *   Wet and dry glass transition temperature (T_g) values for matrix resins and adhesives must be obtained and recorded.

    *   The amount of dimensional change as a result of moisture uptake also must be obtained and recorded.

    *   Coefficients of thermal expansion are required for resins and fibers because experience on many engine cowlings has shown that carbon and Kevlar, the fabrics often used to make hybrid components, become disbonded if the temperature is cycled

between low and high values. The same hybrid structures on helicopters and other low-flying aircraft, where the temperature range is much less, do not have the same problem. The ability to calculate the range of cyclic temperature that various hybrids are likely to be able to accept is necessary, and this requires basic data.

13. Moisture ingress can occur at the trailing edges of tapered parts (e.g., control surfaces). This problem was first investigated when water was found 250 mm (10 in.) from the trailing edge in a spoiler made from aluminum alloy skins and aluminum alloy honeycomb. The first reaction to such information, if not seen personally, would be that this is impossible. However, after removal of one skin and investigation with a magnifying glass, it was easy to see how this could happen. See Fig. 16.13, particularly the sketch of the existing design.

The cause of the problem is twofold: first, the splitting of the honeycomb node bonds during cutting to a feather edge, and second, the fact that no wraparound sealing is provided at the trailing edge.

Proposed Solution: The proposed solution is shown in Figure 16.13(b). Fill the last 50 mm (2 in.) with a good paste epoxy instead of honeycomb and allow it to cure. Sand to profile, and bond the skin or patch. Apply two layers of 0.08 mm (0.003 in.) fiberglass bonded with a toughened two-part paste adhesive around the trailing edge to minimize water ingress. This method has been used successfully during repair work and is applicable to composite and metal parts.

(a)

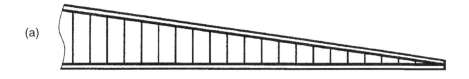

(b)

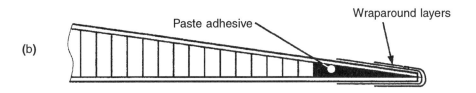

Fig. 16.13 (a) Sketch of the existing design of the trailing edge of tapered control surface panel; (b) Proposed solution is sealing of a trailing edge.

16.4 References

16.1   Armstrong, K.B., "Parts Integration—Advantages and Problems," Carbon Fibers III, 3rd International Conference, October 8–10, 1985, Kensington, London, UK, pp. 15/1–15/6.

16.2   Allen, H.G., *Analysis and Design of Structural Sandwich Panels,* Pergamon, Oxford, UK, 1969.

16.3   Armstrong, K.B., "Aircraft Floor Panel Developments at British Airways (1967–1973)," *Composites,* Vol. 5, No. 4, July 1974.

16.4   Palmer, R.J., "Investigation of the Effect of Resin Material on Impact Damage to Graphite/Epoxy Composites," NASA Contractor Report 165677, McDonnell-Douglas Corporation, Douglas Aircraft Company, Long Beach, CA 90846, 1981.

16.5   Robson, J.E., Matthews, F.L., and Kinloch, A.J., "The Fatigue Behavior of Bonded Repairs to CFRP," 2nd International Conference on Deformation and Fracture of Composites, March 1993, The Institute of Materials.

16.6   Armstrong, K.B., "The Selection of Adhesives and Composite Matrix Resins for Aircraft Repairs," Ph.D. thesis, The City University, London, UK, 1990.

16.7   Baker, A.A. and Jones, R. [eds.], *Bonded Repair of Aircraft Structures*, Martinus Nijhoff, Dordrecht, The Netherlands, 1987.

16.8   Armstrong, K.B., "Effect of Absorbed Water in CFRP Composites on Adhesive Bonding," 33rd Annual Conference on Adhesion and Adhesives, April 1995, Oxford Brookes University, Headington, Oxford, UK, published in *International Journal of Adhesion and Adhesives*, Vol. 16, No. 1, 1996, pp 21–28.

Additional Information

The following is a list of useful Society of Automotive Engineers (SAE) Aerospace Recommended Practices (ARPs) and other documents.

1. IATA DOC.GEN: 3043—Guidance Material for the Design, Maintenance, Inspection, and Repair of Thermosetting Epoxy Matrix Composite Aircraft Structures

2. AIR 4844—Composites and Metal Bonding Glossary

3. ARP 4916—Masking and Cleaning of Epoxy and Polyester Matrix Thermosetting Composite Materials

4. AIR 4938—Composite and Bonded Structure Technician/Specialist Training Document

5. ARP 4977—Drying of Thermosetting Composite Materials

6. ARP 4991 Draft—Core Restoration of Thermosetting Composite Materials

7. ARP 5143 Draft—Vacuum Bagging of Thermosetting Composite Repairs

8. ARP 5144 Draft—Heat Application for the Local Repair of Thermosetting Composite Materials

9. ARP 5089—Composite Repair NDT/NDI Handbook

10. ARP 5256—Resin Mixing

11. AE 27—Design of Durable, Repairable, and Maintainable Aircraft Composites

Other draft ARPs and documents in preparation include the following topics:

- Squeeze-Out/Impregnation
- Machining

For additional information on these documents or a complete list, contact the Society of Automotive Engineers, 400 Commonwealth Drive, Warrendale, PA 15096-0001, USA; phone (412) 776-4841, fax (412) 776-5760, http://www.sae.org.

Index

About the Authors

Dr. Keith Armstrong is well respected for his many years of experience in advanced composite aircraft structures. He is best known for his involvement at British Airways in the development of carbon fiber/Nomex honeycomb floor panels. These panels were the first carbon fiber composites ever to fly on commercial aircraft, and most flooring in modern commercial aircraft meets specifications that evolved from these early composite structures.

Dr. Armstrong earned an M.Sc. in 1978 and a Ph.D. in 1990 in Adhesion Science and Technology from The City University in London, England. His career began in 1948 at Vickers-Armstrong (Aircraft) Ltd. Weybridge as an aviation apprentice and then as a design draftsperson. He subsequently served as a technical officer in the Royal Air Force and later as an experimental officer at the National Physical Laboratory in Teddington.

Dr. Armstrong spent the next 24 years of his career with British Airways. Along with his work on carbon fiber/Nomex honeycomb floor panels, he developed many new methods in the infancy of the industry. He later served as a consultant to Du Pont on composite repairs using Nomex honeycomb and Kevlar, and then as a quality audit engineer for Aerobond in the United Kingdom.

From 1988 to 1991, Dr. Armstrong chaired the IATA Composite Repair Task Force, and he continues to participate in the Training Task Group of the SAE/IATA/ATA Commercial Aircraft Composite Repair Committee. He is a member of SAE International, SAMPE, and ASTM. He also is a Fellow of the Institution of Mechanical Engineers, the Royal Aeronautical Society, and the Institute of Materials.

Dr. Armstrong has written more than 30 technical papers and compiled the *Composites and Metal Bonding Glossary* published by SAE. He is a part-time lecturer at Brunel and Bristol universities in the United Kingdom and has held a private pilot's license for more than 25 years.

Richard T. Barrett has more than 25 years of experience in the aircraft industry. He earned an airframe and powerplant license from the Detroit Institute of Aeronautics in 1974 and an Associates Degree in Engineering from Washtenaw Community College in 1976. Since that time, he has gained significant hands-on training in aircraft maintenance and composite material design and repair techniques, and he is a certified autoclave operator.

Mr. Barrett began his career in 1971 as an aircraft mechanic with Rosenbalm, Zantop, Royal Air. He next worked as an engineering technician and aircraft mechanic for General Motors, Inc. and then performed aircraft maintenance and inspection for Frontier Airlines, Inc. From 1987 to 1995, Mr. Barrett worked in technical management at Continental Airlines. In his role there as composite shop manager, he developed his expertise in performing, managing, and training hands-on repair of damaged composite parts.

From 1992 to 1994, Mr. Barrett chaired the Training Task Group of the SAE/IATA/ATA Commercial Aircraft Composite Repair Committee. He is a member of SAE International, SAMPE, EAA, and Toastmasters International.